AF847917

Axiomatic Design in Large Systems

Amro M. Farid · Nam P. Suh
Editors

Axiomatic Design in Large Systems

Complex Products, Buildings and Manufacturing Systems

Springer

Editors
Amro M. Farid
Thayer School of Engineering
Dartmouth College
Hanover, NH
USA

Nam P. Suh
Department of Mechanical Engineering
Massachusetts Institute of Technology
Cambridge, MA
USA

ISBN 978-3-319-32387-9 ISBN 978-3-319-32388-6 (eBook)
DOI 10.1007/978-3-319-32388-6

Library of Congress Control Number: 2016936680

© Springer International Publishing Switzerland 2016
This work is subject to copyright. All rights are reserved by the Publisher, whether the whole or part of the material is concerned, specifically the rights of translation, reprinting, reuse of illustrations, recitation, broadcasting, reproduction on microfilms or in any other physical way, and transmission or information storage and retrieval, electronic adaptation, computer software, or by similar or dissimilar methodology now known or hereafter developed.
The use of general descriptive names, registered names, trademarks, service marks, etc. in this publication does not imply, even in the absence of a specific statement, that such names are exempt from the relevant protective laws and regulations and therefore free for general use.
The publisher, the authors and the editors are safe to assume that the advice and information in this book are believed to be true and accurate at the date of publication. Neither the publisher nor the authors or the editors give a warranty, express or implied, with respect to the material contained herein or for any errors or omissions that may have been made.

Printed on acid-free paper

This Springer imprint is published by Springer Nature
The registered company is Springer International Publishing AG Switzerland

To Our Supportive Wives,
Inas and Young

Preface

Design is one of the intellectual activities that have played a fundamental role in creating and shaping the modern civilization. Design begins with the identification of the *need* that must be satisfied and ends with an *innovation.* This translation of the need to a final solution is the responsibility and domain of a designer. The goal is to create a design that satisfies the perceived needs as specified by functional requirements (FRs) within specified constraints. How we perform this task well is the main subject addressed in this book.

Since the Industrial Revolution, major scientific discoveries and technological innovations have enabled humans to harness energy, transport people and goods, generate wealth, secure health and security, and explore the universe. All of these advances were possible because of the human ability to design and innovate. Without these advances, the world would not have had the means of supporting its rapidly expanding population and improving the quality of life at the same time. Innovations will continue to be important in the twenty-first century as we now have new challenges: global warming, sustainability, and securing peace throughout the world. Our ability to design well will continue to affect all of these important issues.

Through design, we create things and artifacts. Some designs are purely artistic. Others are technological or organizational or scientific. Some designs may involve large intelligent systems that exceed human capability. Some are in the form of algorithms that perform pre-set routines to deliver solutions to a class of common problems. To solve these challenging problems of the twenty-first century, we need to advance the field of design.

While there are many approaches to design, this book deals with the design from the perspective of Axiomatic Design (AD). The goal is to advance the "science of design" to solve practical problems by creating the foundation for "design thinking". The innovations created based on AD share two common characteristics: simplicity and reliability of the final product. Simplicity and reliability represent the dual nature of creative design. When the design is simple, the development of the product or the system tends to take short time, and the product tends to be more reliable. These are the foundations of the two design axioms.

The idea that we should create *design science* to strengthen the discipline of design and manufacturing came out of necessity. In the late 1970s, when MIT launched a new effort for the field of design and manufacturing, we chose the creation of the science base for design and manufacturing as the ultimate goal of the Laboratory for Manufacturing and Productivity (LMP). With the generous support of the U.S. National Science Foundation (NSF), MIT embarked on AD research. A few years later, NSF created a research support program in design, which generated a cadre of young design specialists, who are now leading scholars and practitioners of the design field. Since then, much has been achieved as indicated by the chapters written by many authors of this book attest, but we still have much more to do in order to merge the scientific quest and the industrial practice so that they are indistinguishable to the scholars and the practitioners of design.

As will be discussed in greater length in Chap. 1, one defining characteristic of twenty-first century challenges is the breadth of their scope. The twentieth century was marked by several transformative technologies such as the generator, telephone, and automobile. Now, the twenty-first century arrives to address the large complex systems which have formed around these individual innovations. Our power grid, communication infrastructure, and transportation systems are all integral parts of daily life and its twenty-first century challenges. Indeed, the trends towards integrating renewable energy and smart buildings, spanning the digital divide, and electrifying transportation are all design efforts to solve today's large-scale problems.

In order to solve these broad scope challenges, design science must continue to advance beyond individual products to large complex systems. This means revisiting and potentially setting aside several limiting constraints and assumptions. First, these large complex systems span the traditional boundaries of individual engineering disciplines and so must design science seek to integrate this knowledge into a consistent framework. Second, these large complex systems rarely have one engaged "customer" but rather a diversity of internal and external stakeholders. Consequently, large complex systems have a mix of soft and hard requirements rather than a single fixed contract.

While these broad twenty-first century challenges will likely remain as such for decades and cannot be addressed within a single volume, this book seeks to advance AD as a design science specifically towards this purpose. To support this goal, the book addresses three application domains of greater scope than traditional products but more tractable than the grand challenges of the century. These three domains are large complex products, buildings, and manufacturing systems. Together, these three domains exhibit many of the previously mentioned characteristics of large complex systems while still remaining grounded as current engineering design problems. Consequently, this book is organized as follows.
Part I of the book provides introductory material common to all the applications found in the book.

- Chapter 1 provides an engineering system's introduction to AD. It introduces the fundamentals of AD within the context of large complex engineering

systems and existing efforts in model-based systems engineering. It also highlights several areas where AD has made many contributions to large complex systems: quantitative measures of life cycle properties, design of cyber-physical systems, and design of hetero-functional networks.
- Chapter 2 provides a mathematical exposition of AD focusing specifically on the quantitative implications of the Independence and Information Axioms.

Part II of the book shifts the focus to address large complex products specifically:

- Chapter 3 addresses new development to guide strategic product design and systematic innovation. In particular, it explains the Linearity Theorem that guides product designers to select design parameters and drive innovation.
- Chapter 4 provides several considerations of information and complexity in AD. In particular, it provides a taxonomy of information and analyzes how each affects the progression of design.
- Chapter 5 provides a novel approach for AD for the environment. It proposes a conceptual framework for a smart eco-design platform that integrates information across AD multiple-design domains.

Part III of the book shifts the focus to address buildings specifically:

- Chapter 6 provides a literature review of the application of AD to the built environment.
- Chapter 7 applies AD to prefabricated buildings. It provides a special attention to the robustness and flexibility in the conceptual design phase.
- Chapter 8 applies AD to temporary housing with the application to refugee populations. It uses a methodology that integrates Quality Function Deployment (QFD) and AD to tie stakeholder requirements to the design solution.

Part IV of the book shifts the focus to address manufacturing systems specifically:

- Chapter 9 provides an AD and an implementation approach for distributed manufacturing systems.
- Chapter 10 seeks to address the multiple competing performance criteria of manufacturing systems. It provides a model for project identification and prioritization.

Finally, the epilogue brings the book to a conclusion. It discusses many of the challenges in designing and implementing large complex systems. It highlights many applications where AD has served to overcome cost overruns and missed project schedules. Together, these 11 chapters serve to demonstrate AD's future role as design science evolves to address the twenty-first century challenges of large complex systems.

January 2016

Amro M. Farid
Nam P. Suh

Contents

Contributors

Gabriele Arcidiacono Department of Innovation and Information Engineering, Marconi University, Rome, Italy

Efren M. Benavides Department of Fluid Mechanics and Aerospace Propulsion, Universidad Politécnica de Madrid, Madrid, Spain

Christopher A. Brown Department of Mechanical Engineering, Worcester Polytechnic Institute, MA, USA

Luca Bucciarelli Business School, Leanprove, Florence, Italy

Darek Ceglarek International Digital Laboratory, WMG, University of Warwick, Coventry, UK

Paolo Citti Dipartimento di Ingegneria dell'Innovazione e dell'Informazione, Università degli Studi Guglielmo Marconi, Rome, Italy

Massimo Delogu Dipartimento di Ingegneria Industriale, Università degli Studi di Firenze, Florence, Italy

Amro M. Farid Department of Engineering, Thayer School of Engineering, Dartmouth College, Hanover, NH, USA; Mechanical Engineering Department, MIT, Cambridge, MA, USA

John E. Fernández Department of Architecture, Massachusetts Institute of Technology, Cambridge, MA, USA

Lindsey R. Gilbert III Engineering Systems and Management, Masdar Institute of Science and Technology, Abu Dhabi, UAE

Alessandro Giorgetti Dipartimento di Ingegneria dell'Innovazione e dell'Informazione, Università degli Studi Guglielmo Marconi, Rome, Italy

Andrea Girgenti Dipartimento di Ingegneria dell'Innovazione e dell'Informazione, Università degli Studi Guglielmo Marconi, Rome, Italy

Marianna Marchesi Faculty of Science and Technology, Free University of Bozen-Bolzano, Bolzano, Italy

Dominik T. Matt Faculty of Science and Technology, Free University of Bozen-Bolzano, Bolzano, Italy

Francesco Melosi Department of Industrial Engineering, University of Bologna, Bologna, Italy

Hilario L. Oh MIT Park Center for Complex Systems, Cambridge, MA, USA

Mohammed A. Omar Engineering Systems and Management, Masdar Institute of Science and Technology, Abu Dhabi, UAE

Erik Puik HU University of Applied Sciences Utrecht, Utrecht, The Netherlands

Erwin Rauch Faculty of Science and Technology, Free University of Bozen-Bolzano, Bolzano, Italy

Joan B. Rodriguez Consultant Engineer—Sides, Madrid, Spain

Nam P. Suh MIT, Cambridge, MA, USA

Reviewers

Mouza Al Kaabi (Masdar Institute, UAE)
Miguel Cavique (Escola Naval, Portugal)
David Cochran (Indiana University, USA)
Amro M. Farid (Dartmouth/MIT, USA)
Sang Gook Kim (MIT, USA)
Taesik Lee (KAIST, South Korea)
Dominik T. Matt (Free University of Bozen-Bolzano, Italy)
James R. Morrison (KAIST, South Korea)
Hilario Oh (MIT, USA)
Erik Puik (HU University of Applied Sciences Utrecht, Netherlands)
Erwin Rauch (Free University of Bozen-Bolzano, Italy)
Luis Ribeiro (Linköpings University, Sweden)
Joan Bruno Rodriguez (Sides, Spain)
Nam Suh (MIT, USA)

Editorial Support

Kathy Di Antonio (Dartmouth, USA)

Part I
Introduction

Chapter 1
An Engineering Systems Introduction to Axiomatic Design

Amro M. Farid

Abstract Since its first publication in 1978, Axiomatic Design has developed to become one of the more commonly applied engineering design theories in the academic literature and industrial practice. In parallel, model-based systems engineering (MBSE) has developed from industrial origins in the aerospace, communications, and defense sectors. As the scope of humanity's engineering efforts grows to include evermore complex engineering systems, the engineering design methodologies that guide these efforts must also develop. These two, now well-established but independently developed, engineering design methodologies now appear well poised to support the synthesis, analysis, and resynthesis of large complex engineering systems. As the first chapter in this book on the application of Axiomatic Design to large complex systems, it introduces the fundamentals of Axiomatic Design within the context of engineering systems and as a conceptual foundation for subsequent chapters. It also relates Axiomatic Design's key concepts and terminology to those found in current MBSE techniques including SysML. The chapter concludes with applications in which Axiomatic Design has served to advance the development of engineering systems including quantitative measures of life cycle properties, design of cyber-physical systems, and design of hetero-functional networks.

A.M. Farid (✉)
Department of Engineering, Thayer School of Engineering, Dartmouth College, 14 Engineering Drive, Hanover, NH 03755, USA
e-mail: amfarid@dartmouth.edu; amfarid@mit.edu

A.M. Farid
Mechanical Engineering Department, MIT, 77 Massachusetts Ave, Cambridge, MA 02139, USA

© Springer International Publishing Switzerland 2016
A.M. Farid and N.P. Suh (eds.), *Axiomatic Design in Large Systems*,
DOI 10.1007/978-3-319-32388-6_1

1.1 Introduction

1.1.1 The Evolution of Axiomatic Design

Since its first publication in 1978 [1], Axiomatic Design [2, 3] has developed to become one of the more commonly applied engineering design theories in the academic literature and industrial practice [4]. It arose from the need to make the field of design more of a science rather than an art [2, 3]. The originator of Axiomatic Design, Prof. Nam P. Suh, believed from his own practical experience as a designer that if design curriculum had a more solid theoretical foundation, then a new generation of engineering designers could be trained to make more effective products and systems in less time and at lower cost. Consequently, Axiomatic Design's most distinguishing characteristic is the use of design axioms which guide the designer through the engineering design process. From these axioms, many theorems and corollaries have been subsequently proven [2, 3]. This theoretical foundation facilitated many subsequent academic works in engineering design [5, 6] without diminishing the practical application of engineering design in industry [4]. In the beginning, Axiomatic Design found applications within Suh's home field: mechanical engineering of products [2]. Since then, Axiomatic Design has expanded to many other disciplines including software and more generally large complex systems in the twenty-first century [7–9]. This successful expansion into new design applications of ever larger system scale has suggested a degree of universality to Axiomatic Design as a theory.

1.1.2 The Evolution of Model-Based Systems Engineering

Meanwhile, the modern systems engineering field developed methodologically from industrial origins in the aerospace, communications, and defense sectors [10].

Definition 1 Systems Engineering [11]: An interdisciplinary approach and means to enable the realization of successful systems. It focuses on defining customer needs and required functionality early in the development cycle, documenting requirements, and then proceeding with design synthesis and system validation while considering the complete problem: operations, cost and schedule, performance, training and support, test, manufacturing, and disposal. SE considers both the business and the technical needs of all customers with the goal of providing a quality product that meets the user needs.

Here, the focus is on complex product and systems where many teams of engineers have to integrate their efforts on complex products from first conception to final decommissioning (i.e., "birth to death") [11]. To support this emerging field, the International Council on Systems Engineering (INCOSE) was founded as a professional organization to develop and disseminate the practice of systems engineering [12]. Consequently, many academic departments were founded [12] and along with several archival journals [13, 14]. INCOSE has also sought to standardize systems engineering knowledge to improve communication and “interoperability” between practitioners [10, 11].

One important aspect of this activity has been the trend toward model-based systems engineering (MBSE).

Definition 2 Model-based systems engineering (MBSE) [11]: The formalized application of modeling to support system requirements, design, analysis, verification, and validation activities beginning in the conceptual design phase and continuing throughout development and later life cycle phases.

While Wayne Wymore is often credited with introducing a mathematical foundation for MBSE [15], much of its development arose only recently from the need to manage system complexity as physical systems integrated more and more control, automation, and information technology [16]. It may be viewed as a trend away from a “document-centric approach” to systems engineering toward a “model-centric” approach integrated into all systems engineering processes [16]. At the heart of this initiative has been the development of several modeling standards, most notably the Systems Modeling Language (SysML) [17, 18]. While these are primarily graphical in nature, they directly support the integration of quantitative models.

1.1.3 The Emergence of Engineering Systems

The maturation of Axiomatic Design and MBSE as engineering design methodologies and theories into the arena of large complex systems is timely. In the twentieth century, individual technology products such as the generator, telephone, and automobile were connected to form many of the large-scale infrastructure networks we know today: the power grid, the communication infrastructure, and the transportation system [19]. Over time, these networked systems developed even more interactions while continuing to incorporate many new technology artifacts (e.g., renewable energy, smart phones, and electric vehicles). Naturally, this meant greater complexity, not just because of the greater interaction within these systems, but also because of the presence of an expanding heterogeneity of functionality. Furthermore, these already large-scale, complex, network systems began to develop interactions between themselves in what is now called *systems-of-systems* [20, 21]. The “smart grid” [22], the energy–water nexus [23], the electrification of transport [24] are all good examples where one network system has fused with another to form a new and much more capable system. This trend is only set to continue. The energy–water–food nexus [25] fuses three such systems, and the recent interest in smart cities [26] provides a platform upon which to integrate all of these efforts. This work classifies such systems as engineering systems:

Definition 3 Engineering system [19]: A class of systems characterized by a high degree of technical complexity, social intricacy, and elaborate processes aimed at fulfilling important functions in society.

As engineering systems have evolved, so too, must the role of the engineer within them [19]. The scope of engineering systems, and in particular systems-of-systems, often spans the traditional borders of individual engineering disciplines (e.g., mechanical, electrical, civil, chemical). Furthermore, as engineering systems become ever more ubiquitous and intertwined with daily life, the requirements that they fulfill must also grow and diversify. Therefore, engineering systems should not be viewed in terms of cost and quality of function alone but also include a full taxonomy of system requirements (see Fig. 1.4). These "requirements" are not just of the traditional type where a single client contractually expects specific line items from the engineer; rather, engineering systems requirements also take the form of policies, regulations, and standards where engineers are one of many public and private stakeholders that help to shape the planning and operation of the engineering system in the present and the future [19]. Engineering design methodologies and theories, at their current stage of development, and when interpreted formally and strictly, are likely inadequate for engineering systems. However, they are likely to provide the mental constructs and models that may serve as foundations for coherent methodological developments.

1.1.4 Classification and Characterization of Engineering Systems

The challenge of developing consistent methodological foundations for engineering systems is formidable. Consider the engineering systems taxonomy presented in Table 1.1 [19]. It classifies engineering systems by five generic functions that fulfill human needs: (1) transform, (2) transport, (3) store, (4) exchange, and (5) control. On another axis, it classifies them by their operands: (1) living organisms (including people), (2) matter, (3) energy, (4) information, and (5) money. This classification presents a broad array of engineering domains that must be consistently treated. Furthermore, these engineering systems are at various stages of development and will continue to do so for decades, if not centuries. And so the field of engineering systems must equally support design synthesis, analysis, and resynthesis while supporting innovation, be it incremental or disruptive. Axiomatic Design and MBSE present themselves as promising engineering design methodologies and theories with the flexibility to address the breadth of different engineering systems.

Table 1.1 A classification of engineering systems by function and operand [19]

Function/Operand	Living organisms	Matter	Energy	Information	Money
Transform	Hospital	Blast furnace	Engine, electric motor	Analytic engine, calculator	Bureau of Printing and Engraving
Transport	Car, airplane, train	Truck, train, car, airplane	Electricity grid	Cables, radio, telephone, and Internet	Banking Fedwire and Swift transfer systems
Store	Farm, apartment complex	Warehouse	Battery, flywheel, capacitor	Magnetic tape and disk, book	US Bullion Repository (Fort Knox)
Exchange	Cattle auction, (illegal) human trafficking	eBay trading system	Energy market	World Wide Web, Wikipedia	London Stock Exchange
Control	US constitution and laws	National highway traffic safety administration	Nuclear regulatory commission	Internet engineering task force	US Federal Reserve

1.1.5 Methodological Challenges in Engineering Systems

Across the broad array of engineering systems applications, several important and recurring themes have emerged as methodological challenges. One of these is the required attention to life cycle properties or "ilities" [19].

Definition 4 Life Cycle Properties ("ilities") [19]: Desired properties of systems, such as flexibility or maintainability (usually but not always ending in "ility"), that often manifest themselves after a system has been put to its initial use. These properties are not the primary functional requirements of a system's performance, but typically concern wider system impacts with respect to time and stakeholders that are embodied in those primary functional requirements. The "ilities" do not include factors that are always present, including size and weight (even if these are described using a word that ends in "ility").

Life cycle properties usually have an emergent nature that cannot be predicted from individual system components. Therefore, understanding the factors that enable these properties is fundamental to engineering systems as they develop over many years.

A second engineering systems challenge is in their cyber-physical nature [10, 19]. Engineering systems, as expected, are largely physical in order to realize their important primary function. In the meantime, by virtue of their size and complexity, they require many decision-making components, be they human or automated control. Designing, planning, and controlling such large-scale cyber-physical systems go well beyond traditional control theory research. It now includes more fundamental questions that balance centralization versus distribution, automation versus human decisions, and authority versus cooperative negotiation bounded within a context of human stakeholders and actors.

Finally, a third engineering systems challenge is managing the integration of hetero-functional systems-of-systems. As mentioned in Sect. 1.1.3, well-known engineering systems such as those that deliver electricity, information, natural gas, water, transportation, and healthcare are fusing [19, 27]. In the meantime, engineering education remains organized into departments along these well-established and often self-reinforcing silos [19]. Very few universities prepare engineers to span two or more integrated engineering systems; even fewer do so while addressing the fundamental questions into life cycle properties and cyber-physical systems. Efforts to address these three challenges require a methodological base founded within engineering design methodologies and theories such as Axiomatic Design and MBSE.

1.1.6 Contribution

As the first chapter in this book on the application of Axiomatic Design to large complex systems, it seeks to introduce the fundamentals of Axiomatic Design as a conceptual foundation for subsequent chapters. These include complex products, buildings, and manufacturing systems. As a group, they contain many of the same challenges found in other application domains for systems research. Therefore, the chapter also seeks to relate Axiomatic Design's key concepts to those found in current MBSE techniques including SysML. As the discussion is of an introductory nature, the chapter draws heavily from several well-established texts in Axiomatic Design [2, 3], MBSE [17, 18, 28, 29], and engineering systems [19]. It, also, seeks to clarify nuances within and between these texts that can cause confusion or misinterpretation. Finally, the chapter returns to the engineering systems discussion provided in this introduction. It concludes with directions in which Axiomatic Design has served to address the methodological challenges facing engineering systems today.

1.1.7 Chapter Outline

The remainder of the chapter proceeds as follows. Section 1.2 introduces Axiomatic Design and its relationship to MBSE in terms of four domains of engineering design: stakeholder requirements, functional architecture, physical architecture, and process domains. Next, Sect. 1.3 focuses specifically on the design synthesis and analysis of the allocated architecture as the mapping between the functional and physical architectures. Next, Sect. 1.4 discusses the relationship between these domains with a focus on Axiomatic Design's Independence and Information

Axioms. Section 1.5 goes on to address how Axiomatic Design manages the complexity of systems via a dual functional and physical system hierarchy. Section 1.6 then highlights potential applications of Axiomatic Design in the development of engineering systems. The chapter is brought to a close in Sect. 1.7.

1.2 Four Domains in the Engineering Design of Systems

From an Axiomatic Design perspective, the engineering design of systems consists of four domains. They are defined here as follows drawing upon consistent definitions from both the MBSE and Axiomatic Design literature.

Definition 5 Stakeholder Requirements Domain [17]: A collection of statements that describe the system properties and behaviors that all stakeholders need to be met.

Definition 6 Functional Architecture Domain [28]: A logical model of a functional decomposition plus the flow of inputs and outputs to which input/output requirements have been traced. It constitutes the system behavior or function.

Definition 7 Physical Architecture Domain [28, 30]: The components of a system and the relationships among them. It constitutes the system form.

Definition 8 Process Architecture Domain [3]: The set of processes and their relationships that characterize how the physical architecture is generated or produced.

These four domains are sequentially mapped one onto the next as shown in Fig. 1.1. Motion from left to right represents an engineer's synthesis activity from "what needs to be achieved" to "how it is to be achieved" [3]. Motion from right to left represents an engineer's analysis activity which supports validation and verification. The first three of these domains are consonant with the "Vee" in traditional

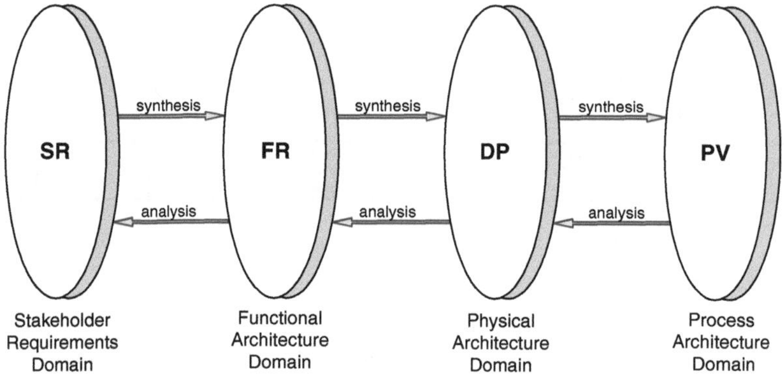

Fig. 1.1 Four domains in the engineering design of systems—an axiomatic design perspective (adapted from [3])

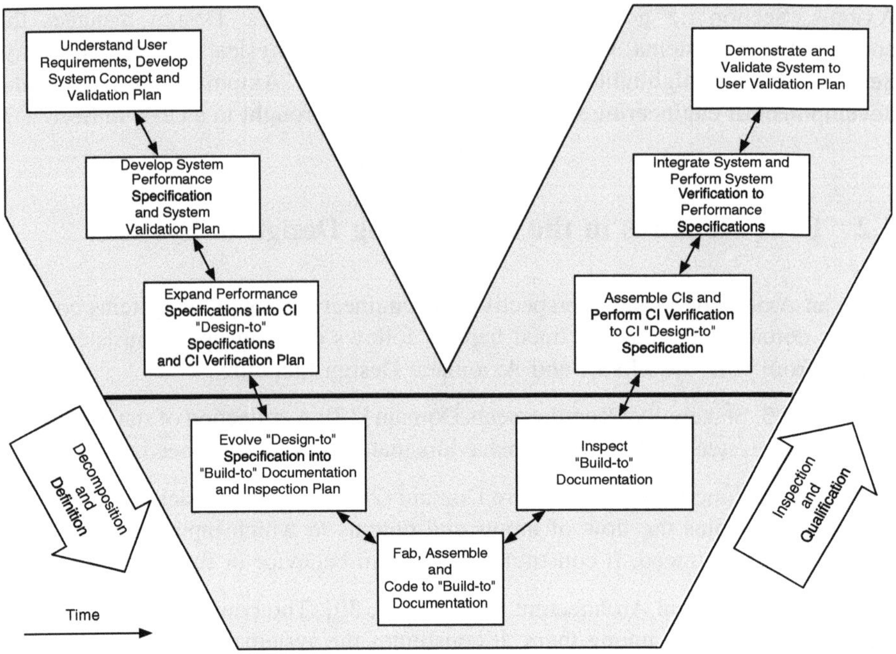

Fig. 1.2 Traditional top-down systems engineering "Vee" (adapted from [28, 31])

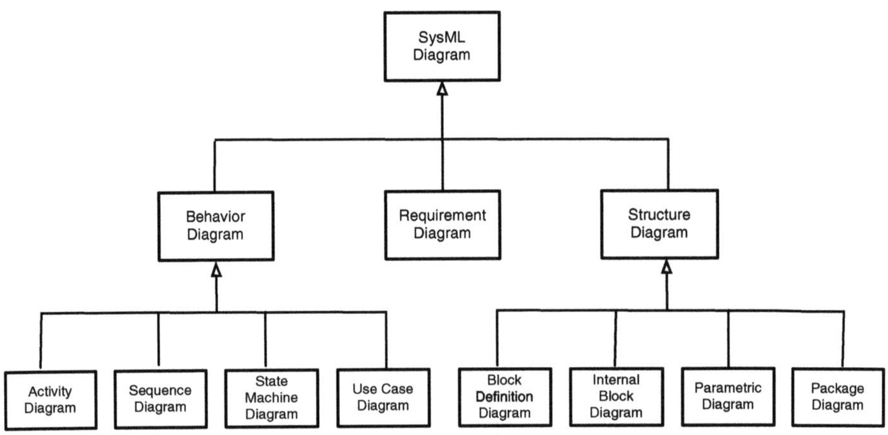

Fig. 1.3 Taxonomy of SysML diagrams (adapted from [17, 18])

top-down systems engineering depicted in Fig. 1.2 [28, 31]. Figure 1.1's synthesis path is consonant with Fig. 1.2's left half describing decomposition and definition. Meanwhile, Fig. 1.1's analysis path is consonant with Fig. 1.2's right half describing integration and qualification. Furthermore, Fig. 1.3 shows that the

SysML supports the engineering design domains with three classes of diagrams: requirements, behavior, and structure. Each of these domains is now described in turn.

1.2.1 Stakeholder Requirements Domain

In Axiomatic Design, the stakeholder requirements domain is more often called the customer domain in recognition of AD's roots in the engineering design of products with customers as sole stakeholders. The elements of the domain are called customer needs **CN** [3]. Here, the term stakeholder requirements domain is used instead to address engineering systems' multiple stakeholders. Similarly, the domain is populated with stakeholder requirements **SR** [28].

Definition 9 Stakeholder Requirements [28]: Statements by the stakeholders about the system's capabilities that define the constraints and performance parameters within which the system is to be designed. These stakeholders' requirements focus on the boundary of the system in the context of these mission requirements, are written in the stakeholders' language, are produced in conjunction with the stakeholders of the system, and are based upon the operational needs of these stakeholders.

The main challenge with the stakeholder requirements domain is that the "voice" of the stakeholder is not that of the engineer. Therefore, the engineer must work with all stakeholders to determine a complete set of stakeholder requirements [3, 28]. These are then used to "translate" and derive a set of system requirements $\mathcal{SR}$ in an engineering language.

Definition 10 System Requirements [28]: A translation (or derivation) of the originating requirements into engineering terminology.

This requirements engineering process is usually completed before the rest of the synthesis path in Fig. 1.1 can continue. That said, subsystem and component requirements may be derived at a later stage to support internal delegation or external subcontracting [28]. The interested reader is referred to several dedicated texts on requirements engineering [32–35]. Buede provides a relatively concise treatment that consists of seven steps [28]:

1. Develop the operational concept
2. Define the system boundary
3. Develop the system objectives hierarchy
4. Develop, analyze, and refine requirements (stakeholders and system)
5. Ensure requirements feasibility
6. Define the qualification system requirements
7. Obtain approval of system documentation

Definition 11 Operational Concept [28]: A vision for what the system is (in general terms), a statement of mission requirements, and a description of how the system will be used. The shared vision is based on the perspective of the system's stakeholders of how the system will be developed, produced, deployed, trained, operated and maintained, refined, and retired to overcome some operational problem and achieve the stakeholders' operational needs and objectives. The mission requirements are stated in terms of measures of effectiveness. The operational concept includes a collection of scenarios (one or more for each group of stakeholders in each relevant phase of the system's life cycle).

Definition 12 Objectives Hierarchy [28]: A hierarchy of objectives that are important to the system's stakeholders in a value sense, that is, the stakeholders would (should) be willing to pay to obtain increased performance (or decreased cost) in any one of these objectives. It is also the definition of the natural subsets of the fundamental objective into a collection of performance requirements.

Stakeholder and system requirements may be classified as shown in Fig. 1.4. Using a requirements classification structure serves to organize requirements (especially in large systems) so as to avoid conflicts and/or duplication. The system requirements include the functional requirements as a subset which also appear later in the functional architecture domain (Sect. 1.2.2). Note that in Axiomatic Design,

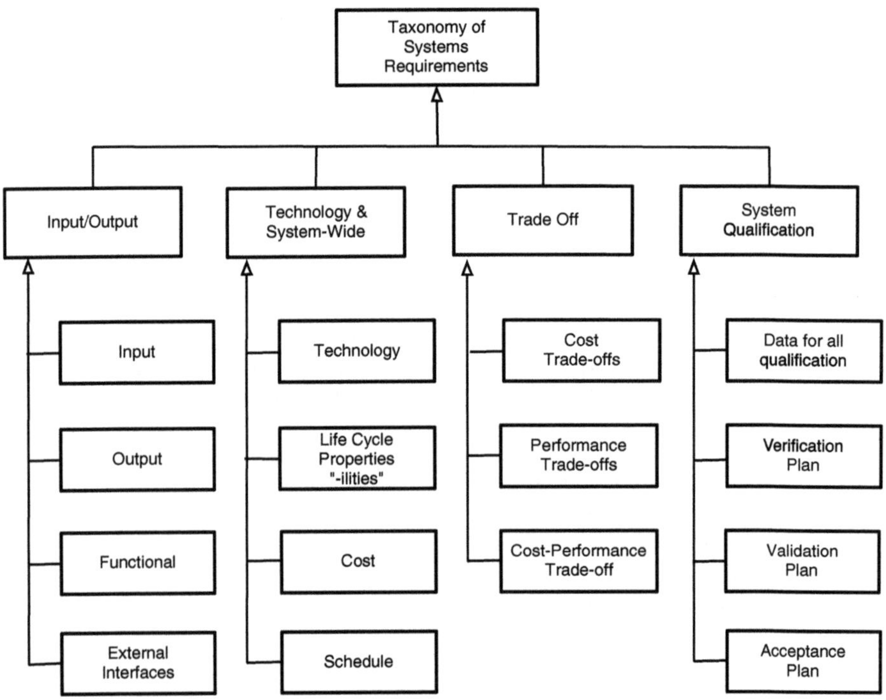

Fig. 1.4 Requirements classification in model-based systems engineering (derived from [28])

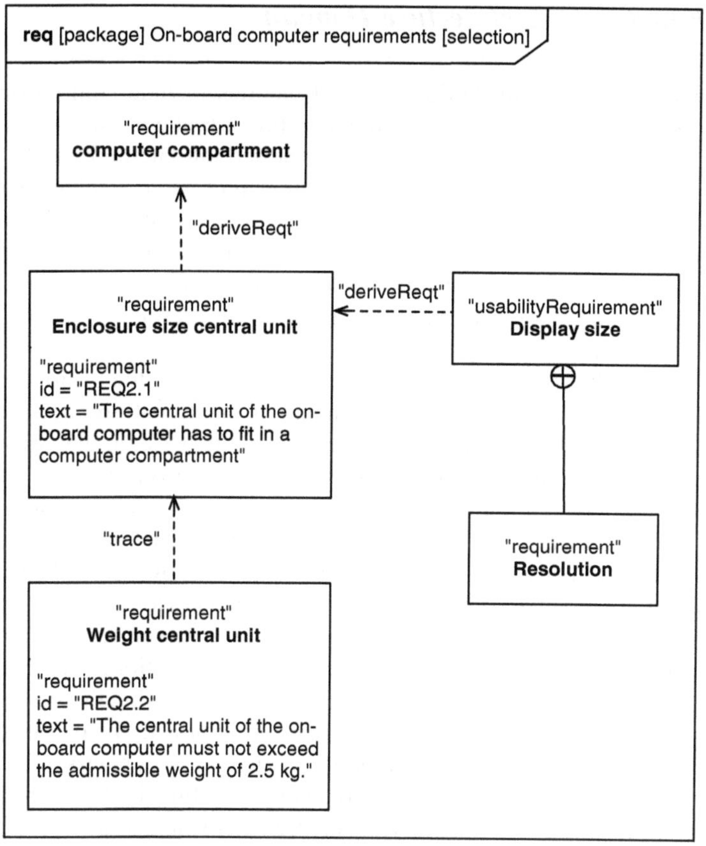

Fig. 1.5 An example requirements diagram in SysML

this requirements taxonomy is traditionally reduced to customer needs **CN**, functional requirements **FR**, and constraints **C**. More recently, Thompson adds non-functional requirements **nFR**, selection criteria **SC**, and optimization criteria **OC** to the taxonomy [36] and highlights common errors in their misclassification [37].

Requirements engineering documentation is well supported by the requirements diagram in SysML [17] (see Fig. 1.5). Derived requirements serve during the synthesis path as the primary relationship within the requirements domain up to and including the functional requirements. This is similar to the "House of Quality" in Quality Function Deployment methodologies [38–40]. During the analysis path, the "verified relationship" links requirements to functions (in test cases) and the "satisfied relationship" links requirements to components.

As expected, the stakeholder requirements domain is primarily described in text with some numerical specifications. It is only after several steps of engineering synthesis and modeling can more mathematical treatments begin to be applied.

1.2.2 Functional Architecture Domain

Most engineering design methodologies and theories include some form of functional architecture domain [4, 29]. As shown in Fig. 1.6, it consists of functions that are arranged in serial or in parallel and may be nested into hierarchies. As described in Sect. 1.1.4, these functions may transform, transport, store, exchange, or control their operands which in turn may be classified as material, energy, information, money, or people [19]. By convention, each function is defined as a transitive verb stated in the third person singular followed by its associated object/operand. It must also be defined in a solution-neutral way that does not presuppose the technologies within the physical architecture [3, 28].

There is considerable variation in nomenclature across the Axiomatic Design and MBSE literature in regard to the elements of functional architecture domain. Generically, they are called functions. In Axiomatic Design, the functions are instead called *functional requirements*. Both of these conventions are used in this chapter. The AD convention serves several logical purposes. First, it emphasizes that what the system must do, its system function, is not defined in a vacuum but rather is the logical consequence of the stakeholder requirements identified previously. Second, it recognizes that functional requirements viewed at a high level are just as binding as when they are decomposed to a lower level. Third, in not distinguishing between a functional requirement and a function, Axiomatic Design is not distinguishing between what the system must do and what the system does (as designed). After all, they should be the same. In contrast, Buede considers that functional requirements have a dual purpose: first as a subclass of the system requirements and second as (only) the top level of the functional architecture [28]. This serves to link the two domains with common elements but distinguish between functional requirements and the rest of the system functionality. To complicate matters, the term "process" is sometimes used in place of function in MBSE [19, 27,

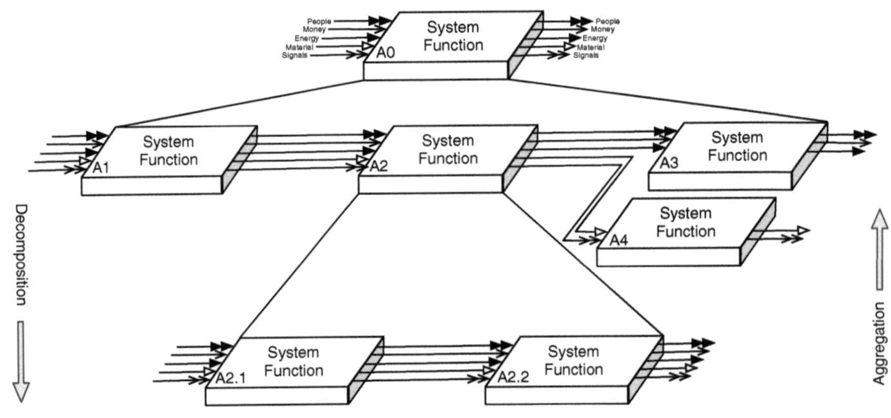

Fig. 1.6 A functional architecture with parallel, serial, and nested interactions

29, 41, 42]. While this practice is common, it ultimately confuses the differences between the functional architecture domain and the process architecture domain in the Axiomatic Design framework in Fig. 1.1.

Generation of the functional architecture is very much a synthetic—"forward-engineering"—activity. In order to be successful, it requires that the set of functions be mutually exclusive and collectively exhaustive [3]. There is a general consensus in engineering design that overlapping system functions will cause downstream design errors [3, 28]. Meanwhile, generating an exhaustive set of functions begins from the previously defined operational concept and high-level functional requirements [28]. Several notable functionality templates have been developed over the years to spur designer creativity and prevent unintentional omission of functionality [43–45]. Again, the identification of solution-neutral functions supports maximally innovative physical designs downstream [3].

That said, it is not uncommon to generate the functional architecture as an analytical—"reverse engineering"—activity. Often, in an engineering systems context, a part or a whole of the system has already been built [19] and the development of the functional architecture is required to determine how to best "evolve" the system to the next stage of development. Furthermore, well-known functions have tried-and-true physical solutions which may be reimplemented successfully as part of design patterns or in novel configurations. It is often unnecessary to "reinvent the wheel." Therefore, using a reverse engineering analytical mind-set, functions can be identified as an abstraction of existing components in the physical architecture. For example, I-beams in buildings support weight, and railways transport trains. In reality, however, the functional architecture, at its multiple levels of decomposition, must be developed in parallel with the physical architecture via the allocated architecture [3, 28] and will be discussed in detail from an Axiomatic Design perspective over several sections.

The development of the functional architecture is well supported in SysML [17, 18]. As shown in Fig. 1.3, this includes four diagrams that provide complementary views of the overall system behavior at different levels of engineering design detail. These include activity, state machine, use case, and sequence diagrams. While all of these are useful in detailed design, the first three have common applications in conceptual design prior to the synthesis of the physical architecture. These include:

- **Activity diagrams**—support general purpose functional modeling that closely resemble Fig. 1.6. Functions in this diagram are called actions.
- **State machine diagrams**—support the organization of functionalities into modes of operation. Functions in this diagram are implicit to what happens during a particular operating state.
- **Use case diagrams**—support the interactions with external entities such as people and organizations. Functions in this diagram are called use cases.

While each of these diagrams has formal semantics, their appropriate use is often daunting for novice systems engineers beginning a conceptual design. As a partial alternative, the object process methodology supports system function models in a manner that resembles simplified activity diagrams [41, 42]. In both models, it is important to distinguish the flow of power from the other types of operands. This is because the directionality of power flow does not fully coincide with the direction of *dynamic causality* [46]. For example, a voltage source will impose a voltage on downstream functions (e.g., loads), but they in return (e.g., by their impedance) will impose the required current.

The functional architecture domain also very much lends itself to mathematical description. From a static perspective, functional elements (at any given level of abstraction) can be organized into a directed graph and its associated adjacency matrix [47]. Alternatively, adjacency matrices have been called "N^2" diagrams or design structure matrices within the engineering systems literature [48].

Definition 13 Directed Graph (digraph) [47]: D consists of a collection nodes B and a collection of arcs E, for which we write $D = (B, E)$. Each arc $e = \langle b_1, b_2 \rangle$ is said to join node $b_1 \in B$ to another (not necessarily distinct) node b_2. Vertex b_1 is called the tail of e, whereas b_2 is its head.

Definition 14 Adjacency matrix [47]: A is binary and of size $\sigma(B) \times \sigma(B)$, and its elements are given by:

$$A(y_1, y_2) = \begin{cases} 1 & \text{if } \langle b_{y_1}, b_{y_2} \rangle \text{ exists} \\ 0 & \text{otherwise} \end{cases} \tag{1.1}$$

where the $\sigma()$ gives the size of a set.

Here, the functions would represent the nodes and would be interconnected with the lines found in activity diagrams.

Example 1 Consider the second level of abstraction of the functional architecture in Fig. 1.6 as a directed graph. It's (functional architecture) adjacency matrix is:

$$A_f = \begin{bmatrix} 0 & 1 & 0 & 0 \\ 0 & 0 & 1 & 1 \\ 0 & 0 & 0 & 0 \\ 0 & 0 & 0 & 0 \end{bmatrix} \tag{1.2}$$

From a dynamic perspective, and perhaps one of the central tasks in traditional engineering effort, functional elements can be replaced by their mathematical function equivalents called device models. Given inputs $\mathbf{U}$, outputs $\mathbf{Y}$, and state variables $\mathbf{X}$, device models with algebraic equations take the form $\mathbf{Y} = g(\mathbf{X}, \mathbf{U})$. Device models with differential equations take the form $\dot{\mathbf{X}} = f(\mathbf{X}, \mathbf{U})$ where the output $\mathbf{Y}$ is algebraically related to the states and inputs $\mathbf{Y} = g(\mathbf{X}, \mathbf{U})$ [49]. Device models with difference equations take the form $\mathbf{X}_{k+1} = f(\mathbf{X}_k, \mathbf{U}_k)$ where the output $\mathbf{Y}_k$ is algebraically related to the states and inputs $\mathbf{Y}_{k+1} = g(\mathbf{X}_k, \mathbf{U}_k)$ [50]. These device models are then aggregated via the functional architecture and may then be simulated to quantitatively understand aggregate *system behavior* and overall system performance.

1.2.3 Physical Architecture Domain

The physical architecture domain embodies the engineering system and is made up of mutually connected components. Again, there is considerable variation in nomenclature across the Axiomatic Design and MBSE literature in regard to the elements of the physical architecture domain. Generically, they are called components which may be aggregated into modules, resources, and subsystems. Furthermore, they may be characterized by attributes such as size, shape, and color. In Axiomatic Design, both attributes and their associated components at any level of aggregation are called design parameters **DP**. The rationale for the AD convention is discussed later in Sect. 1.4. Both of these conventions are used in this chapter.

In traditional top-down systems engineering and Axiomatic Design, the generation of the physical architecture proceeds as a synthesis activity in concert with the allocated architecture [3, 28] to be discussed in Sect. 1.3. In contrast, bottom-up design methodologies generate the physical architecture as an analytic activity presupposing the set of components and their associated technologies [28].

The development of the physical architecture is well supported in SysML [17, 18]. As shown in Fig. 1.3, this includes four diagrams that provide complementary views of the overall system structure at different levels of engineering design detail. These include the block definition, internal block, parametric, and package diagrams. While all of these are useful in detailed design, the block definition diagram is most often applied in conceptual design before detailed analytical equations are available. Figure 1.7 shows an example of a SysML block diagram which includes many of the systems thinking concepts related to the conceptual design of physical architectures. Association links represent interconnected components. They would ultimately realize function output as material, energy, information, people, or money. Composition links represent whole–part relationships. Classification links represent generalization–specialization relationships. It is important to note that the SysML block definition diagram either models the generic or the instantiated physical architecture but not both at the same time.

Definition 15 Generic Physical Architecture [28]: A description of the partitioned elements of the physical architecture without any specification of the performance characteristics of the physical resources that comprise each element.

Definition 16 Instantiated Physical Architecture [28]: A generic physical architecture to which complete definitions of the performance characteristics of the resources have been added (including the number of each type of resource).

For example, a block definition diagram can represent generic relationships between roles in an organization chart or they can instantiate those roles to the specific people that hold them. In the precursor to SysML, the UML 2.0 specification reserved class diagrams for the generic physical architecture and the object diagram for the instantiated physical architecture [52].

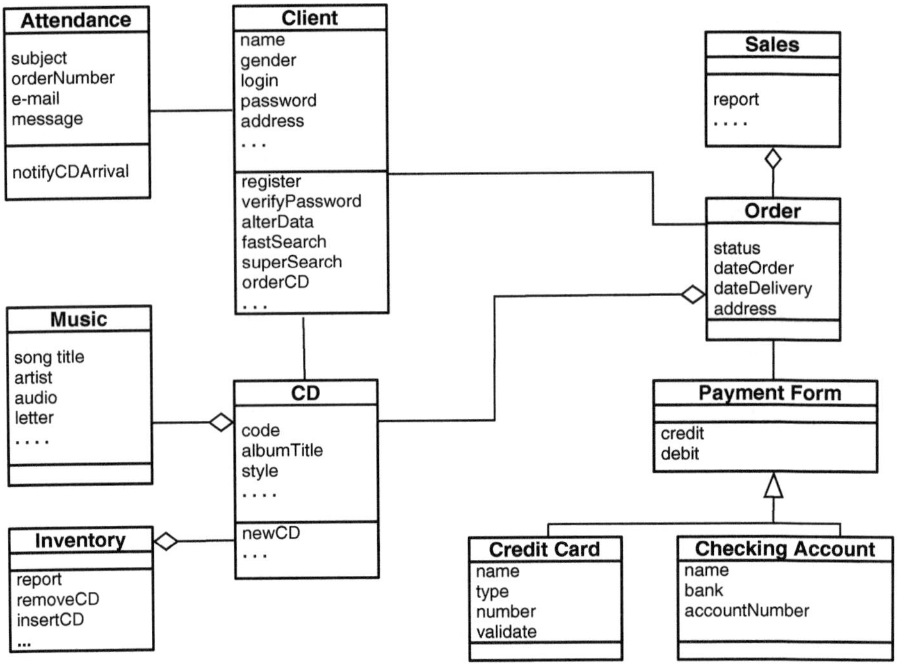

Fig. 1.7 An example SysML block definition diagram (adapted from [51])

Much like the functional architecture domain, the physical architecture domain also lends itself to mathematical description. From a static perspective, physical elements (at any given level of abstraction) can be organized into a directed graph and its associated adjacency matrix [47].

Example 2 Consider the second level of abstraction of the physical architecture in Fig. 1.7 as a graph. It's (physical architecture) adjacency matrix is:

$$A_p = \begin{bmatrix} 0 & 1 & 0 & 0 & 0 \\ 1 & 0 & 1 & 1 & 0 \\ 0 & 1 & 0 & 0 & 0 \\ 0 & 1 & 0 & 0 & 1 \\ 0 & 0 & 0 & 1 & 0 \end{bmatrix} \tag{1.3}$$

From a dynamic perspective, the evolution of a physical architecture remains a subject of cutting edge research.

1.2.4 *Process Architecture Domain*

Recalling Definition 8: The process architecture domain is composed of the set of processes and their relationships that characterize how the physical architecture is to be produced. In many ways, it strongly resembles the functional architecture domain [53]. Each process is defined as a transitive verb stated in third person singular followed by its associated object/operand. Indeed, recent work on the Axiomatic Design of manufacturing systems (independent of the products that they produce) treats manufacturing processes as elements of the manufacturing system's functional architecture domain [54–56]. Therefore, there is significant similarity in how the two domains are modeled in general as well as in SysML.

Despite the similarities between the two domains, their respective roles in MBSE and Axiomatic Design are fundamentally different. In Axiomatic Design, as shown in Fig. 1.1, the process domain occurs as a synthesis *after* the physical architecture has been developed. In contrast, MBSE does not explicitly treat downstream "manufacturability" as a fourth domain. It is possible to include manufacturing requirements and constraints as part of the requirements engineering process in the stakeholder requirements domain. However, such an approach assumes that the physical architecture, its required manufacturing processes, and the stakeholders that own them are already known to some degree in advance.

1.2.5 Multi-domain Mapping in the Engineering Design of System

With Axiomatic Design's four domains introduced, it becomes clear that the engineering design of large systems is particularly complex. As shown in Figs. 1.1 and 1.2, not only must engineers proceed sequentially from one domain to the next to synthesize the system, they must also retrace those steps backwards to analytically validate and verify the original intent of synthesis. This is a tremendous task of information management and requires the three activities identified in Table 1.2.

In order to conceptualize this undertaking, graph theory again proves to be useful. The literature has proposed the engineering systems matrix (ESM) (Fig. 1.8) as a form of multi-domain matrix [57, 58]. In addition to the four domains of engineering design in Fig. 1.1, it adds the system drivers domain and the stakeholder domain.

Definition 17 System Drivers Domain [57, 58]: A representation of the non-human portion of the environmental domain is composed of the set of all non-human components that act or are acted on by the system. The system drivers can include

Table 1.2 Information management tasks in the engineering design of systems

• Element information: All of the elements in all of the domains must be systematically identified
• Intra-domain information: The links between elements within a given domain must be systematically identified
• Inter-domain Information: The links between elements across two domains must be systematically identified

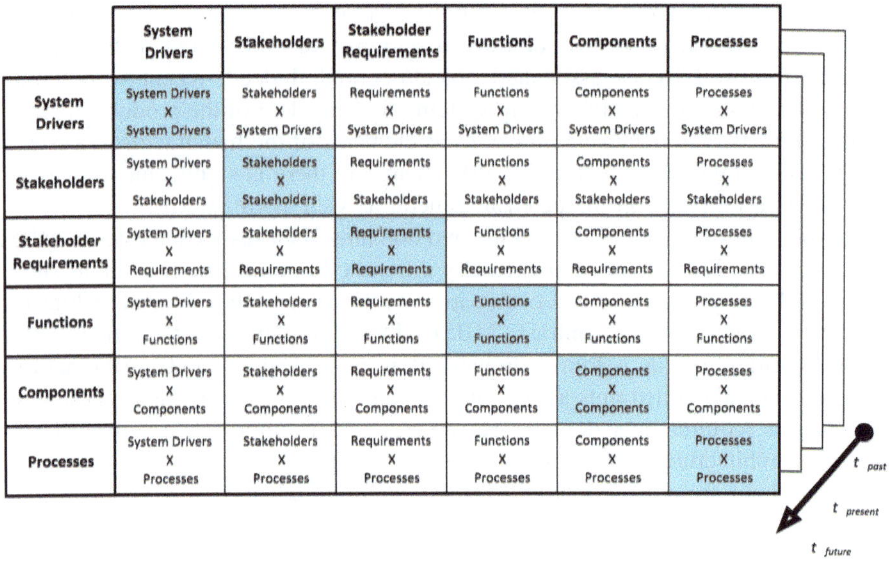

Fig. 1.8 Engineering systems multiple-domain matrix [28, 31]

the economic, political, and technical influences that constrain, enable, or alter the characteristic of components in the system.

Definition 18 Stakeholder Domain [57, 58]: A social network of stakeholders in an engineering system which may be classified as external or internal. The external stakeholders constitute the remaining portion of the environmental domain and consist of the human entities that affect or are affected by the system but that do not control components within the system boundary. Likewise, internal stakeholders are the human entities that contribute to the goals of the system and control components within the system.

This addition serves to recognize that an engineering system exists within a context in which multiple system drivers are influencing its conception, planning, and operation. It also recognizes the presence of multiple stakeholders which may indeed pose conflicting or misaligned stakeholder requirements. The elements of these six domains combined essentially form the nodes of the underlying engineering systems graph. The nonzero elements of the ESM indicate the presence of links between them, be they within a given domain or across multiple domains. The ESM in Fig. 1.8 is naturally highly sparse, but it nevertheless serves as a tool to manage the information highlighted in Table 1.2. Finally, it is important to recognize that the ESM can be viewed as "snapshots" in time, either as the system is designed, or as it is operated.

From the lens of the ESM, Fig. 1.1 now appears highly structured. It addresses the bottom four blocks of the main block diagonal. Its mappings address their first off-block diagonals immediately above and below the main block diagonal. The other interactions are intentionally omitted so as to avoid needless complexity in the engineering design process. In other words, an engineering design process that specifically eliminates needless interactions is one that can allow the system to be designed, planned, and operated more efficiently.

1.3 The Allocated Architecture: Design Synthesis and Analysis

Much of the focus of Axiomatic Design has gone specifically to the mapping between the functional and the physical domains [3]. In MBSE, this is often called the allocated architecture [28].

Definition 19 Allocated Architecture [28]: A complete description of the system design, including the functional architecture allocated to the physical architecture; derived input/output; technology, system-wide, trade-off, and qualification requirements for each component; an interface architecture that has been integrated as one of the components; and complete documentation of the design and major design decisions.

There are several ways to model the allocated architecture in model-based systems engineering. Two of these are described here. The first has already been shown in Fig. 1.7 where a method can be executed as a function allocated to a given class. For example, "verify password" is such a method for the "Client" class. The second approach is shown in Fig. 1.9. Here, an activity diagram has been partitioned into "swim lanes" so that a given action is allocated to the physical

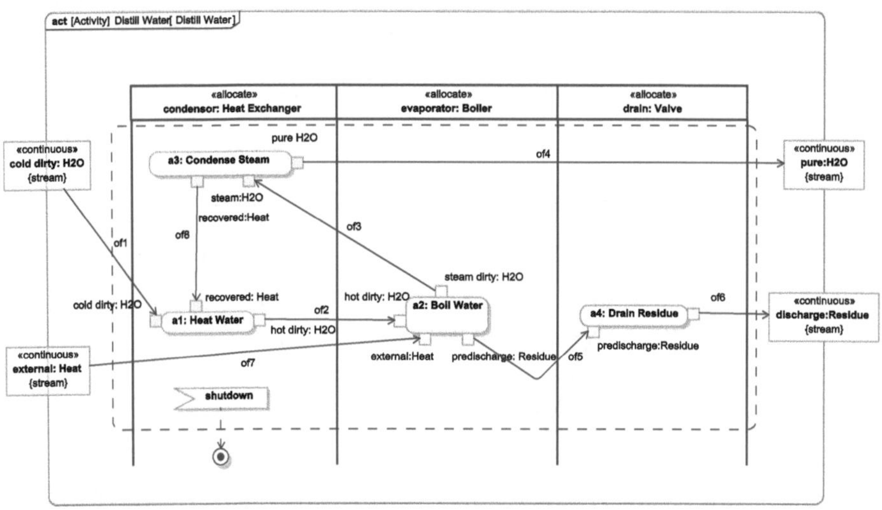

Fig. 1.9 For example, system processes, resources, and allocation [27]

component that executes it. In such a way, the functional and physical architecture domains are closely tied.

While the other activities in the engineering design of systems are certainly not to be neglected, there are several reasons for Axiomatic Design's specific focus on the allocated architecture. First, the allocation of function to form represents in engineering design the "*the moment of synthesized embodiment.*" In other words, prior to that moment, the design only had a set of functional requirements, but afterward, the design now includes a set of physical elements or design parameters **DP** that now describe a physically embodied way to achieve these functions. Second, this allocation of form to function is done *quantitatively* (rather than graphically) to the level of mathematical detail that is available at the time. Third, the nature of this allocation, as later sections describe, ultimately drives many aspects of an engineering system including its life cycle properties. The successful transition from the functional to the physical domain requires effective design synthesis and analysis.

1.3.1 Design Synthesis

Engineering discussions on design synthesis are often neglected. Casually speaking, a designer's "creativity" is engaged and "voila" innovation happens! However, a rigorous understanding of design synthesis must root itself into the formal foundations of philosophy, logic, and linguistics. After all, it is a process which brings a *system model* $\mathcal{M}$ into being from the mind(s) of its designer(s). In this regard, the Ullmann triangle [59] shown in Fig. 1.10 proves to be a useful construct. It derives from fundamental works [60, 61] upon which much of modern linguistics is based. In the left-hand triangle, a *domain conceptualization* C is an immaterial entity that only exists in the mind of a community of users of a *language* $\mathcal{L}$ [62]. As such, it is a mental abstraction of a *real domain* $\mathcal{D}$ (i.e., as it is observed in the natural sciences) [62]. Furthermore, the language $\mathcal{L}$ is composed of set of modeling

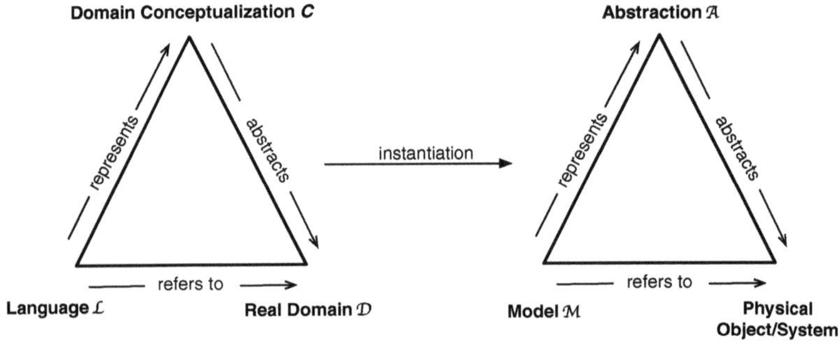

Fig. 1.10 The role of the Ullman triangle in design synthesis

primitives which collectively represent the domain conceptualization $\mathcal{C}$ [62]. The right-hand triangle instantiates the one the left. The abstraction $\mathcal{A}$ is an instance of the domain conceptualization $\mathcal{C}$ [62] and now abstracts a system model $\mathcal{M}$ as the output of a design process. Such a process is not direct. It must return to the domain conceptualization $\mathcal{C}$, its representing language $\mathcal{L}$, and its associated modeling primitives. The system model $\mathcal{M}$ then follows as an instance of the language $\mathcal{L}$.

One practical challenge in the engineering systems field is that modeling primitives are domain specific. For example, the topic of motion in machine design is often treated with primitives such as linkages, cams, and gear trains [63]. Similarly, the design of dynamic systems across multiple energy domains has lead to primitives such as generalized capacitors, inductors, resistors, transformers, and gyrators [46]. In business dynamics, stocks and flows are often used as primitives [64]. More broadly, the engineering systems literature has recently developed simple but encompassing taxonomies of function and form [19]. The object process modeling language, as the name suggests, uses objects and processes as primitives [42].

In all cases, the *domain appropriateness* and *comprehensibility* of a language can be formally assessed. Guizzardi writes [62]: "In order for a model $\mathcal{M}$ to faithfully represent an abstraction $\mathcal{A}$, the modeling primitives of the language $\mathcal{L}$ used to produce $\mathcal{M}$ should faithfully represent the domain conceptualization $\mathcal{C}$ used to articulate the represented abstraction $\mathcal{A}$." A formal assessment of a language $\mathcal{L}$ yields the properties of *soundness*, *completeness*, *lucidity*, and *laconicity* which are graphically depicted in Fig. 1.11 and formally defined [65].

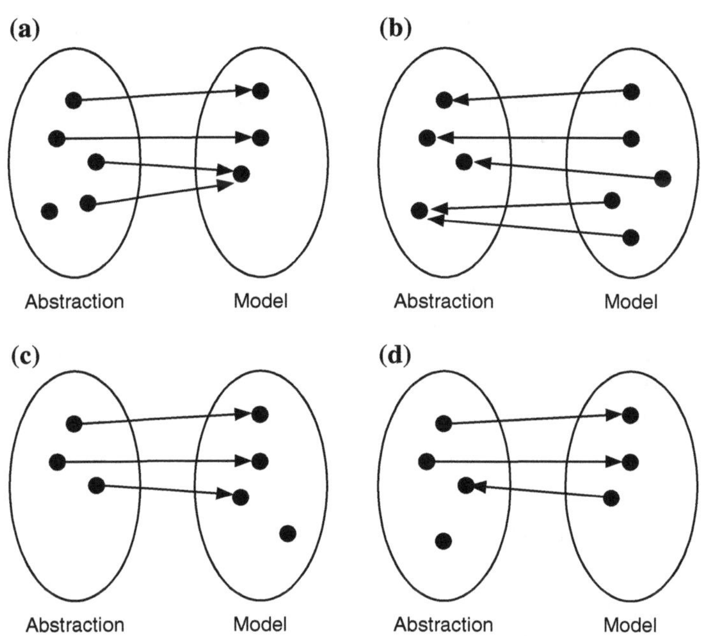

Fig. 1.11 Graph theoretical representation of mapping between a model and its abstraction: **a** soundness, **b** completeness, **c** lucidity, and **d** laconicity [65]

Definition 20 Soundness [65]: A language $\mathcal{L}$ is sound w.r.t. to a domain $\mathcal{D}$ iff every modeling primitive in the language has an interpretation in the domain abstraction $\mathcal{A}$.

Definition 21 Completeness [65]: A language $\mathcal{L}$ is complete w.r.t. to a domain $\mathcal{D}$ iff every concept in the domain abstraction $\mathcal{A}$ of that domain is represented in a modeling primitive of that language.

Definition 22 Lucidity [65]: A language $\mathcal{L}$ is lucid w.r.t. to a domain $\mathcal{D}$ iff every modeling primitive in the language represents at most one domain concept in $\mathcal{A}$.

Definition 23 Laconicity [65]: A language $\mathcal{L}$ is laconic w.r.t. to a domain $\mathcal{D}$ iff every concept in the abstraction $\mathcal{A}$ of that domain is represented at most once in the model of that language.

The absence of these properties violates conversational maxims that assume thought is "relevant, clear, unambiguous, brief, not overly informative, and true" [66]. Interestingly, UML [67] has been assessed relatively positively in the context of Figs. 1.10 and 1.11 [62]. Perhaps, this result may serve to provide a theoretical reason for the successful adoption of SysML/UML as an integral part of design synthesis in MBSE.

While the process of conceptualization in Fig. 1.10 is necessary to define design synthesis, it is ultimately insufficient. After all, it must be reconciled with the constrained mapping presented in Fig. 1.1. To this end, the term synthesis may be defined in its philosophical sense.

Definition 24 Synthesis [68]: The third stage of argument in a dialectic which reconciles the mutually contradictory first two propositions of thesis and antithesis.

Therefore, design synthesis can be defined as follows:

Definition 25 Design Synthesis: a synthesis process which reconciles the conceptualization of a set of design parameter primitives $\mathbb{DP}$ (as a thesis) with the satisfaction of a set of functional requirements **FR** (as an antithesis). Mathematically,

$$\mathbf{DP} = f_s(\mathbf{FR}, \mathbb{DP}) \tag{1.4}$$

Furthermore, it is understood that these design parameter primitives are domain appropriate and comprehensible and effectively represent a conceptualization of the designer's experience. Two distinct designers may generate different sets of design parameters **DP** given that they may retain different design parameter primitives $\mathbb{DP}$ in their mental conceptualization.

1.3.2 Design Analysis

Unlike design synthesis, engineering discussions on design analysis are given significantly greater attention. Perhaps this is because, the inputs of design analysis are design parameters. As entities, they are well described, often quantitatively, in the natural sciences which form the roots of the modern engineering science. In contrast, design parameter primitives exist in the ontological sciences which draw from philosophy, logic, and linguistics. Furthermore, while design synthesis requires the reconciliation of design parameter primitives with functional requirements to identify a set of design parameters, design analysis takes the previously identified design parameter information to determine whether they satisfy the functional requirements. In a sense, design synthesis defines the nature of an engineering system/artifact and design analysis refines it.

Axiomatic Design describes design analysis with a design equation:

$$\mathbf{FR} \$ f_a(\mathbf{DP}) \tag{1.5}$$

where $f_a()$ retains the "function of" meaning, and the relatively new symbol \$ means "satisfies" when read from right to left. When the design parameters and functional requirements quantitatively represent the physical quantities of an engineering system, then $f_a()$ comes to represent its associated laws of physics. In such a way, Eq. 1.5 can be rewritten as:

$$\mathbf{FR} = f_a(\mathbf{DP}) \tag{1.6}$$

whose first derivative gives:

$$\Delta\mathbf{FR} = [B]\Delta\mathbf{DP} \tag{1.7}$$

where now the nonzero elements of the *design matrix* $B(i,j)$ highlight the existence of a dependence between an arbitrary $\mathbf{FR}_i$ and an arbitrary design parameter $\mathbf{DP}_j$.[1] It is important to very clearly distinguish $f_s()$ and $f_a()$; while the former describes the designers' mental process of generating the design parameters, the latter describes the laws of physics that relate the now already existing design parameters to their functional requirements. Axiomatic Design does not require the designer(s) to have full knowledge of the mathematical form of $f_a()$ during design synthesis. As Sect. 1.5 later discusses, the knowledge of these mathematical forms may not be fully available during early-stage conceptual design. Instead, its axioms only require the designer(s) to have knowledge of the existence of nonzero elements in B and act accordingly. Graphically, the designer needs only have the intent of allocating a functional element to a physical one as depicted in Fig. 1.9.

[1]Note that many works on Axiomatic Design, including later chapters in this book, simply write $\mathbf{FR} = [B]\mathbf{DP}$ to concisely convey the meaning of Eqs. 1.5–1.7. While this notational shorthand is often sufficient to properly implement Axiomatic Design, it does cloud the small but meaningful differences between the three equations. Furthermore, such a shorthand suggests that the $f()$ in Eq. 1.6 is a linear matrix equation consisting of real numbers when indeed no such restriction is formally required.

1.4 The Independence and Information Axioms

Axiomatic Design was developed out of a need to make the field of design more scientific [2, 3]. In 2001, Suh writes: "The goal of Axiomatic Design is manifold: to make human designers more creative, to reduce the random search process, to minimize iterative trial-and-error process, to determine the best designs among those proposed, and to endow the computer with creative power through the creation of a scientific base for the design field." [3]. These lofty goals brought about a highly intensive and *empirical* research process in which the common elements of "*good*" designs were identified [2, 3]. These common elements were ultimately distilled into Axiomatic Design's two axioms. The interested reader is referred to [2] for further details on the research process used to develop Axiomatic Design. These two axioms, stated today, are as follows:

Axiom 1 The Independence Axiom [2, 3]: This maintains the independence of the functional requirements (FRs).

Axiom 2 The Information Axiom [2, 3]: This minimizes the information content of the design.

Consequently, these axioms have lead to the development of Axiomatic Design's many theorems and corollaries summarized for convenience in the Appendix of this book. Each of these is now discussed conceptually.

1.4.1 The Independence Axiom

The Independence Axiom is a statement that applies as equally to design synthesis as design analysis. Its interpretation in the former requires that the set of functional requirements be *mutually exclusive and collectively exhaustive* [2, 3]. In other words, the requirements engineering process that produces the functional requirements may be viewed as an ontology development activity that produces part of the system's design language. Furthermore, it is very difficult to conceive any synthesis function $f_s()$ that retains its nature as a function when its input domain is neither mutually exclusive nor collectively exhaustive. This agrees with the discussion in Sect. 1.2.2, which made this requirement to avoid downstream design errors.

The Independence Axiom is applied in design analysis through the use of Eq. 1.6 and more specifically the matrix properties of the design matrix B. When B is a diagonal matrix, then the system is said to be *uncoupled.* When B is either a lower triangular matrix or may be converted into a lower triangular matrix by row swapping operations, then the system is said to be *decoupled.* When B does not have either of these two forms, then the system is said to be *coupled.* Uncoupled designs are preferred over decoupled ones. And coupled designs are said to not comply with the Independence Axiom. Therefore, the application of the Independence Axiom has a component that allows the synthesis function f_s to exist, and then, it guides the designer(s) through an analysis step to verify if the resulting laws of physics describe an uncoupled or coupled system.

Researchers, educators, and practitioners often experience several misconceptions as they convey the Independence Axiom to their peers. The most notable of these misconceptions is in the concept of coupling. As Sect. 1.2.2 has described, the functional domain contains couplings that occur from the sequential relationship between functions. The MBSE literature often calls these couplings *interactions* [69]. They are formally modeled by the existence of nonzero elements in the functional domain's adjacency matrix. Similarly, and as Sect. 1.2.3 has described, the physical domain contains couplings that occur from the sequential relationship between components. The MBSE literature often calls these couplings *interfaces* [69]. They are formally modeled by the existence of nonzero elements in the physical domain's adjacency matrix. Both of these are examples of *intradomain* information.[2] In analysis, the Independence Axiom exclusively addresses the *interdomain* information with the design matrix that describes the allocation architecture. Intradomain coupling is not relevant.

Another concern that emerges over the Independence Axiom is its statement in the imperative rather than more traditionally as a declarative (e.g., $1 * X = X$—multiplicative identity axiom) or a conditional (e.g., if $x = y$, then $y = x$—reflexivity axiom) statement. Here, again, it is important to recall that design is both synthesis and analysis. A statement in the imperative is conducive in the practical sense to the process of design synthesis. In other words, Suh's Independence Axiom is directed to a design synthesis practitioner rather than a design analyst audience. The Independence Axiom can be recast as a declarative statement as follows.

Axiom 3 Independence Axiom (Recast): Maintaining the independence of the functional requirements **FR** during design synthesis yields "good" designs.

Another misconception arises when Suh [3] speaks of *good* designs being synthesized as a result of the application of Axiomatic Design. Here, the criticism is directed to the term "good" as a statement of subjective value rather than a quantifiable scientific measure. The Independence Axiom's mathematical statement of a diagonal (or lower triangular) design matrix is matched to the qualitative notion of a "good design" by extensive empirical observation. Methodologically, and logically, this is an extension of the corresponding concept in ontological science. The formal mathematical definitions of soundness, completeness, lucidity, and laconicity yield the qualitatively and widely held conversational maxims of "relevance and clarity." Both statements are built upon extensive empirical observation relating a qualitative conceptual idea to a formal definition in the corresponding analytical model. There is no difference in the nature of the logical reasoning.

---

[2]Note that the flows of matter, energy, information, money, and people within interfaces and interactions are collectively the same artifacts. However, their representation need not be the same in the two domains. Indeed, it is easy to prove that they are same if and only if the design matrix is square and diagonal.

1.4.2 The Information Axiom

The Information Axiom introduced at the beginning of the section is applied in design analysis once the Independence Axiom has been applied. It calls for the minimization of a design's information content I, which is defined in terms of the probabilities P_i of satisfying each of the functional requirements FR_i [3].

$$I = -\sum_{i}^{\sigma(FR)} \log_2 P_i \tag{1.8}$$

These probabilities may be understood practically by Fig. 1.12. Each functional requirement may be specified as a design range. In practice, however, the true value of the functional requirements falls within a probability density function that is characterized by a system range. The area under the probability density function that falls within the design range provides a measure of the probability of satisfying a given functional requirement $A_{cr_i} = P_i$ [3]. A deeper discussion of the Information Axiom and its applications is provided in Chap. 2.

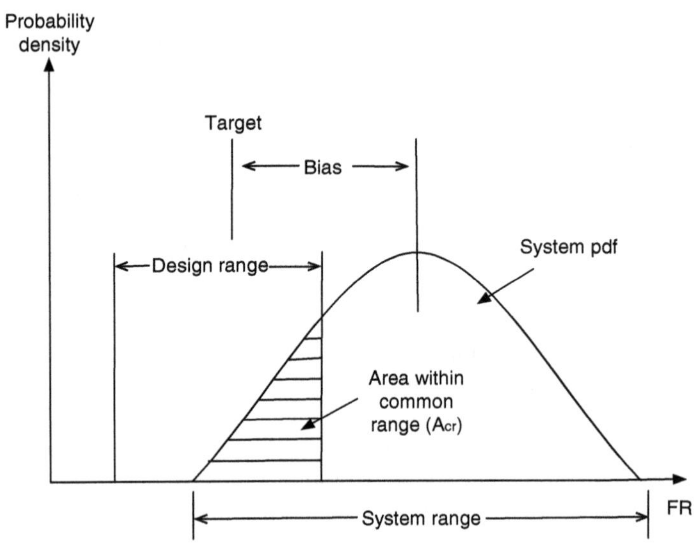

Fig. 1.12 A practical understanding of the Information Axiom: design range, system range, and probability density function of a functional requirement

1.4.3 Axiomatic Design's Theorems and Corollaries

These two axioms form the foundation of Axiomatic Design. Over several decades, many theorems and corollaries have been proven from these two axioms. The interested reader can find many of these summarized in the appendix of this book with citations to their original references and corresponding proofs.

1.5 Functional and Physical System Hierarchy in Large Systems

At this point, a careful reader would recognize that the previous section's treatment of Axiomatic Design was at a single level of decomposition and hence is insufficient to address the functional and physical system hierarchy in large systems as represented in Figs. 1.6 and 1.7. This section now expands the discussion of the previous one to address the systems thinking concepts of decomposition and specialization in large fixed and large flexible engineering systems.

Definition 26 Large Fixed Engineering System [3]: An engineering system with a large set of functional requirements which do not evolve over time and whose components also do not change over time.

Definition 27 Large Flexible Engineering System [3, 27]: An engineering system with many functional requirements that not only evolve over time, but also can be fulfilled by one or more design parameters.

1.5.1 Large Fixed Engineering Systems

A discussion on large fixed engineering systems follows from those provided in continue to follow the Axiomatic Design discussions provided in Sects. 1.3 and 1.4. A synthesis function $f_s()$ is used to conceptually represent a designers' generation of a set of design parameters **DP**, and an analysis function $f_a()$ following the laws of physics is assessed to determine adherence to the Independence and Information Axioms. For small systems (i.e., those with a very few functional requirements), such a process is relatively straightforward.

For large fixed engineering systems, however, such an approach is impractical for two reasons. The first issue is in the size of **FR** and **DP** in $f_s()$. In 1956, as a psychologist, Miller [70] noted that human short-term memory is limited to 7 ± 2 elements. Therefore, the synthesis function $f_s()$ is ill-defined beyond this size. Instead, the functional requirements must be *aggregated* into this manageable size, and design parameters must be synthesized *conceptually* at a corresponding *level of* *abstraction*. For example, the design parameters can now be whole subsystems such as whole drivetrains, buildings, or organizations. This brings about the second practical issue which is in the nature of **FR** and **DP** in Eq. 1.4. **FR** and **DP** are no longer real numbers and so $f_a()$ is no longer well-defined as an algebraic or differential equation. In practice, designers may not know the exact impact of a given design parameter on a given functional requirement, and yet they must continue to synthesize engineering systems in spite of this. Inevitably, this causes a profound intellectual conflict between the mathematical rigor of engineering analysis and the creativity of engineering synthesis. It appears most vividly early on in the conceptualization of an engineering system where interestingly engineering design decisions have the greatest impact.

Axiomatic Design resolves this conflict by allowing design analysis to occur, albeit with a less precise form of mathematics. At higher levels of abstraction, early on in the conceptualization of an engineering system, **FR** and **DP** represent elements not numbers. Therefore, Eq. 1.3 must be represented using graph and set theory. In large fixed engineering systems, Eq. 1.3 becomes

$$\mathbf{FR}\$(B \circledast \mathbf{DP}) \tag{1.9}$$

where the aggregation operation $\circledast$ is defined as follows:

Definition 28 Aggregation Operator $\circledast$ [54, 71]: Given boolean matrix A and sets B and C, $C = A \circledast B$ is equivalent to:

$$C(i) = \bigcup_j a(i,j) \wedge b(j) \tag{1.10}$$

The \$ in Eq. 1.9 is often replaced with a simple = as a matter of notational convenience without change in the underlying meaning.

$$\mathbf{FR} = B \circledast \mathbf{DP} \tag{1.11}$$

Note that, B now comes to represent an (undirected) incidence matrix between the sets **FR** and **DP**.

Definition 29 Incidence matrix [47] M of size $\sigma(B) \times \sigma(E)$ is given by:

$$M(i,j) = \begin{cases} -1 & \text{if } b_i \text{ is the head of arc } e_j \\ 1 & \text{if } b_i \text{ is the tail of arc } e_j \\ 0 & \text{otherwise} \end{cases} \tag{1.12}$$

As mentioned previously, the presence of nonzero elements in the design matrix B is graphically represented in SysML as an allocation of a functional element to a physical element as shown in Fig. 1.9. Axiomatic Design collates these graphical interactions to highlight their underlying mathematical form. Furthermore, Eq. 1.11

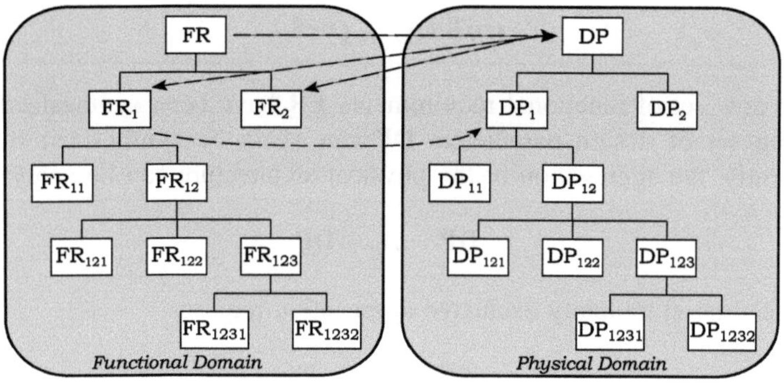

Fig. 1.13 Synthesis paths in simultaneous hierarchical physical and functional decomposition

also implies that Eq. 1.6 remains true and consequently the Independence Axiom can still be applied without change.

The introduction of graph and set theory into the discussion now allows a formal understanding of how Axiomatic Design manages system complexity and multiple layers of abstraction. Figure 1.13 shows Axiomatic Design's dual hierarchy of the functional and physical architecture domains. It represents the full allocated architecture of the system and is generated along the depicted synthesis arrows by what Suh calls a "Zig-Zag" approach. The highest level of design parameters is synthesized from the highest level of functional requirements by Eq. 1.4 and then analyzed by Eq. 1.11 for adherence to the Independence Axiom. At this point, a new set of decomposed functional requirements $\underline{\mathbf{FR}}$ must be synthesized based upon the designer's knowledge of the higher-level functional requirements **FR** and design parameters **DP**.

$$\underline{\mathbf{FR}} = f_s(\mathbf{FR}, \mathbf{DP}) \tag{1.13}$$

As with Eq. 1.4, Eq. 1.13 describes the designer's mental process of synthesis as an abstract mathematical function. Again, two distinct designers may produce the decomposition entirely differently depending on their knowledge of the design parameters **DP** as an abstract model of the system in real life. The result of the decomposition can be analyzed using the aggregation operation [72].

$$\mathbf{FR} = A_f \circledast \underline{\mathbf{FR}} \tag{1.14}$$

where A_f is a binary functional aggregation matrix that describes to which high-level functional requirement, each low-level functional requirement pertains [72]. Strict mutually exclusive aggregation places a constraint on the nature of the aggregation matrix.

$$\mathbf{1}^{\sigma(\mathbf{FR})T} A_f = \mathbf{1}^{\sigma(\underline{\mathbf{FR}})T} \tag{1.15}$$

Once the new set of functional requirements $\underline{\mathbf{FR}}$ have been synthesized, the corresponding set of design parameters $\underline{\mathbf{DP}}$ can again be synthesized by Eq. 1.4. Consequently, the aggregation of the physical architecture can be analyzed [72].

$$\mathbf{DP} = A_p \circledast \underline{\mathbf{DP}} \tag{1.16}$$

where again, strict mutually exclusive aggregation requires

$$1^{\sigma(\mathbf{DP})T} A_f = 1^{\sigma(\underline{\mathbf{DP}})T} \tag{1.17}$$

Consequently,

$$B * A_p = A_f \underline{B} \tag{1.18}$$

where B is the higher-level design matrix and $\underline{\mathrm{B}}$ is the lower-level design matrix. Equation 1.18 may be solved when the left and right inverses of B and $\underline{\mathrm{B}}$, respectively, exist. Furthermore, when they are identity matrices (e.g., the Independence Axiom is fulfilled), the aggregation in the functional and physical architectures becomes the same. $A_f = A_p$.

1.5.2 Large Flexible Engineering Systems

The Axiomatic Design of large flexible engineering systems was first mentioned by Suh in his 2001 text [3] and has since been significantly developed [27, 54, 55, 72–74]. Large flexible engineering systems typically require attention at higher levels of abstraction but are otherwise similar to large fixed systems. Equation 1.4 describes design synthesis, and Eq. 1.11 describes design analysis. Recalling Definition 27: The distinguishing feature of flexibility is achieved by a strict adherence to the Independence Axiom. Therefore, $\mathbf{B} = \mathbf{I}^n$, where n equivalently represents the number of design parameters or functional requirements. Conceptually, this is because a non-identity design matrix would imply that either a single design parameter affects more than one functional requirement or vice versa. Consequently, when it comes time to *reconfigure* the engineering system with an addition or removal of a functional or physical element, other changes would need to be made as well. In contrast, adherence to the Independence Axiom enables a "plug & play" engineering system where functional and physical elements can be added or removed at will [56, 71].

By Definition 27, large flexible engineering systems have functional requirements that can be fulfilled by potentially many design parameters. An identity design matrix does not show this. Therefore, in order to reveal this functional redundancy, the set of functional requirement *instances* **FR** must be distinguished from the set of functional requirement *classes* $\mathbb{FR}$.[3] A new design equation can then be written to relate $\mathbb{FR}$ to **DP**.

$$\mathbb{FR} = \mathbf{J} \odot \mathbf{DP} \tag{1.19}$$

where **J** represents the system knowledge base and $\odot$ represents matrix boolean multiplication.

Definition 30 System Knowledge Base [27, 54, 55, 72–74]: a binary matrix **J** of size $\sigma(\mathbb{FR}) \times \sigma(\mathbf{DP})$ whose element $\mathbf{J}(w, v) \in \{0, 1\}$ is equal to one when action e_{wv} (in the SysML sense)[4] exists as a functional requirement $\mathbb{FR}_w$ being executed by a design parameter $\mathbf{DP}_v$.

Definition 31 Matrix Boolean Multiplication $\odot$ [27, 54, 55, 72–74]: Given sets or boolean matrices B and C and boolean matrix A, $C = A \odot B$ is equivalent to:

$$C(i, k) = \bigvee_j A(i, j) \wedge B(j, k) \tag{1.20}$$

Interestingly, it is equally valid to replace the set of design parameter instances **DP** with the set of design parameter classes $\mathbb{DP}$. In such a case, Axiomatic Design addresses the design of generic or *reference architectures* rather than specific, instantiated or system architectures [75–77].

By Definition 27, large flexible engineering systems have functional requirements that can evolve over time. To that effect, the Axiomatic Design literature introduces a system constraints matrix.

Definition 32 System Constraints Matrix [27, 54, 55, 73, 74]: a binary matrix **K** of size $\sigma(\mathbb{FR}) \times \sigma(\mathbf{DP})$ whose element $\mathbf{K}(w, v) \in \{0, 1\}$ is equal to one when a constraint eliminates action e_{wv} from the action set.

A *reconfiguration process* is said to change the value of the system constraints matrix [78]. Therefore, the system knowledge base contains information on the *existence* of capabilities in the engineering system. Meanwhile, the constraints matrix contains information of their *availability* [77, 79]. Quantitatively, keeping track of these capabilities is done via the system's structural degrees of freedom as a measure [27, 54, 55, 72–74].

[3] Note that many works on Axiomatic Design do not make this distinction between functional requirement instances and functional requirement classes because it is rarely needed within a single design work. Here, the distinction is made in order to maintain the conceptual link between large fixed and large flexible engineering systems and the universality of the Independence Axiom in both cases.

[4] The word "action" is meant in the technical sense of allocated functional elements in SysML's activity diagram. See Fig. 1.9 for details. These actions represent capabilities in the engineering system.

Definition 33 LFES Sequence-Independent Structural Degrees of Freedom [27, 54, 55, 72–74]: The set of independent actions $\mathcal{E}_S$ that completely defines the available processes in a LFES. Their number is given by:

$$\text{DOF} = \sigma(\mathcal{E}) = \sum_{w}^{\sigma(\mathbb{FR})} \sum_{v}^{\sigma(\mathbf{DP})} [\mathbf{J} \ominus \mathbf{K}](w, v) \tag{1.21}$$

Consequently, the redundancy of functional requirement $\mathbf{FR}_w$ is [27, 54, 72]:

$$\mathcal{R}_w = \sum_{v}^{\sigma(\mathbf{DP})} [\mathbf{J} \ominus \mathbf{K}](w, v) \tag{1.22}$$

The flexibility of the design parameter $\mathbf{DP}_v$ is [27, 54, 72]:

$$\mathcal{F}_v = \sum_{w}^{\sigma(\mathbb{FR})} [\mathbf{J} \ominus \mathbf{K}](w, v) \tag{1.23}$$

These measures are important because redundancy and flexibility are important enabling properties for many life cycle properties [27, 56, 72].

Large flexible engineering systems require a careful discussion of the Axiomatic Design dual hierarchy [71, 72]. Fundamentally, this is because functional and physical elements can be added or removed. Consequently, their respective hierarchies must be allowed to change as well. Developing the Axiomatic Design dual hierarchy for large flexible engineering systems, downward in the direction of design synthesis, proceeds in the same way as for large fixed engineering systems. The system is viewed in terms of functional requirement *instances* rather than classes. Because the Independence Axiom has been strictly maintained, each structural degree of freedom can be designed as previously described as if it were its own system. The engineering design problem is separable. Therefore, the addition or removal of a structural degree of freedom adds or removes all of the associated lower branches in the dual hierarchy.

It is also useful to consider the dual hierarchy of a large flexible engineering system upward in the direction of design analysis. Here, it is no longer required to aggregate the physical and functional hierarchies simultaneously [27, 72, 74, 77]. It is particularly common in bottom-up design to aggregate only the physical hierarchy into higher-level design parameters or *resources*. A corresponding functional aggregation may not occur. This is because physical aggregation and functional aggregation do not have the same meaning and do not necessarily imply each other [41, 42]. Consider, for example, five tasks as functional requirements and five individuals as design parameters each of whom completes one task. This is a large flexible engineering system that fulfills the Independence Axiom. The five individuals may be aggregated into a resource called a team without making any statement about the five tasks. They may not be related in any way (i.e., share any functional interaction).

Similarly, the five tasks may be aggregated into a project without making any statement about the five individuals who complete them. They may have never met (i.e., share any physical interface). Physical aggregation is particularly interesting because it yields resources with high flexibility. An addition or removal of a design parameter yields the corresponding change in a resource's flexibility. In contrast, the functional aggregation of a large flexible engineering system may result a rigid top-down structure. Any time the set of functional requirements changes, the functional hierarchy would need to change as well. In a project, the elimination of a single task causes the elimination of the project as a whole.

Thus far, the two systems thinking concepts of instantiation and aggregation/decomposition have been discussed as a means of managing system complexity. The discussion now turns to the last such concept: specialization/generalization. The Axiomatic Design for large flexible engineering system literature has addressed this topic implicitly in several works [27, 54, 55, 73, 77]. More explicitly, bottom-up generalization is a form of conditional aggregation.

Definition 34 Generalized Design Parameter: A generalized design parameter $\widetilde{\mathbf{DP}}_i$ is an aggregation of a set of design parameters $\mathbf{DP}$ if any $\mathbf{DP}_k \in \mathbf{DP}$ is capable of doing any of the common functional requirements $\mathbf{cFR} \subseteq \mathbf{FR}$.

$$\widetilde{\mathbf{DP}}_i = A_g \circledast \mathbf{DP} \tag{1.24}$$

where $A_g(i,k) = 1$ iff $J(j,k) = 1$ for any $FR_j \in \mathbf{cFR}$.

Note that the definition of a generalized design parameter requires the identification of a set of common functional requirements that can be done by the low-level design parameters $\mathbf{DP}$ as well as its generalization $\widetilde{\mathbf{DP}}$. Also, note that unlike a regular aggregration, specialization does not require constraint in Eq. 1.17.

1.5.3 An Illustrative Example

To summarize the discussion on Axiomatic Design for large flexible engineering systems, consider the following example.

Example 3 Consider the manufacturing system depicted in Fig. 1.14. It consists of a drill press and milling machine. The former is able to drill a hole, and the latter is

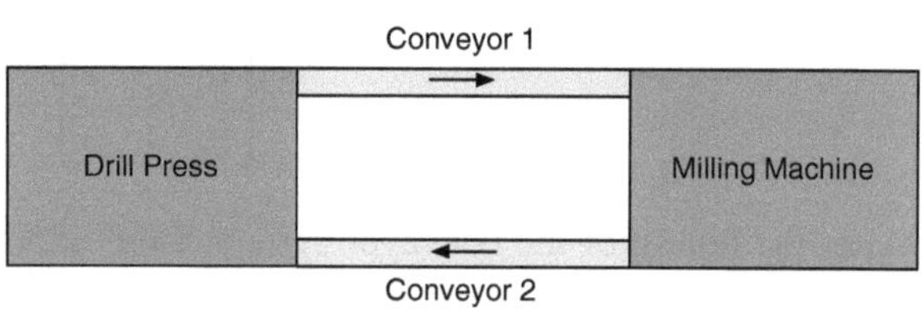

Fig. 1.14 A simple manufacturing system with one drill press, one milling machine, and two conveyors

able to do the same and mill surfaces. Each contains its respective fixture. It also has two one-way conveyors between them.

A large fixed engineering system analysis yields:

FR = {drill hole, drill hole, mill surface, store the part at point A, transport part from point A to point B, transport part from point B to point A, store the part at B}.
DP = {drill press, milling machine drill, milling machine end mill, drill press fixture, conveyor 1, conveyor 2, milling machine fixture}.
The design matrix $\mathbf{B} = \mathbf{I}^7$. The Independence Axiom is satisfied.

For a large flexible engineering system analysis, the functional requirement classes are viewed instead of their instances.

$\mathbb{FR}$ = {drill hole, mill surface, store the part at A, transport part from point A to point B, transport part from point B to point A, store the part at B}.

An aggregation matrix is applied so that the drill press, milling machine, and conveyor system appear as single resources.

$\overline{\mathbf{DP}}$ = {drill press, milling machine, conveyor system}.

$$J = \begin{bmatrix} 1 & 1 & 0 \\ 0 & 1 & 0 \\ 1 & 0 & 0 \\ 0 & 0 & 1 \\ 0 & 0 & 1 \\ 0 & 1 & 0 \end{bmatrix} = \left[\begin{array}{c|c} J_M & \mathbf{0} \\ \hline \multicolumn{2}{c}{J_H} \end{array}\right] \tag{1.25}$$

That partitioning of the system knowledge base into J_M and J_H comes from generalization. J_M represents structural degrees of freedom that have a transformational function. J_H represents structural degrees of freedom that have a transportational function.

Resource flexibility: The three resources have flexibilities of 2, 3, and 2 structural degrees of freedom, respectively.
Functional redundancy: All the functional requirements have a functional redundancy of 1 except "drill hole."

The failure of the conveyor system would appear as two constraints in the constraints matrix

$$K = \begin{bmatrix} 0 & 0 & 0 \\ 0 & 0 & 0 \\ 0 & 0 & 0 \\ 0 & 0 & 1 \\ 0 & 0 & 1 \\ 0 & 0 & 0 \end{bmatrix} \tag{1.26}$$

After the failure of the conveyor system, DOF = 5.

1.6 Engineering Systems Applications of Axiomatic Design

With a solid Axiomatic Design foundation in place, the chapter can now return to the engineering systems discussion initiated in the introduction. This section highlights the potential applications of Axiomatic Design in the development of engineering with regard to three specific challenges: (1) a quantitative understanding of life cycle properties, (2) a treatment of cyber-physical systems, and (3) a treatment of hetero-functional networks.

1.6.1 Quantitative Understanding of Life Cycle Properties

The subject of life cycle properties in engineering systems is an expansive one [10] with potentially whole textbooks devoted to a single property (e.g., resilience [80]). Consequently, a detailed discussion cannot be provided here. Nevertheless, the Axiomatic Design, MBSE, and engineering systems concepts provided thus far can serve to provide a guiding structure to the subject. A quantitative formulation of life cycle properties first requires a qualitative understanding of which engineering domains in Fig. 1.1 or more generally the ESM in Fig. 1.8 pertain to that specific life cycle property. Furthermore, the life cycle property must be classified as a description of system structure or system behavior [71].

Therefore, Table 1.3 presents a first-pass classification of life cycle properties. As mentioned at the end of Sect. 1.2.2, the central focus of traditional engineering effort is often devoted to understanding system behavior from quantitative models of system function [28]. Sustainability, when viewed in the sense of the provision of a certain level of product or service while limiting the quantities of input resources and by-product emissions, may be similarly classified [84–86]. Many life cycle properties, however, depend on an explicit—often graph theoretic—description of system structure. Modularity [82, 83] and centrality [81] are two such life cycle properties that depend on the form of a graph's adjacency matrix, be it in

Table 1.3 A preliminary classification of life cycle properties in engineering systems

	System structure	System behavior
Functional architecture domain	Centrality [81], modularity [82, 83]	Stability [49, 50], sustainability [84–86]
Physical architecture domain	Centrality [81], modularity [82, 83]	Not applicable
Allocated architecture domain	Flexibility [54], redundancy [54], reconfigurability [56, 71], static resilience [27]	Dynamic resilience [87, 88], stability/synchronization [89], sustainability [84–86]

the functional or physical architecture domains. One may argue that perhaps one of the great benefits of MBSE (e.g., through SysML) is that it can abstract details of system behavior to provide a clear view of system structure and its associated life cycle properties.

Still other life cycle properties emerge from the allocated architecture. It is here that the Axiomatic Design matrix B and knowledge base J, as different types of incidence matrices, are quite valuable in developing a quantitative treatment. Section 1.5.2 already provided measures for two relatively simple life cycle properties of system structure: flexibility [54] and redundancy [54]. More complex life cycle properties such as reconfigurability [56, 71] and static resilience [27] often require that a new adjacency matrix A_ρ be constructed from the system's structural degrees of freedom [27, 74].

$$A_\rho = [J \ominus K]^V [J \ominus K]^{VT} \ominus K_\rho \tag{1.27}$$

where K_ρ is a constraints matrix that imposes continuity relations between the individual structural degrees of freedom. Interestingly, reconfigurability clearly differentiates between large fixed and large flexible engineering systems [27, 54, 55, 74]. As expected, engineering systems that adhere to the Independence Axiom are fundamentally more reconfigurable than systems that do not [56, 71].

Finally, many life cycle properties require an understanding of the relationship between the allocated architecture and the system behavior. Dynamic resilience—in particular the capacity to "bounce back" to a certain system performance after a disruption—depends equally on the system's constituent device models [76] as on flexibility of its resources and their redundancy [87, 88]. Synchronization of engineering systems with coupled oscillators (e.g., the electric power grid, swarms/fleets of moving vehicles) utilizes many of the techniques required to analyze stability but add further steps that consider the physical architecture's adjacency matrix [89]. Finally, when the prior view of sustainability is expanded to also include cost performance, it must balance the performance of the functional architecture to the cost of the physical architecture via the allocated architecture.

1.6.2 Treatment of Cyber-Physical Systems

Axiomatic Design sheds light on many of the architectural questions related to cyber-physical systems. Consider the four simple control theory examples shown in Fig. 1.15. Figure 1.15a depicts an open-loop physical system C0P1. The second column shows its corresponding SysML activity diagram. The system, as a whole, transforms an input $\mathbf{U}$ into an output $\mathbf{Y}$. A single action $\mathcal{G}_1$ achieves this activity, and it is allocated to the physical system G_1. The distinction between the functional element $\mathcal{G}_1$ and the physical element G_1 is critical. The third column shows the corresponding system behavior as a transfer function involving $\mathcal{G}_1$. Meanwhile, the fourth column shows the corresponding allocated architecture as an Axiomatic Design equation involving G_1. A single functional requirement is placed on the output **Y**, and the single resource G_1 is designed to achieve it, and consequently, the Independence Axiom is fulfilled with an identity design matrix.

It is important to note that from a mathematical perspective, the two equations in Fig. 1.15a are indeed equivalent, albeit very differently arranged. At this level of abstraction, the functional form of $\mathcal{G}_1$ is hidden away. Similarly, G_1 hides away all of its constituent (design) parameters; the same one would expect to find in $\mathcal{G}_1$. Proving their equivalence requires two steps. First, $\mathcal{G}_1$ is written explicitly and then differentiated with respect to each of the design parameters in G_1 so that it takes the form of Eq. 1.7. Similarly, G_1 is decomposed down to an "atomic" level of design parameters represented by real numbers and then differentiated. Although these two equations are equivalent, one focuses the designer on a system's behavior, and the other focuses the designer's attention to its allocated architecture.

Figure 1.15b depicts a closed-loop cyber-physical system. The SysML diagram depicts two components: a cyber component K_1 and a physical component G_1. The former realizes the action (i.e., transfer function) $\mathcal{K}_1$, while the latter realizes the action $\mathcal{G}_1$. An output feedback loop is introduced. The third column shows the overall closed-loop transfer function $\mathcal{G}_{cp}$ as a top level of abstraction or equivalently one level of abstraction down in terms of $\mathcal{K}_1$ and $\mathcal{G}_1$. At the highest level of abstraction, the Axiomatic Design equation resembles the open-loop system and the Independence Axiom is fulfilled. However, one level of abstraction down, the design equation reveals a "redundant design."[5] This coupled design does not adhere to the Independence Axiom and requires the physical system to be fixed first before the controller can be designed. Not surprisingly, many feedback control design methods require iterative tuning.

Figure 1.15c now depicts a closed-loop cyber-physical system with one centralized controller and n resources. As in Fig. 1.15a, b, this system may be viewed as an open-loop system fulfilling the Independence Axiom. However, one level of abstraction down, the coupled and redundant design matrix reappears as expected. This is unfortunate from the perspectives of reconfigurability and resilience. Although the four physical systems are mathematically uncoupled, the failure or "hack" of the centralized control affects the performance of all of the functional requirements [54–56, 71]. Therefore, from an Axiomatic Design perspective, centralized controllers are to be avoided.

Finally, Fig. 1.15d depicts a closed-loop cyber-physical system with n controllers matched to n resources. If the n controllers are entirely independent, the system now fully adheres to the Independence Axiom supporting the case for distributed control. Furthermore, from a reconfigurability and resilience perspective, there exists no single point of failure [54–56, 71]. This is a very special and rare case, however. Instead, much research on multi-agent control systems [90–95]

---

[5]In the Axiomatic Design of large fixed systems, redundant designs have more design parameters than functional requirements [3].

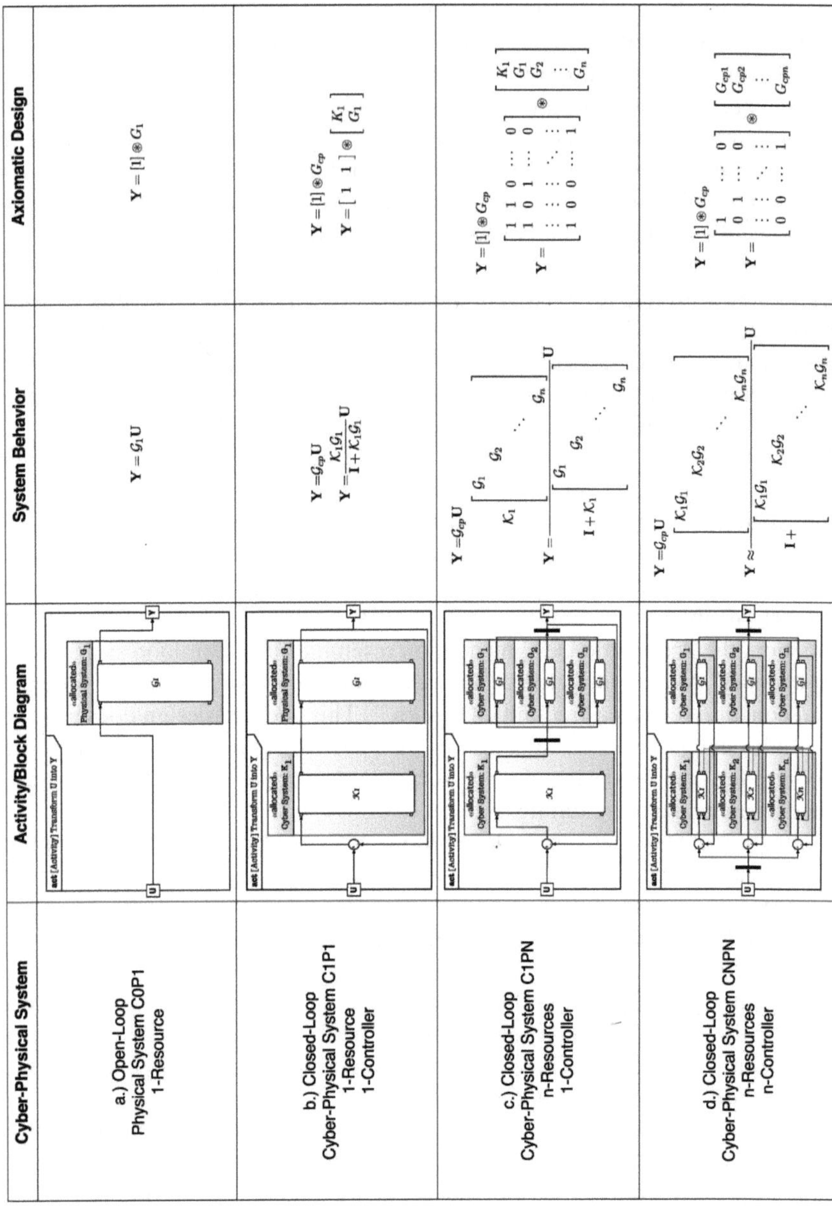

Fig. 1.15 Cyber-physical systems from the perspective of SysML, transfer functions, and Axiomatic Design

introduces communication between the n agent controllers to achieve greater coordination between the n physical resources. If this interagent communication algorithm is either 5× faster or slower than the physical system's dynamics, then the system's transfer function is approximately timescale separable and the Independence Axiom continues to be supported [79]. Furthermore, the performance of each physical resource can be enhanced with the addition of a single fast but local controller for each physical system. These Axiomatic Design principles have been used to develop multi-agent control system architectures for production [77] and power systems [79].

1.6.3 Treatment of Hetero-Functional Networks

As engineering systems integrate together to form hetero-functional networks, they pose several challenges to existing approaches to engineering design and modeling. As has been previously mentioned in Sects. 1.2.2 and 1.2.3, adjacency matrices are typically used to provide abstract graph theoretic models of either the functional or the physical architecture. Furthermore, the most common applications of graph theory are homo-functional in nature [27, 81]. Artifacts (of some kind) are transported along edges between physical locations represented as nodes. This is sufficient for individual engineering systems. For example, in transportation systems, the nodes often physically represent intersections and stations, while edges/arcs represent roads, rails, or transportation routes [96–101]. Meanwhile, in power systems, the nodes often physically represent generators, substations, and loads, while the edges represent the power lines [102–105]. The integration of two or more engineering systems, however, requires a richer approach because the nodes and edges have completely different physical meanings. Alternatively, bond graphs [46] and linear graphs [106] are promising techniques to quantitatively model continuous-time physical systems across multiple energy domains. Their current level of development, however, lacks the systems thinking abstractions mentioned throughout this chapter. Furthermore, they have limited capability to handle systems with discrete-event dynamics and consequently offer limited support for dynamics and decision-making driven by people, be they individuals or organizations.

In contrast, Axiomatic Design enables the study of engineering systems as they integrate together to form hetero-functional networks. Production systems, due to their hetero-functional nature, have been proven to be an excellent application domain for advancing Axiomatic Design. In Example 3, Axiomatic Design for large flexible engineering systems was used to model the physical part of a production system's allocated architecture at multiple levels of abstraction. Later chapters in this book will demonstrate Axiomatic Design's application to decision-making processes as the cyber-part of production systems. More explicitly, Eq. 1.27 allows the system knowledge base to be converted into a hetero-functional graph with structural degrees of freedom as nodes [27, 74]. Such a graph based upon the

Axiomatic Design knowledge base was later used to directly derive a production system's discrete-event dynamics [107]. Similarly derived discrete-event dynamics were demonstrated for transportation systems as an engineering system with no transformation functions [108]. Meanwhile, the Axiomatic Design knowledge base was used with device models to derive the continuous-time dynamics of power systems [79]. With these methodological developments in place, Axiomatic Design has been used to develop full simulations of the energy–water nexus [75, 109], electrified transportation systems [110], and microgrid-enabled production systems [107] as truly hetero-functional and integrated engineering systems. The broad diversity of these applications demonstrates the utility of Axiomatic Design to engineering systems as a field.

1.7 Conclusion

In the twenty-first century, engineers are facing engineering challenges of increasingly greater scope. These include many large complex products and systems described later in this book but even more generally whole engineering systems. This chapter has introduced Axiomatic Design within this larger engineering systems context. It began by identifying Axiomatic Design and MBSE as two engineering design methodologies and theories that when appropriately developed have the potential to address the methodological challenges of engineering systems. The chapter introduced Axiomatic Design and its relationship to MBSE in terms of four domains of engineering design: stakeholder requirements, functional architecture, physical architecture, and process domains. It also discussed a system's allocated architecture with special care given to differentiate its synthesis and analysis. Here, Axiomatic Design's ability to quantify the allocated architecture was highlighted in terms of its Independence and Information Axioms. At that point, the chapter generalized these concepts with several hierarchical techniques to manage system complexity. This allowed the discussion to return to the three methodological challenges mentioned in the introduction: quantification of life cycle properties, design of cyber-physical systems, and design of hetero-functional networks. Taken together, the chapter details the essentials of Axiomatic Design, relates it to MBSE, and highlights its potential applications in the field of engineering systems.

References

1. N.P. Suh, A. Bell, D. Gossard, On an axiomatic approach to manufacturing and manufacturing systems. J. Manuf. Sci. Eng. **100**(2), 127–130 (1978)
2. N.P. Suh, *The Principles of Design* (Oxford University Press, Oxford, 1990)
3. N.P. Suh, *Axiomatic Design: Advances and Applications* (Oxford University Press, Oxford, 2001)

4. T. Tomiyama, P. Gu, Y. Jin, D. Lutters, C. Kind, F. Kimura, Design methodologies: industrial and educational applications, in *CIRP Annals—Manufacturing Technology*, vol. 58, no. 2 (2009), pp. 543–565. (Online) Available: http://www.sciencedirect.com/science/article/pii/S000785060900170X
5. N.P. Suh, *Complexity: Theory and Applications* (Oxford University Press, New York, 2005)
6. E.M. Benavides, *Advanced Engineering Design: An Integrated Approach* (Elsevier, Amsterdam, 2011)
7. S.J. Kim, N.P. Suh, S.G. Kim, Design of software systems based on axiomatic design, in *Robotics and Computer-Integrated Manufacturing*, vol. 8, no. 4 (1991)
8. N.P. Suh, Design and operation of large systems. J. Manuf. Syst. **14**(3) (1995)
9. S.H. Do, N.P. Suh, Systematic OO programming with axiomatic design. Computer **32**(10), 121–124 (1999)
10. BKCASE Editorial Board, *The Guide to the Systems Engineering Body of Knowledge (SEBoK)*, v1.3 edn (The Trustees of the Stevens Institute of Technology, Hoboken, NJ, 2014)
11. SE Handbook Working Group, *Systems Engineering Handbook: A Guide for System Life Cycle Processes and Activities*. International Council on Systems Engineering (INCOSE), 2015
12. A. Pyster, D.H. Olwell, T.L. Ferris, N. Hutchison, S. Enck, J.F. Anthony Jr., D. Henry, A. Squires, Graduate reference curriculum for systems engineering (grcse) version 1.0, The Trustees of Stevens Institute of Technology, Tech. Rep., 2012
13. V. Piuri, A.G. Aghdam, S. Nahavandi, Editorial. IEEE Syst. J. **7**(1), 2–3 (2013)
14. O.L. de Weck, A vision for the future of the journal systems engineering. Syst. Eng. **16**(4), 379–380 (2013)
15. A.W. Wymore, *Model-based systems engineering*, vol. 3 (CRC press, Boca Raton, 1993)
16. MBSE Initiative Working Group, Model-based systems engineering (mbse) wiki, INCOSE, Technical Report, 2015. (Online) Available: http://www.omgwiki.org/MBSE/doku.php
17. T. Weilkiens, *Systems Engineering with SysML/UML Modeling, Analysis, Design* (Morgan Kaufmann, Burlington, 2007)
18. S. Friedenthal, A. Moore, R. Steiner, *A Practical Guide to SysML: The Systems Modeling Language*, 2nd edn. (Morgan Kaufmann, Burlington, 2011)
19. O.L. De Weck, D. Roos, C.L. Magee, *Engineering Systems: Meeting Human Needs in a Complex Technological World* (MIT Press, Cambridge, 2011). (Online) Available: http://www.knovel.com/knovel2/Toc.jsp?BookID=4611, http://mitpress-ebooks.mit.edu/product/engineering-systems
20. M. Amin, Toward secure and resilient interdependent infrastructures. J. Infrastruct. Syst. **8**(3), 67–75 (2002)
21. M. Amin, System-of-systems approach, in *Intelligent Monitoring, Control, and Security of Critical Infrastructure Systems* (Springer, Berlin, 2015), pp. 317–354
22. A.M. Annaswamy, M. Amin, C.L. Demarco, T. Samad, J. Aho, G. Arnold, A. Buckspan, A. Cadena, D. Callaway, E. Camacho, M. Caramanis, A. Chakrabortty, A. Chakraborty, J. Chow, M. Dahleh, A.D. Dominguez-Garcia, D. Dotta, A.M. Farid, P. Flikkema, D. Gayme, S. Genc, M.G.i. Fisa, I. Hiskens, P. Houpt, G. Hug, P. Khargonekar, H. Khurana, A. Kiani, S. Low, J. McDonald, E. Mojica-Nava, A.L. Motto, L. Pao, A. Parisio, A. Pinder, M. Polis, M. Roozbehani, Z. Qu, N. Quijano, J. Stoustrup, in *IEEE Vision for Smart Grid Controls: 2030 and Beyond* eds. by A.M. Annaswamy, M. Amin, C.L. Demarco, T. Samad (IEEE Standards Association, New York, 2013). (Online) Available: http://www.techstreet.com/ieee/products/1859784
23. G. Olssonn, *Water and Energy: Threats and Opportunities* (IWA Publishing, London, 2012)
24. W. Su, H. Rahimi-eichi, W. Zeng, M.-Y. Chow, A survey on the electrification of transportation in a smart grid environment. IEEE Trans. Industr. Inf. **8**(1), 1–10 (2012)
25. Stockholm Environment Institute, Understanding the Nexus: Background paper for The Water, Energy and Food Security Nexus Conference, Stockholm Environment Institute, Bonn, Technical Report, Nov 2011

26. B. Murgante, G. Borruso, Smart cities in a smart world, in *Future City Architecture for Optimal Living* (Springer, Berlin, 2015), pp. 13–35
27. A.M. Farid, Static resilience of large flexible engineering systems: Axiomatic design model and measures. IEEE Syst. J. (99), 1–12 (2015). (Online) Available: http://amfarid.scripts.mit.edu/resources/Journals/IES-J19.pdf
28. D.M. Buede, *The Engineering Design of Systems: Models and Methods*, 2nd edn. (Wiley, Hoboken, 2009)
29. SE Handbook Working Group, *Systems Engineering Handbook: A Guide for System Life Cycle Processes and Activities* (International Council on Systems Engineering (INCOSE), 2011)
30. D.W. Oliver, T.P. Kelliher, J.G. Keegan, *Engineering complex systems with models and objects* (McGraw-Hill, New York, 1997)
31. K. Forsberg, H. Mooz, The relationship of systems engineering to the project cycle. Eng. Manage. J. **4**(3), 36–43 (1992)
32. K. Pohl, *Requirements Engineering: Fundamentals, Principles, and Techniques* (Springer, Berlin, 2010)
33. B. Berenbach, D. Paulish, J. Kazmeier, A. Rudorfer, *Software & Systems Requirements Engineering: In Practice* (McGraw-Hill, Inc., New York, 2009)
34. E. Hull, K. Jackson, J. Dick, *Requirements Engineering* (Springer Science & Business Media, 2010)
35. P.A. Laplante, *Requirements Engineering for Software and Systems* (CRC Press, Boca Raton, 2013)
36. M.K. Thompson, Improving the requirements process in axiomatic design theory. CIRP Ann. Manuf. Technol. **62**(1), 115–118 (2013)
37. M.K. Thompson, A classification of procedural errors in the definition of functional requirements in axiomatic design theory, in *Proceedings of the 7th International Conference on Axiomatic Design (ICAD'13)*, Worcester, MA, June 2013
38. L.-K. Chan, M.-L. Wu, A systematic approach to quality function deployment with a full illustrative example. Omega **33**(2), 119–139 (2005)
39. L.-K. Chan, M.-L. Wu, Quality function deployment: a literature review. Eur. J. Oper. Res. **143**(3), 463–497 (2002)
40. L.-K. Chan, M.-L. Wu, Quality function deployment: a comprehensive review of its concepts and methods. Qual. Eng. **15**(1), 23–35 (2002)
41. D. Dori, *Object-Process Methodology: A Holistics Systems Paradigm* (Springer, Berlin, 2002)
42. D. Dori, *Object-Process Methodology: A Holistic Systems Paradigm* (Springer Science & Business Media, 2013)
43. I. Pirbhai, D. Hateley, *Strategies for Real-Time System Specification* (Dorset House, New York, 1987)
44. S.M. McMenamin, J.F. Palmer, *Essential Systems Analysis* (Yourdon Press, 1984)
45. J.G. Miller, *Living Systems* (Mcgraw-Hill, New York, 1978)
46. D. Karnopp, D.L. Margolis, R.C. Rosenberg, *System Dynamics: A Unified Approach*, 2nd ed (Wiley, New York, 1990). (Online) Available: http://www.loc.gov/catdir/enhancements/fy0650/90012110-t.html, http://www.loc.gov/catdir/enhancements/fy0650/90012110-b.html, http://www.loc.gov/catdir/enhancements/fy0650/90012110-d.html
47. M. van Steen, *Graph Theory and Complex Networks: An Introduction* (Maarten van Steen, 2010), no. January
48. S.D. Eppinger, T.R. Browning, *Design Structure Matrix Methods and Applications* (MIT Press, Cambridge, 2012)
49. B. Friedland, *Control System Design: An Introduction to State-space Methods* (McGraw-Hill, New York, 1986)
50. K. Ogata, *Discrete-Time Control Systems*, 2nd edn. (Prentice Hall, Englewood Cliffs, 1994)
51. J. Castro, F. Alencar, G. Cysneiros, Closing the gap between organizational requirements and object oriented modeling. J. Braz. Comput. Soc. **7**(1), 05–16 (2000)

52. J. Rumbaugh, I. Jacobson, G. Booch, *The Unified Modeling Language Reference Manual* (Addison-Wesley, Reading, 2005)
53. C.A. Brown, Axiomatic design of manufacturing processes considering coupling, in Proceedings of ICAD2014 the Eighth International Conference on Axiomatic Design, 2014
54. A.M. Farid, D.C. McFarlane, Production degrees of freedom as manufacturing system reconfiguration potential measures. Proc. Inst. Mech. Eng. Part B (J. Eng. Manuf.) (invited paper) **222**(B10), 1301–1314 (2008). (Online) Available: http://amfarid.scripts.mit.edu/resources/Journals/IEM-J05.pdf
55. A.M. Farid, Product degrees of freedom as manufacturing system reconfiguration potential measures. Int. Trans. Syst. Sci. Appl. (invited paper) **4**(3), 227–242 (2008). (Online) Available: http://amfarid.scripts.mit.edu/resources/Journals/IEM-J04.PDF
56. A.M. Farid, Measures of reconfigurability & its key characteristics in intelligent manufacturing systems. J. Intell. Manuf. **1**(1), 1–26 (2014). (Online) Available: http://dx.doi.org/10.1007/s10845-014-0983-7
57. J.E. Bartolomei, Qualitative knowledge construction for engineering systems: extending the design structure matrix methodology in scope and procedure, Massachusetts Institute of Technology Engineering Systems Division, Technical Report, 2007
58. J.E. Bartolomei, D.E. Hastings, R. de Neufville, D.H. Rhodes, Engineering systems multiple-domain matrix: an organizing framework for modeling large-scale complex systems. Syst. Eng. **15**(1), 41–61 (2012)
59. S. Ullmann, *Semantics: An Introduction to the Science of Meaning* (1979)
60. C.K. Ogden, I.A. Richards, B. Malinowski, F.G. Crookshank, *The Meaning of Meaning* (Kegan Paul London, 1923)
61. F. De Saussure, W. Baskin, R. Harris (trans.), *Course in General Linguistics* (Open Court Publishing Company, 1986 (original 1916))
62. G. Guizzardi, *Ontological Foundations for Structural Conceptual Models* (CTIT, Centre for Telematics and Information Technology, 2005)
63. J.E. Shigley, C.R. Mischke, T.H. Brown, *Standard Handbook of Machine Design*, 3rd edn. (McGraw-Hill, New York, 2004)
64. J.D. Sterman, *Business Dynamics: Systems Thinking and Modeling for a Complex World*, vol. 19 (Irwin/McGraw-Hill Boston, 2000)
65. G. Guizzardi, On ontology, ontologies, conceptualizations, modeling languages, and (meta) models. Front. Artif. Intell. Appl. **155**, 18 (2007)
66. H.P. Grice, Logic and conversation, in *Syntax and Semantics*, vol. 3 (Academic Press, New York, 1970), pp. 43–58
67. M. Blaha, J. Rumbaugh, M. BlBlaha, Object-oriented modeling and design with UML (2005), pp. xvii, 477 p.
68. Anonymous, Synthesis, Dictionary.com, Technical Report, 2015. (Online) Available: http://dictionary.reference.com/browse/synthesis?s=t
69. A. Kossiakoff, W.N. Sweet, Knovel (Firm), *Systems Engineering Principles and Practice* (Wiley-Interscience, Hoboken, 2003). (Online) Available: http://www.knovel.com/knovel2/Toc.jsp?BookID=1430
70. G.A. Miller, The magical number seven, plus or minus two: some limits on our capacity for processing information. Psychol. Rev. **63**(2), 81 (1956)
71. A.M. Farid, Reconfigurability Measurement in Automated Manufacturing Systems, Ph.D. Dissertation, University of Cambridge Engineering Department Institute for Manufacturing, 2007. (Online) Available: http://amfarid.scripts.mit.edu/resources/Theses/IEM-TP00.pdf
72. I.S. Khayal, A.M. Farid, Axiomatic design based volatility assessment of the Abu Dhabi Healthcare Labor Market. J. Enterp. Transform. **5**(3), 162–191 (2015)
73. A.M. Farid, An axiomatic design approach to non-assembled production path enumeration in reconfigurable manufacturing systems, in 2013 IEEE International Conference on Systems Man and Cybernetics, Manchester, UK, 2013, pp. 1–8. (Online) Available: http://dx.doi.org/10.1109/SMC.2013.659

74. A. Viswanath, E.E.S. Baca, A.M. Farid, An axiomatic design approach to passenger itinerary enumeration in reconfigurable transportation Systems. IEEE Trans. Intell. Transp. Syst. **15** (3), 915–924 (2014). (Online) Available: http://amfarid.scripts.mit.edu/resources/Journals/TES-J08.pdf
75. W.N. Lubega, A.M. Farid, A reference system architecture for the energy-water nexus. IEEE Syst. J. (99), 1–11 (2014). (Online) Available: http://amfarid.scripts.mit.edu/resources/Journals/EWN-J11.pdf
76. S. Rivera, A.M. Farid, K. Youcef-Toumi, Chapter 15—A multi-agent system coordination approach for resilient self-healing operations in multiple microgrids, in *Industrial Agents*, ed. by P.L. Karnouskos (Morgan Kaufmann, Boston, 2015), pp. 269–285. (Online) Available: http://amfarid.scripts.mit.edu/resources/Books/SPG-B03.pdf
77. A.M. Farid, L. Ribeiro, An axiomatic design of a multi-agent reconfigurable mechatronic system architecture. IEEE Trans Ind. Inform. **11**(5), 1142–1155 (2015). (Online) Available: http://dx.doi.org/10.1109/TII.2015.2470528
78. A.M. Farid, W. Covanich, Measuring the effort of a reconfiguration process, in *IEEE International Conference on Emerging Technologies and Factory Automation, 2008. ETFA 2008,* Hamburg, Germany, 2008 (pp. 1137–1144). (Online) Available: http://dx.doi.org/10.1109/ETFA.2008.4638540
79. A.M. Farid, Multi-agent system design principles for resilient coordination & control of future power systems. Intell. Ind. Syst. **3**(1), 225–269 (2015). (Online) Available: http://amfarid.scripts.mit.edu/resources/Journals/SPG-J17.pdf
80. E. Hollnagel, D.D. Woods, N. Leveson, *Resilience Engineering: Concepts and Precepts, kindle*, edi edn. (Ashgate Publishing Limited, Aldershot, 2006)
81. M. Newman, *Networks: An Introduction* (Oxford University Press, Oxford, 2009). (Online) Available: http://books.google.ae/books?id=LrFaU4XCsUoC
82. K. Gershenson, G.J. Prasad, Y. Zhang, Product modularity: definitions and benefits. J. Eng. Des. **14**(3), 295–313 (2003)
83. J.K. Gershenson, G.J. Prasad, Y. Zhang, Product modularity: measures and design methods. J. Eng. Des. **15**(1), 33–51 (2004). (Online) Available: http://www.tandf.co.uk/journals
84. D. Mebratu, Sustainability and sustainable development: historical and conceptual review. Environ. Impact Assess. Rev. **18**(6), 493–520 (1998)
85. P. Glaviĉ, R. Lukman, Review of sustainability terms and their definitions. J. Clean. Prod. **15** (18), 1875–1885 (2007)
86. C. Böhringer, P.E. Jochem, Measuring the immeasurable—A survey of sustainability indices. Ecol. Econ. **63**(1), 1–8 (2007)
87. A. Madni, S. Jackson, Towards a conceptual framework for resilience engineering. IEEE Syst. J. **3**(2), 181–191 (2009)
88. R. Bhamra, S. Dani, K. Burnard, Resilience: the concept, a literature review and future directions. Int. J. Prod. Res. **49**(18), 5375–5393 (2011)
89. A. Arenas, A. Diaz-Guilera, J. Kurths, Y. Moreno, C. Zhou, Synchronization in complex networks. Phys. Rep. **469**(3), 93–153 (2008). (Online) Available: http://dx.doi.org/10.1016/j.physrep.2008.09.002
90. W. Shen, D. Norrie, Agent-based systems for intelligent manufacturing: a state-of-the-art survey. Knowl. Inf. Syst. Int. J. **1**(2), 129–156 (1999)
91. P. Leitao, Agent-based distributed manufacturing control: a state-of-the-art survey. Eng. Appl. Artif. Intell. **22**(7), 979–991 (2009). (Online) Available: http://dx.doi.org/10.1016/j.engappai.2008.09.005
92. R. Babiceanu, F. Chen, Development and applications of holonic manufacturing systems: a survey. J. Intell. Manuf. **17**, 111–131 (2006)
93. V. Marik, M. Fletcher, M. Pechoucek, O. Stepankova, H. Krautwurmova, M. Luck, Holons and agents: recent developments and mutual impacts, in *Multi-Agent Systems and Applications II: Lecture Notes in Artificial Intelligence* (Springer, Berlin, 2002), pp. 233–267
94. D.C. McFarlane, S. Bussmann, Developments in holonic production planning and control. Prod. Plann. Control **11**(6), 522–536 (2000)

95. D. McFarlane, S. Bussmann, S.M. Deen, Holonic manufacturing control: rationales, developments and open issues, in *Agent-Based Manufacturing* (Springer, Berlin, 2003), pp. 303–326
96. W.H. Ip, D. Wang, Resilience and friability of transportation networks: evaluation, analysis and optimization. IEEE Syst. J. **5**(2), 189–198 (2011)
97. V. Pillac, M. Gendreau, C. Gueret, A.L. Medaglia, A review of dynamic vehicle routing problems. Eur. J. Oper. Res. **225**(1), 1–11 (2013). (Online) Available: http://www.sciencedirect.com/science/article/pii/S0377221712006388
98. K.G. Zografos, K.N. Androutsopoulos, Algorithms for itinerary planning in multimodal transportation networks. IEEE Trans. Intell. Transp. Syst. **9**(1), 175–184 (2008)
99. K.G. Zografos, K.N. Androutsopoulos, V. Spitadakis, Design and assessment of an online passenger information system for integrated multimodal trip planning. IEEE Trans. Intell. Transp. Syst. **10**(2), 311–323 (2009)
100. L. Hame, H. Hakula, Dynamic journeying in scheduled networks. IEEE Trans. Intell. Transp. Syst. **14**(1), 360–369 (2013)
101. L. Häme, H. Hakula, Dynamic journeying under uncertainty. Eur. J. Oper. Res. **225**(3), 455–471 (2013)
102. R.D. Zimmerman, C.E. Murillo-Sanchez, R.J. Thomas, MATPOWER: steady-state operations, planning, and analysis tools for power systems research and education. IEEE Trans. Power Syst. **26**(1), 12–19 (2011). (Online) Available: http://dx.doi.org/10.1109/TPWRS.2010.2051168
103. J. Ash, D. Newth, Optimizing complex networks for resilience against cascading failure. Phys. A Stat. Mech. Appl. **380**, 673–683 (2007). (Online) Available: http://www.sciencedirect.com/science/article/pii/S0378437107002543
104. P. Holme, B. Kim, C. Yoon, S. Han, Attack vulnerability of complex networks. Phys. Rev. E **65**(5), 56101–56109 (2002)
105. R. Albert, H. Jeong, A.-L. Barabási, Error and attack tolerance of complex networks. Nature **406**(6794), 378–382 (2000)
106. D. Rowell, D.N. Wormley, *System Dynamics: An Introduction* (Prentice Hall, Upper Saddle River, 1997)
107. W. Schoonenberg, A.M. Farid, A dynamic production model for industrial systems energy management, in 2015 IEEE International Conference on Systems Man and Cybernetics, Hong Kong, 2015, pp. 1–7
108. A. Viswanath, A.M. Farid, A hybrid dynamic system model for the assessment of transportation electrification, in American Control Conference 2014, Portland, Oregon, 2014, pp. 1–7. (Online) Available: http://dx.doi.org/10.1109/ACC.2014.6858810
109. W.N. Lubega, A.M. Farid, Quantitative engineering systems model & analysis of the energy-water nexus. Appl. Energy **135**(1), 142–157 (2014). (Online) Available: http://dx.doi.org/10.1016/j.apenergy.2014.07.101
110. A.M. Farid, Electrified transportation system performance: conventional vs. online electric vehicles, in *Electrification of Ground Transportation Systems for Environment and Energy Conservation*, Chap. 22, eds. by N.P. Suh, D.H. Cho (MIT Press, 2015), pp. 1–25

Chapter 2
Mathematical Exposition of the Design Axioms

Hilario L. Oh

Abstract Axiomatic design (AD) offers designers two fundamental principles to follow for a successful design: (1) identify and define the design objectives, i.e., functional requirements (FRs), in such a way that they are inherently independent; and (2) conceive solutions for the FRs that comply with two design axioms: the independent axiom and the information axiom. In the previous chapter, the rationale and origin for the axiomatic nature of the design axioms were provided. In this chapter, the two axioms are given a deeper mathematical understanding, thereby strengthening their value. Starting with the formal definition of functional independence, the criterion for functional independence of FRs in a design is derived as the Jacobian determinant $|J| \neq 0$. Since $| J | \neq 0$ implies independence of FRs and existence of design solutions, the $|J|$ criterion corroborates the declaration of independence axiom that a good design must "maintain the independence of the functional requirements." The $|J|$ criterion further reveals that AD criterion for functional independence—design with single input–single output—is only a sufficient condition. For rigor and completeness, the $|J|$ criterion is shown to be necessary and sufficient. In implementing information axiom, AD assessment of uncertainty in design should cover a larger extent than it currently does. AD has not and should begin to recognize and identify the sources of variability and the countermeasures to them. The chapter ends with a summary of implementation steps in AD expressed in mathematical terms.

2.1 Introduction

Axiomatic design (AD) offers designers two fundamental principles to follow for a successful design: (1) define the design objectives, i.e., functional requirements (FRs), in such a way that they are inherently independent; and (2) conceive solutions in terms of design parameters (DPs) that maintain the independence of FRs as

H.L. Oh (✉)
MIT Park Center for Complex Systems, Cambridge, MA, USA
e-mail: ohlarry@mit.edu

© Springer International Publishing Switzerland 2016
A.M. Farid and N.P. Suh (eds.), *Axiomatic Design in Large Systems*,
DOI 10.1007/978-3-319-32388-6_2

originally intended and have the highest probability of achieving FRs' targets. AD uses independent axiom to check for functional independence and information axiom to assess probability of success. What follows are mathematical expositions of the two axioms. Chapter 1 discusses the dual application of the two axioms in design synthesis and design analysis. In this chapter, we will consider design analysis only.

2.2 Mathematical Exposition of Independence Axiom

Independent axiom in AD declares a criterion to check whether a conceived solution in terms of DPs maintains the functional independence of FRs. We develop another criterion derived from formal definition of functional independence. In the sections to follow, we discuss these two criteria. The discussion is confined to no more than 3 FRs and 3 DPs. However, the logic behind the discussion can be extended to n FRs and n DPs, $n > 3$.

2.2.1 AD Criterion for Functional Independence

Per independence axiom, "in an acceptable design, the DPs and the FRs are related in such a way that specific DP can be adjusted to satisfy its correspondent FR without affecting other functional requirements," p. 48 [1]. In other words, AD criterion for functional independence is that adjustment Δ, of one and only one DP should affect only correspondent FR but not other FRs. It implies a single input–single output (SISO) relationship of FRs to DPs. A mathematical representation of the criterion is as follows:

$$\begin{aligned}\Delta \mathrm{FR}_1 &= \Delta \mathrm{FR}_1(\Delta \mathrm{DP}_1)\\ \Delta \mathrm{FR}_2 &= \Delta \mathrm{FR}_2(\Delta \mathrm{DP}_2)\\ \Delta \mathrm{FR}_3 &= \Delta \mathrm{FR}_3(\Delta \mathrm{DP}_3)\end{aligned}$$

An alternative representation is with a design matrix (DM). A DM is indexed row-wise by FR_j and column-wise by DP_k. If DP_k has an effect on FR_j, the cell DM (j, k) is marked "X". If it has no effect, the cell is marked "O".

In DM representation, AD's SISO criterion for independence is a diagonal DM. Such a design is called an uncoupled design.

$$\begin{bmatrix}\Delta \mathrm{FR}_1\\ \Delta \mathrm{FR}_2\\ \Delta \mathrm{FR}_3\end{bmatrix} = \begin{bmatrix}X & O & O\\ O & X & O\\ O & O & X\end{bmatrix}\begin{bmatrix}\Delta \mathrm{DP}_1\\ \Delta \mathrm{DP}_2\\ \Delta \mathrm{DP}_3\end{bmatrix}$$

Another representation that also satisfies the SISO criterion is as follows:

$$\begin{aligned}
\Delta \mathrm{FR}_1 &= \Delta \mathrm{FR}_1(\Delta \mathrm{DP}_1);\\
\Delta \mathrm{FR}_2 &= \Delta \mathrm{FR}_2(\Delta \mathrm{DP}_1,\ \Delta \mathrm{DP}_2)\\
\Delta \mathrm{FR}_3 &= \Delta \mathrm{FR}_3(\Delta \mathrm{DP}_1,\ \Delta \mathrm{DP}_2,\ \Delta \mathrm{DP}_3).
\end{aligned}$$

In the above representation, $\Delta \mathrm{FR}_k$ can be made a function solely of $\Delta \mathrm{DP}_k$ if the adjustment $\Delta \mathrm{DP}_k$ to satisfy the corresponding $\Delta \mathrm{FR}_k$ follows the sequence: $k = 1$ firstly—so that $\Delta \mathrm{FR}_1$ becomes a constant in subsequent equation for $\Delta \mathrm{FR}_2$—followed by $k = 2$ secondly, and so on:

$$\begin{aligned}
\Delta \mathrm{FR}_1 &= \Delta \mathrm{FR}_1(\Delta \mathrm{DP}_1);\\
\Delta \mathrm{FR}_2 &= \Delta \mathrm{FR}_2(\Delta \mathrm{FR}_1,\ \Delta \mathrm{DP}_2);\\
\Delta \mathrm{FR}_3 &= \Delta \mathrm{FR}_3(\Delta \mathrm{FR}_1,\ \Delta \mathrm{FR}_2,\ \Delta \mathrm{DP}_3).
\end{aligned}$$

This adjustment sequence is known as forward substitution; an algorithm used in solving lower triangular linear systems [2]. If the adjustment adheres to the sequence, the above shows $\Delta \mathrm{FR}_k$ is a function exclusively of $\Delta \mathrm{DP}_k$. The SISO rule is thereby fulfilled. The DM is triangular. The design is called a decoupled design.

$$\begin{Bmatrix} \Delta \mathrm{FR}_1 \\ \Delta \mathrm{FR}_2 \\ \Delta \mathrm{FR}_3 \end{Bmatrix} = \begin{bmatrix} X & O & O \\ X & X & O \\ X & X & X \end{bmatrix} \begin{Bmatrix} \Delta \mathrm{DP}_1 \\ \Delta \mathrm{DP}_2 \\ \Delta \mathrm{DP}_3 \end{Bmatrix}$$

In AD, only uncoupled and decoupled designs are acceptable. Since they are SISO, the FRs are obviously functionally independent of one another.

$$\text{SISO} \Rightarrow \text{functional independence}$$

Any other design with DM that is neither diagonal nor triangular cannot satisfy SISO criterion. Such designs are called coupled designs. Per AD, they should be avoided since it is not obvious that the associated FRs are functionally independent of one another. We will show later that FRs in a design that does not satisfy SISO can still be functionally independent. In other words,

$$\text{Functional independence} \neg \Rightarrow \text{SISO}$$

Accordingly, the independence axiom with SISO criterion is only a sufficient condition for functional independence.

To recap, per independent axiom, there are three categories of design: uncoupled, decoupled, and coupled. Within the coupled design, there are three subcategories.

One subcategory is designed with cyclic interaction: DP_1 affects FR_2, DP_2 affects FR_3, and DP_3 affects FR_1:

$$\begin{bmatrix} \Delta FR_1 \\ \Delta FR_2 \\ \Delta FR_3 \end{bmatrix} = \begin{bmatrix} X & O & X \\ X & X & O \\ O & X & X \end{bmatrix} \begin{bmatrix} \Delta DP_1 \\ \Delta DP_2 \\ \Delta DP_3 \end{bmatrix}.$$

Another subcategory is redundant design with more DPs than FRs:

$$\begin{bmatrix} \Delta FR_1 \\ \\ \Delta FR_2 \end{bmatrix} = \begin{bmatrix} X & X & X \\ & & \\ X & X & X \end{bmatrix} \begin{bmatrix} \Delta DP_1 \\ \Delta DP_2 \\ \Delta DP_3 \end{bmatrix}.$$

A third subcategory is design with insufficient DPs:

$$\begin{bmatrix} \Delta FR_1 \\ \Delta FR_2 \\ \Delta FR_3 \end{bmatrix} = \begin{bmatrix} X & X \\ X & X \\ X & X \end{bmatrix} \begin{bmatrix} \Delta DP_1 \\ \\ \Delta DP_2 \end{bmatrix}.$$

We next show examples of various categories and subcategories of design.

2.2.1.1 Examples of Various Categories of Design

Water Faucet Illustrating Coupled and Uncoupled Design

An example frequently used to illustrate the uncoupled and coupled design has the two alternative designs of a water faucet shown in Fig. 2.1. In the figure, Q subscripted h and c are respectively the flow rate of the hot and cold water. Both faucet designs **a** and **b** have the same functional requirements:

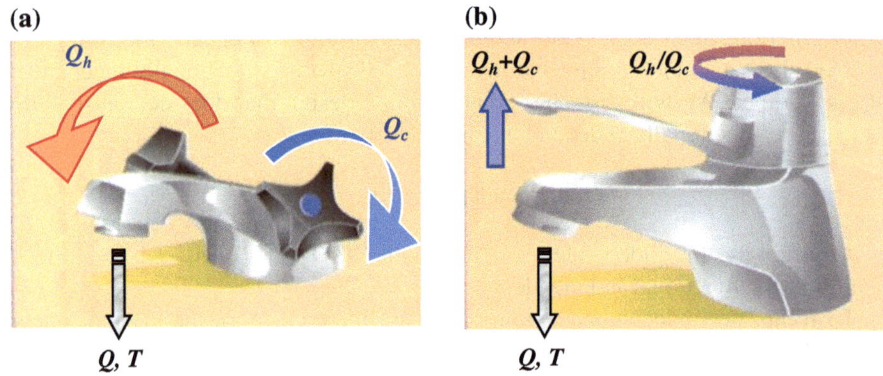

Fig. 2.1 Alternative designs **a** and **b** of a water faucet

$$\mathrm{FR}_1 = \text{control flow rate } Q;$$
$$\mathrm{FR}_2 = \text{control flow temperature } T.$$

For Faucet **a**, we choose as DP_1: the left knob controlling Q_h and as DP_2, the right knob controlling Q_c (see Fig. 2.1a). By assessing the effect of DPs on FRs, we arrive at the first category of coupled DM exhibiting cyclic interaction below.

$$\begin{bmatrix} \mathrm{FR}_1 \\ \mathrm{FR}_2 \end{bmatrix} = \begin{bmatrix} X & X \\ X & X \end{bmatrix} \begin{bmatrix} \mathrm{DP}_1 \\ \mathrm{DP}_2 \end{bmatrix}$$

Similarly for Faucet **b**, we choose as DP_1, the up/down of the lever to control total flow rate ($Q_\mathrm{h} + Q_\mathrm{c}$); and as DP_2, the clockwise/counterclockwise of the lever to control the ratio of the hot/cold water flow rate ($Q_\mathrm{h}/Q_\mathrm{c}$), see Fig. 2.1b. Again, by considering the effect of DPs on FRs, we obtain the uncoupled DM below that is acceptable per AD's SISO criterion for functional independence.

$$\begin{bmatrix} \mathrm{FR}_1 \\ \mathrm{FR}_2 \end{bmatrix} = \begin{bmatrix} X & O \\ O & X \end{bmatrix} \begin{bmatrix} \mathrm{DP}_1 \\ \mathrm{DP}_2 \end{bmatrix}$$

The examples above illustrate a practical aspect of AD. That is, by simply considering the effects of DPs on FRs with (X, O), we can differentiate an acceptable design from an unacceptable design without going into the details of the physics. This practicality is very useful at the concept selection stage of design.

Projector Illustrating Redundant Design

Projector has two FRs:

$$\mathrm{FR}_1 = \text{magnify the image;}$$
$$\mathrm{FR}_2 = \text{focus the image on the projection plane.}$$

Figure 2.2a shows a projector; and Fig. 2.2b, the associated ray tracing of the light beam from the object plane, through the lens, and to the projection plane.

From the similar triangles shown in Fig. 2.2b, we have

$$\mathrm{FR}_1 = \frac{\text{image height}}{\text{object height}} = \frac{D}{d};$$

Also per camera equation, the image is focused whenever

$$\mathrm{FR}_2 = \frac{1}{D} + \frac{1}{d} + \frac{1}{f} = 0.$$

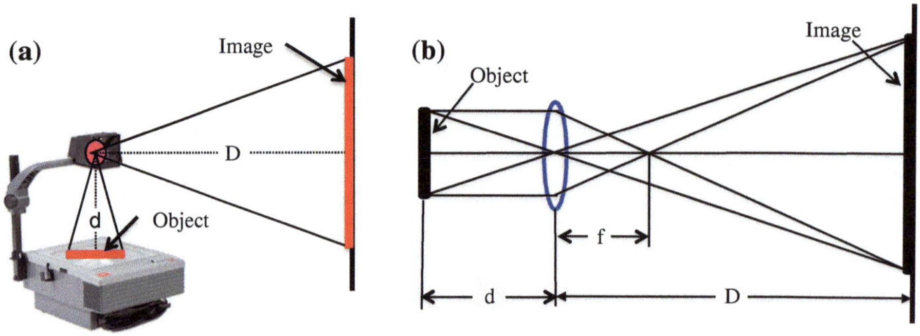

Fig. 2.2 Schematic **a** of a projector and its ray-tracing **b**

Thus, we have a redundant design with 2 FRs and 3 DPs as shown below.

$$\begin{bmatrix} FR_1 \\ \\ FR_2 \end{bmatrix} = \begin{bmatrix} X & X & O \\ \\ X & X & X \end{bmatrix} \begin{bmatrix} DP_1 \\ DP_2 \\ DP_3 \end{bmatrix}.$$

In the above, DP_1 is D, the distance of the lens from the screen aka the throw of the projector; DP_2 is d, the distance of the lens from the object, and DP_3 is f, the focal length of the lens.

A redundant design with more DPs than FRs cannot satisfy SISO criterion unless we fix the extra DPs. For example in the type of overhead projector shown in Fig. 2.2a, the focal length of the lens DP_3 is fixed. Hence, we have a design with equal number of FRs and DPs shown below. In this case, the design is coupled. With this type of projector, it would take several trial and errors to get the right magnification of the image focused at a given throw.

$$\begin{bmatrix} FR_1 \\ FR_2 \end{bmatrix} = \begin{bmatrix} X & X \\ X & X \end{bmatrix} \begin{bmatrix} DP_1 \\ DP_2 \end{bmatrix}$$

By contrast, in a portable projector used in a variety of room sizes that requires various throws, the distance of the object from the lens DP_2 is fixed. A zoom lens with varying focal length DP_3 is used. Thus we have a decoupled design:

$$\begin{bmatrix} FR_1 \\ FR_2 \end{bmatrix} = \begin{bmatrix} X & O \\ X & X \end{bmatrix} \begin{bmatrix} DP_1 \\ DP_3 \end{bmatrix}.$$

With a portable projector, to attain a certain magnification focused at a certain throw is easy: set the throw for magnification then adjust the zoom lens for focus.

Disbursement Algorithm Illustrating a Redundant Design

Let us say we have an ATM that has in it the following bank notes denomination: \$20, \$10, \$5, and \$1. Three demands are made of the ATM as follows:

FR_1: disburse bills that sum up to \$Total.
FR_2: disburse number of bills that totals to N_{Total}.
FR_3: ensure the number of small bills is twice that of large bills.

There are four design parameters (DPs):

DP_1: number of \$20 bills, $N_{\$20}$.
DP_2: number of \$10 bills, $N_{\$10}$.
DP_3: number of \$5 bills, $N_{\$5}$.
DP_4: number of \$1 bills, $N_{\$1}$.

The 4 DPs would satisfy the three FRs as follows:

$$
\begin{aligned}
FR_1 &= \$20N_{\$20} + \$10N_{\$10} + \$5N_{\$5} + \$1N_{\$1} = \$\text{Total} \\
FR_2 &= N_{\$20} + N_{\$10} + N_{\$5} + N_{\$1} = N_{\text{Total}} \\
FR_3 &= 2N_{\$20} - N_{\$1} = 0.
\end{aligned}
$$

Above is a redundant design with 3 FRs and 4 DPs with a DM as shown below:

$$
\begin{bmatrix} FR_1 \\ FR_2 \\ FR_3 \end{bmatrix} = \begin{bmatrix} X & X & X & X \\ X & X & X & X \\ X & O & O & X \end{bmatrix} \begin{bmatrix} N_{\$20} \\ N_{\$10} \\ N_{\$5} \\ N_{\$1} \end{bmatrix}
$$

If we were to fix an extra DP to get equal number of FRs and DPs, we would have 4 (=$_4C_3$) possible DM solutions as follows:

$$
\begin{bmatrix} FR_1 \\ FR_2 \\ FR_3 \end{bmatrix} = \begin{bmatrix} \$20 & \$10 & \$5 \\ 1 & 1 & 1 \\ 2 & 0 & -1 \end{bmatrix} \begin{bmatrix} N_{\$20} \\ N_{\$10} \\ N_{\$5} \end{bmatrix}; \tag{2.1a}
$$

$$
\begin{bmatrix} FR_1 \\ FR_2 \\ FR_3 \end{bmatrix} = \begin{bmatrix} \$20 & \$10 & \$1 \\ 1 & 1 & 1 \\ 2 & 0 & -1 \end{bmatrix} \begin{bmatrix} N_{\$20} \\ N_{\$10} \\ N_{\$1} \end{bmatrix}; \tag{2.1b}
$$

$$
\begin{bmatrix} FR_1 \\ FR_2 \\ FR_3 \end{bmatrix} = \begin{bmatrix} \$20 & \$5 & \$1 \\ 1 & 1 & 1 \\ 2 & 0 & -1 \end{bmatrix} \begin{bmatrix} N_{\$20} \\ N_{\$5} \\ N_{\$1} \end{bmatrix}; \tag{2.1c}
$$

$$\begin{bmatrix} FR_1 \\ FR_2 \\ FR_3 \end{bmatrix} = \begin{bmatrix} \$10 & \$5 & \$1 \\ 1 & 1 & 1 \\ 2 & 0 & -1 \end{bmatrix} \begin{bmatrix} N_{\$10} \\ N_{\$5} \\ N_{\$1} \end{bmatrix}. \tag{2.1d}$$

Note in solution (2.1a), $FR_3 = (FR_1 - \$10 \times FR_2)/\$5$. So that $FR_3 = f(FR_1, FR_2)$ is no longer independent as originally planned. In fact, design Eq. (2.1a) is itself a redundant design with 2 FRs and 3 DPs. In short, fixing extra DPs in a redundant design to obtain a square DM does not guarantee a solution. It may induce coupling and destroy the functional independence originally planned for.

Hubcap Illustrating Insufficient DPs

Figure 2.3a shows the front of GM 1986–88 Pontiac 6000 hubcap. Figure 2.3b shows the wheel rim with a circumferential ledge, shown white, onto which the hubcap snapped on. The diameter of the ledge is D_{rim}. Figure 2.3c shows three pairs of clips at the back of the hubcap. The pair at the 4 o'clock position is shown enlarged in Fig. 2.3d. The clips are cantilevers fixed on a post. As seen in Fig. 2.3c, the three pairs of clips are spaced 120° apart such that the 6 clips form a circle of diameter D_{clip}, larger than D_{rim}. As the hubcap is snapped on to the rim, the rim ledge catches the cantilever clips. Wheel retention is developed through interference fit = $k\delta$; where k is the spring rate of the cantilever clips and δ is the interference = $(D_{clip} - D_{rim})/2$.

The are two FRs for the hubcap design:

FR_1 = retain the hubcap over road bumps and on cornering;
FR_2 = ease the removal of hubcap during a flat tire repair.

There is only one design parameter:

DP_1 = interference, the larger the better for retention; the smaller the better for removal.

Obviously the design is flawed since it has insufficient DPs: one DP, the interference, cannot satisfy two conflicting FRs, retention, and removal. It violates the AD's SISO criterion of one DP affecting only one FR.

$$\begin{bmatrix} FR_1 \\ FR_2 \end{bmatrix} = \begin{bmatrix} X \\ X \end{bmatrix} [DP_1]$$

The consequence was 25 % of hubcaps fell off as the car corners or hits bumps or potholes. And some customers have difficulty removing the hubcap for a flat repair. The solution [3] back then was to implement robust design optimization to find a clip spring rate k that reduces the design sensitivity to the variation in interference. The solution had limited success, as performance of an ill-conceived design cannot be improved through subsequent optimization.

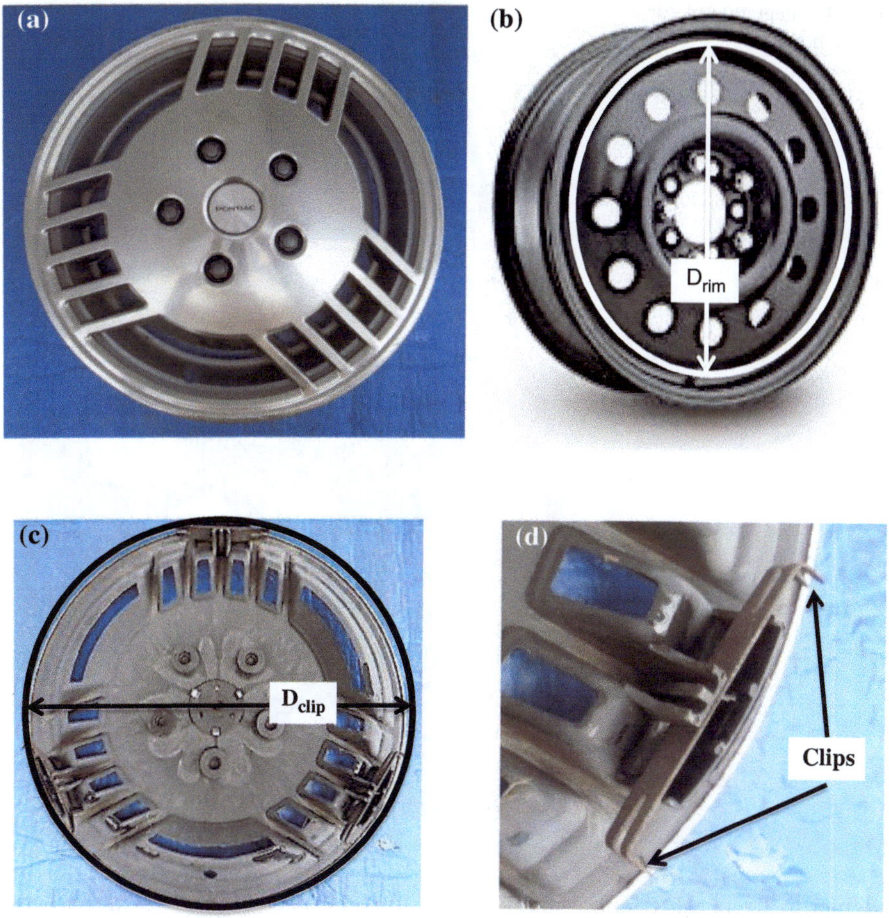

Fig. 2.3 Attachment of hubcap to wheel rim, **a** Hubcap front, **b** Wheel rim, **c** Hubcap back, **d** Cantilever clips

2.2.1.2 Car Door-to-Body Integration—Coupling in Large System

Thus far, we have illustrated the use of AD's SISO criterion to accept or reject various categories of design. These illustrations involve designs with small number of FRs. We now apply SISO to designs with large number of FRs; specifically to a car door-to-body integration. Figure 2.4 shows the car door integrated to the body opening.

Problems in car door-to-body integration, e.g., poor fit of door with neighboring panels; noisy ride and water leak; high opening and closing effort, are typical system problems. They appear only after the system is assembled since only then are couplings triggered. Fixing them is like playing a whack-a-mole game. As one solves a problem in one area, new problems pop up in other areas. This is because

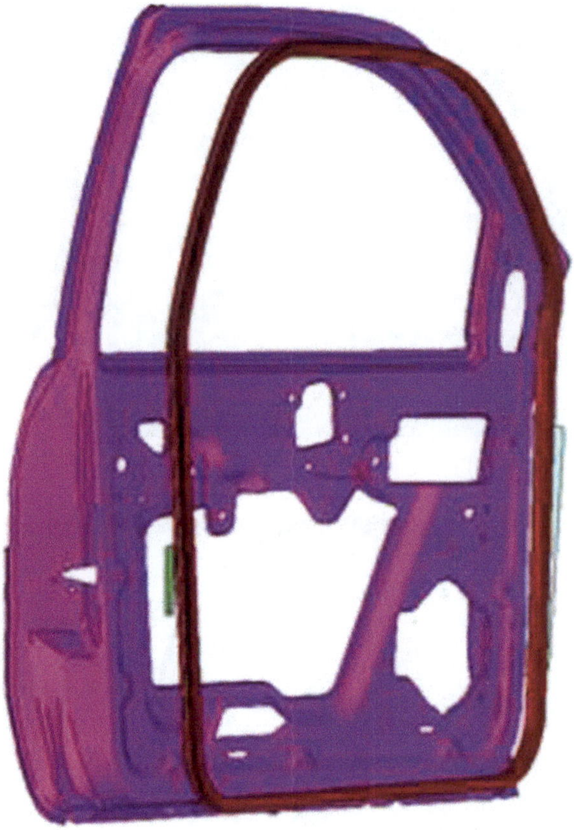

Fig. 2.4 Integration of car door to body

attempt to fix one FR failure inadvertently triggers other FR failures due to coupling. These types of system failures are not detectable by the traditional recursive design/build/test of components since they cannot capture the inter-dependence of FRs among subsystems and components.

AD takes a top-down approach. It starts with system-level FRs and decomposes them down through the subsystems until they reach the levels where known, implementable solutions exist. These levels reached are called leaf-level FRs, and the corresponding solutions, leaf-level DPs. Along the way, AD examines the interrelationships across the subsystems and components. In this way, functional couplings are understood and captured.

The AD top-down approach is as follows. First we decompose the system-level FRs down the design hierarchy to the leaf-level design parameters. From the decomposition, we construct the DM that shows the relationship among the leaf-level FR–DP pairs. We then successively remove those FR–DP pairs that are not part of the coupling. What remains is a reduced DM that contains all the couplings of the original DM. Retracing the leaf-level couplings up the design hierarchy reveals the roots of the system-level functional interactions.

Capturing Functional Couplings with the Design Matrix

Appendix A1 shows the FR decomposition and Appendix A2 the corresponding DP decomposition of the car door–to–body system. The decomposition starts with three subsystems: FR_1: fit door to neighboring panels; FR_2: keep interior quite and intrusion free; and FR_3: ensure door opens and closes properly. The subsystem FRs are then decomposed down the hierarchy, zigzagging between FRs and DPs, i.e., between Appendices A1 and A2, until the leaf-level DPs marked by "+" are reached. These leaf-level DPs, together with the corresponding FRs they satisfy, are listed in Table 2.1. For succinct presentation, Table 2.1 uses serial notations 1–28 in place of hierarchical notations in Appendix A. For example, FR_1 in Table 2.1 refers to $FR_{1.1.1}$ in Appendix A1. Note that FR_1 thru FR_7 are leaf-level FRs that flow from the first subsystem; FR_8 thru FR_{18} are leaf-level FRs, from the second subsystem and FR_{19} thru FR_{28}, from the third subsystem.

Table 2.1 Leaf-level FR–DP of car door-to-body system

FR_1 achieve uniform gap on both edges	DP_1 hinge tip in x–z plane
FR_2 balance leading and trailing edge gaps	DP_2 fore/aft position of hinge axis
FR_3 align feature lines	DP_3 vertical position of hinge datum
FR_4 achieve flushness at leading edge	DP_4 in/out position of hinge axis
FR_5 achieve flushness along both edges	DP_5 hinge tip in y–z plane
FR_6 achieve flushness at trailing edge	DP_6 in/out position of striker
FR_7 achieve flushness above beltline	DP_7 header over bent
FR_8 maintain adequate seal margin	DP_8 position of door interior surface
FR_9 maintain adequate seal height	DP_9 a system to maintain uniform seal height
FR_{10} maintain seal footprint	DP_{10} contour of contacting surfaces
FR_{11} divert away water	DP_{11} channel slope
FR_{12} detune seal from noise transmission.	DP_{12} modal property of seal section
FR_{13} dissipate noise energy	DP_{13} seal damping characteristic
FR_{14} eliminate seal itch	DP_{14} lubricant, substrate loss modulus
FR_{15} prevent gap-induced turbulence	DP_{15} gap filler
FR_{16} stop flushness-induced turbulence	DP_{16} header stiffness
FR_{17} control leakage across seals	DP_{17} sealing energy as barrier to intrusion
FR_{18} maintain mass flow rate of inlet air	DP_{18} fan
FR_{19} ensure reaction force > gravity	DP_{19} stiffness and preloads of check link spring
FR_{20} bar opening door swing thru stops	DP_{20} site, depth and climb of check link valleys
FR_{21} let closing door swing thru stops	DP_{21} site, depth and descent of check link valleys
FR_{22} eliminate resistance to swing	DP_{22} hinge axes aligned with axis of rotation
FR_{23} lower KE to surmount latch misalign	DP_{23} up-down adjustable striker
FR_{24} lower KE to compress seal	DP_{24} area under weather strip CLD
FR_{25} lower KE to deflect header	DP_{25} area under header load-deflection curve
FR_{26} lower KE to overcome air bind	DP_{26} pressure relief valve
FR_{27} store spring energy from opening	DP_{27} preloaded check link torsional spring
FR_{28} reduce effort to unlatch	DP_{28} mechanism to relieve reaction at latch

From Table 2.1, we construct the DM that captures the functional interdependencies among the 28 leaf-level FR–DP pairs, see Fig. 2.5. Row-wise are leaf-level FRs; column-wise are associated leaf-level DPs. Leaf-level FR_1 thru FR_7 and their associated DPs, which flow from the first subsystem, are labeled to the left and top of the DM, not shaded.

The leaf-level FR_8 thru FR_{18} and their associated DPs, which flow from the second subsystem, are shown lightly shaded green. And FR_{19} thru FR_{28}, which flow from the third subsystem are heavily shaded blue. For each cell DM (j, k) of the 28 × 28 cells, assessment is made whether the DP_k has an effect on FR_j. If it has an effect (no effect), the cell is marked with an "X" (blank). While cell-by-cell evaluation is tedious, it is crucial because functional inter-dependencies so obtained among the leaf-level FR–DP decide how the design functions at the system level.

Reducing the Design Matrix to Uncover Functional Coupling

When a DM is sparse or small, we can check for coupling by inspection. When it is large, the task becomes difficult. For DM with $n \times n = 28 \times 28$ and a total off-diagonal elements of $m = 88$ as in Fig. 2.5, the number of possible couplings equals $2^{m-n+1} - 1 = 2.306\text{E}18$ [4]. It is prohibitive to check for couplings among this

	DP1	DP2	DP3	DP4	DP5	DP6	DP7	DP8	DP9	DP10	DP11	DP12	DP13	DP14	DP15	DP16	DP17	DP18	DP19	DP20	DP21	DP22	DP23	DP24	DP25	DP26	DP27	DP28
FR1	**X**																											
FR2	X	X																										
FR3	X		X																				X					
FR4				**X**																								
FR5				X	X																							
FR6					X	X																						
FR7				X	X	X	X	X	X							X								X	X			
FR8				X	X	X	X	X															X	X				
FR9									**X**																			
FR10				X	X	X	X	X		X													X					
FR11											**X**																	
FR12												**X**																
FR13													**X**															
FR14				X	X	X	X	X	X	X			X	**X**														
FR15	X	X	X												**X**								X					
FR16				X		X			X							X									X			
FR17				X	X	X	X	X	X	X						X	**X**							X	X			
FR18																		**X**										
FR19	X				X														X								X	
FR20	X				X														X	X							X	
FR21	X				X														X		**X**						X	
FR22	X				X															X		**X**					X	
FR23	X																						X					
FR24				X	X	X	X	X	X							X							X	X				
FR25				X	X	X	X	X								X							X	X	X			
FR26																										**X**		
FR27																											**X**	
FR28				X	X	X		X															X					**X**

Fig. 2.5 Design matrix relating the leaf-level FR–DP

large number of candidates. Thus, we reduce the dimension of DM by isolating the submatrix that contains the couplings from the rest. We do this by successively relocate rows (columns) whose row-wise (column-wise) entries are all zeros but the diagonal element. For example in Fig. 2.5, we find five zero-rows: FR_1, FR_4, FR_9, FR_{13}, and FR_{27}; six zero-columns: DP_{14}, DP_{15}, DP_{17}, DP_{21}, DP_{22}, and DP_{28} and four combined (zero-rows, zero-columns): (FR_{11}, DP_{11}), (FR_{12}, DP_{12}), (FR_{18}, DP_{18}), and (FR_{26}, DP_{26}). A zero-row corresponds to an FR that is not affected by other DPs; a zero-column corresponds to a DP that does not affect any other FRs; and a combined (zero-row, zero-column) corresponds to an FR–DP pair that does not affect nor be affected by other FRs and DPs. All the three categories do not belong to any coupling loop. They are thus moved from their original locations. Namely through row and column interchange, all combined (zero-row, zero-column) are relocated to the upper left corner of the DM; all zero-columns (zero-rows) and their associated rows (columns) are relocated to the lower right (upper left) corner of the DM. What remains is a 13 × 13 submatrix outlined in thick border as shown in Fig. 2.6.

	DP11	DP12	DP18	DP26	DP1	DP4	DP9	DP13	DP27	DP2	DP3	DP5	DP6	DP7	DP8	DP10	DP16	DP19	DP20	DP23	DP24	DP25	DP14	DP15	DP17	DP21	DP22	FR28
FR11	X																											
FR12		X																										
FR18			X																									
FR26				X																								
FR1					X																							
FR4						X																						
FR9							X																					
FR13								X																				
FR27									X																			
FR2					X					X																		
FR3					X						X									X								
FR5						X						X																
FR6												X	X															
FR7						X	X					X	X	X	X		X				X	X						
FR8						X						X	X	X	X					X	X							
FR10						X						X	X	X	X	X				X								
FR16						X	X						X				X					X						
FR19					X				X			X						X										
FR20					X				X			X						X	X									
FR23					X															X								
FR24						X	X					X	X	X	X		X			X	X							
FR25						X						X	X	X	X		X			X	X	X						
FR14						X	X	X				X	X	X	X	X							X					
FR15					X					X	X									X				X				
FR17						X	X					X	X	X	X	X	X				X	X			X			
FR21					X				X			X						X								X		
FR22					X				X			X							X								X	
FR28						X						X	X		X					X								X

Fig. 2.6 Relocating zero-rows and zero-columns

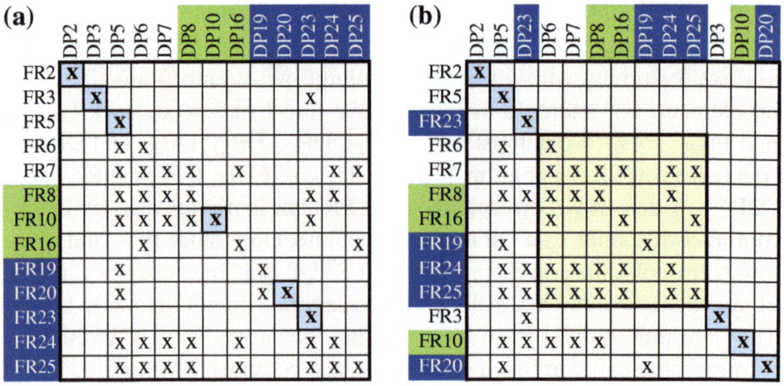

Fig. 2.7 Further detection **a** and relocation **b** of zero-row and zero-column

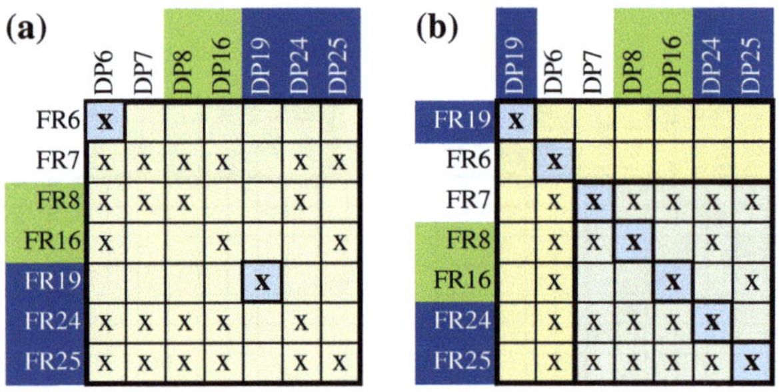

Fig. 2.8 Final detection **a** and relocation **b** of zero-row and zero-column

Further examination of the 13×13 submatrix reveals 3 more zero-rows: FR_2, FR_5 and FR_{23}; and 3 more zero-columns: DP_3, DP_{10} and DP_{20}, as shown in Fig. 2.7a. We repeat the successive relocation of these zero-rows and zero-columns and arrive at a further reduced 7×7 submatrix outlined in thick border in Fig. 2.7b.

Continuing the search for zero-rows and zero-columns in the 7×7 reduced matrix, we found 1 zero-row FR_6 and 1 zero-column DP_{19} (see Fig. 2.8a). Upon relocation of these two, we finally obtain a 5×5 matrix that containing neither zero-row nor zero-column as outlined in thick border, Fig. 2.8b. All the couplings in the original decomposed DM are now isolated and condensed into this 5×5 DM.

Implications from the Reduction of DM

Figure 2.9a shows DM as decomposed juxtaposed with Fig. 2.9b DM as condensed. The as-condensed DM shows three submatrices: a (4×4) uncoupled submatrix in the upper left; a (24×24) decoupled submatrix in the lower right; and a protruding (5×5) coupled submatrix within the decoupled submatrix. These

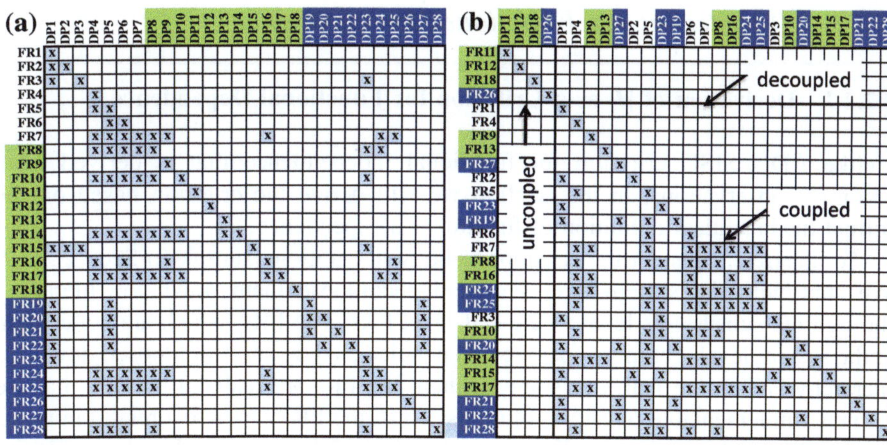

Fig. 2.9 Design matrix, **a** as decomposed and **b** as condensed

results reflect the reduction algorithm: relocating the combined (zero-rows, zero-columns) to DM upper left produces the uncoupled submatrix; relocating the zero-rows to DM upper left plus the zero-columns to DM lower right produce the decoupled submatrix. What remains is a protruding coupled submatrix within the decoupled submatrix. As indicated on top of DM Fig. 2.9b, the algorithm also produces a mingling of leaf-level DPs from the three subsystems. The implications of these results are as follows.

The protruding (5 × 5) coupled submatrix is the source of the whack-a-mole type of failures. The couplings need to be resolved first and foremost by effectively identify and eliminate functional couplings following for example a graph theory-based method described in [4].

Once couplings in the (5 × 5) coupled matrix are resolved, what remains is a (24 × 24) lower triangular DM. The lower triangular DM serves as a road map and provides a sequence to follow in satisfying ΔFRs: adjust ΔDP_j, $j = 1 \ to \ i$, to satisfy/fix ΔFR_i. Without the road map, we will still be fighting the whack-a-mole type of failure.

We must recognize which FR–DP pair falls unto the uncoupled submatrix and take advantage of the information, as they are the easiest to fix and satisfy.

Engineers in a door group are typically tasked with specific leaf-level functions of the door. As indicated in Fig. 2.9b, there is a mingling of the leaf-level DP that forms the triangular DM. The engineers must be made aware of this interdependency, i.e., mingling, of functions since their tasks must conform to the sequence dictated by the triangular DM.

The implications described above hold for any assembly of subsystems and components that form a large system.

2.2.2 |J| Criterion for Functional Independence

2.2.2.1 Derivation of |*J*| as a Criterion for Functional Dependence

Unlike AD's SISO criterion for functional independence that is derived from empirical observations, in this section we will derive the criterion based on the formal mathematical definition of functional independence.

To illustrate, consider the car door-to-body integration example in Sect. 2.2.1.2. We start with the system-level FRs and decompose them down the hierarchy through the subsystems level to the leaf-level FRs. The FRs, which are conceptual at the system level, get more specific and detailed as they are decomposed down the hierarchy. When the leaf levels are reached, the FRs are realized by DPs that are known, implementable physical solutions. Thus, the FRs can be expressed in terms of the DPs through physics. For example, Table 2.1 relates 28-leaf-level FRs to corresponding leaf-level DPs. We denote these relationships as follows:

$$\mathrm{FR}_1 = f_1(\mathrm{DP}_1, \ldots, \mathrm{DP}_m)$$
$$\vdots$$
$$\mathrm{FR}_n = f_n(\mathrm{DP}_1, \ldots, \mathrm{DP}_m)$$

Or in vector notation,

$$\mathbf{FR} = \mathbf{f}(\mathbf{DP});$$

In the above and hereafter, a bolded quantity denotes a vector, a bracketed quantity denotes a matrix, and **f(•)** denotes vector valued functions.

The vector equation above is Eq. (2.6) of Chap. 1, with $\mathbf{f}(\mathbf{DP}) \equiv \mathbf{f}_a(\mathbf{DP})$. The vector **DP** represents the physical quantities of the design, and the vector valued function **f(DP)** represents the laws of physics relating **FR** to **DP**. Since **f(DP)** is drawn from laws of physics, it may be assumed as continuous. So that we may expand **f(DP)** in a Taylor series about a design point $\mathbf{DP}^*$:

$$\mathbf{f}(\mathbf{DP}) = \mathbf{f}(\mathbf{DP}^*) + [J](\mathbf{DP} - \mathbf{DP}^*) + \mathbf{o}(||\mathbf{DP} - \mathbf{DP}^*||)$$
$$\approx \mathbf{f}(\mathbf{DP}^*) + [J](\mathbf{DP} - \mathbf{DP}^*)$$

Thus: $\mathbf{FR} - \mathbf{FR}^* \approx [J](\mathbf{DP} - \mathbf{DP}^*)$

where $[J]$ is the Jacobian matrix whose element $J_{ij} = \partial \mathrm{FR}_i / \partial \mathrm{DP}_j$ evaluated at the design point $\mathbf{DP}^*$ is a constant.

We recognize the Jacobian $[J]$ above is in fact the design matrix $[A]$ in AD, Eq. (3.3) in [1]. Thus, we may rewrite the vector equation as follows:

$$\Delta\mathbf{FR} = [A]\Delta\mathbf{DP}. \tag{2.2}$$

If **FRs** are linear functions of **DP**, the above differential vector equation reduces:

$$\mathbf{FR} = [A]\mathbf{DP}. \tag{2.2a}$$

Equation (2.2a) is known as a "design equation" and appears on page 55 of the first axiomatic design text [1]. It has since been used extensively for conceptual applications in the AD literature. However, it is important to recognize that in most cases Eq. (2.2a) is only a notation to convey a relation between **FR** and **DP**. In actuality, the equation to solve for is the differential form in Eq. (2.2).

Expanding the differential vector equation for the special case of $i, j = 1, 2$:

$$\Delta \mathrm{FR}_1 = A_{11}\Delta \mathrm{DP}_1 + A_{12}\Delta \mathrm{DP}_2; \tag{2.3}$$

$$\Delta \mathrm{FR}_2 = A_{21}\Delta \mathrm{DP}_1 + A_{22}\Delta \mathrm{DP}_2. \tag{2.4}$$

Equation (2.3) × A_{22} minus Eq. (2.4) × A_{12} to eliminate $\Delta \mathrm{DP}_2$ gives:

$$A_{22}\Delta \mathrm{FR}_1 - A_{12}\Delta \mathrm{FR}_2 = (A_{22}A_{11} - A_{12}A_{21})\Delta \mathrm{DP}_1 \tag{2.5}$$

Note that $(A_{22}\ A_{11} - A_{12}\ A_{21}) = \left(\frac{\partial \mathrm{FR}_2}{\partial \mathrm{DP}_2}\right)\left(\frac{\partial \mathrm{FR}_1}{\partial \mathrm{DP}_1}\right) - \left(\frac{\partial \mathrm{FR}_1}{\partial \mathrm{DP}_2}\right)\left(\frac{\partial \mathrm{FR}_2}{\partial \mathrm{DP}_1}\right)$ is the determinant $|A|$ of the DM. It is known as $|J|$, the Jacobian determinant in vector calculus.

$$|J| = \begin{vmatrix} \frac{\partial \mathrm{FR}_1}{\partial \mathrm{DP}_1} & \frac{\partial \mathrm{FR}_1}{\partial \mathrm{DP}_2} \\ \frac{\partial \mathrm{FR}_2}{\partial \mathrm{DP}_1} & \frac{\partial \mathrm{FR}_2}{\partial \mathrm{DP}_2} \end{vmatrix}.$$

Thus, if $(A_{22}\ A_{11} - A_{12}\ A_{21}) = |J| = 0$ in Eq. (2.5), then $\Delta \mathrm{FR}_2 = A_{22}\Delta \mathrm{FR}_1/A_{12}$. Or $\mathrm{FR}_2 = \mathrm{FR}_2^* + A_{22}\ (\mathrm{FR}_1 - \mathrm{FR}_1^*)/A_{12}$. Namely, FR_2 is dependent on FR_1. Hence, $|J| = 0$ implies functional dependence.

$$\text{Proof A:} \quad (|J| = 0) \Rightarrow \text{functional dependence}$$

We next prove the converse is true. Namely if FR_2 is functionally dependent on FR_1, then $|J| = 0$. We start with the formal definition of functional dependency. Namely, FR_2 is dependent on FR_1 if it is a function of FR_1:

$$\mathrm{FR}_2 = \mathrm{FR}_2(\mathrm{FR}_1)$$

Applying the chain rule for differentiation on above equation, we have

$$\frac{\partial \mathrm{FR}_2}{\partial \mathrm{DP}_1} = \frac{\partial \mathrm{FR}_2}{\partial \mathrm{FR}_1}\frac{\partial \mathrm{FR}_1}{\partial \mathrm{DP}_1}$$
$$\frac{\partial \mathrm{FR}_2}{\partial \mathrm{DP}_2} = \frac{\partial \mathrm{FR}_2}{\partial \mathrm{FR}_1}\frac{\partial \mathrm{FR}_1}{\partial \mathrm{DP}_2}.$$

In the above, multiply 1st equation by $\frac{\partial FR_1}{\partial DP_2}$ and the 2nd by $\frac{\partial FR_1}{\partial DP_1}$. Subtract one resulting equation from the other to eliminate $\left(\frac{\partial FR_2}{\partial FR_1}\right)\left(\frac{\partial FR_1}{\partial DP_2}\right)\left(\frac{\partial FR_1}{\partial DP_1}\right)$, we have:

$$\left(\frac{\partial FR_1}{\partial DP_2}\right)\left(\frac{\partial FR_2}{\partial DP_1}\right) - \left(\frac{\partial FR_1}{\partial DP_1}\right)\left(\frac{\partial FR_2}{\partial DP_2}\right) \equiv |J| = 0$$

Hence, functional dependence implies $|J| = 0$. This is the proof of the converse:

Proof B: Functional dependence $\Rightarrow (|J| = 0)$.

Combining both proofs A and B, we have

Functional dependence $\Leftrightarrow (|J| = 0)$.

Namely, $|J| = 0$ is a necessary and sufficient condition for FR_2 to be functionally dependent on FR_1. Likewise, FR_2 is functionally independent of FR_1 if and only if (iff) $|J| \neq 0$. Thus by formal definition of functional dependency, we have derived the criterion: FRs are functionally independent iff $|J| \neq 0$; dependent iff $|J| = 0$.

2.2.2.2 Implications of $|J|$ as a Criterion for Functional Independence

The differential form of Eq. (2.2) may be rewritten as follows.

$$\mathbf{\Delta FR} = [J]\mathbf{\Delta DP}$$

So that the adjustments $\mathbf{\Delta DP}$ necessary to bring **FR** to its target **FR*** is,

$$\mathbf{\Delta DP} = [J]^{-1}\mathbf{\Delta FR}$$

Note that if **FRs** of the design are functionally dependent, then $|J| = 0$ and its inverse $|J|^{-1}$ does not exist. In mathematical lingua, the design has a "singularity" in its first derivative and is non-differentiable. Consequently, no adjustments in $\mathbf{\Delta DP}$ can bring the design to its target value **FR***.

AD's independence axiom declares that a good design must "maintain the independence of the functional requirements (**FR**)." The $|J|$ criterion corroborates this declaration since $|J| \neq 0$ implies independence of **FR** and it guarantees a design solution. Therefore, the $|J|$ criterion provides formidable theoretical evidence that a violation of the independence axiom will impede the design from finding a final value **DP*** that fulfills the design equation and meets all functional requirements **FR***. While the AD independence axiom was established through extensive empirical study to yield "good" designs, the $|J|$ criterion shows that these "good" designs not only fulfill all their functional requirements but also can be found through well established analytical and numerical methods.

AD further proposes SISO as the criterion for independence. Since determinant of a diagonal or triangular DM—a SISO design—is the product of all the diagonal elements none of which is zero, their $|J| \neq 0$. This validates SISO as a criterion for functional independence. However, SISO criterion is only a sufficient condition. Namely, SISO implies functional independence but functional independence does not imply SISO:

$$\text{SISO} \Rightarrow \text{Functional independence;}$$
$$\text{Functional independence} \neg \Rightarrow \text{SISO.}$$

While SISO criterion is more conservative, it does not detract from its utility. It remains as a sufficiency condition. Furthermore, during design synthesis it is relatively easy to mentally keep track of a diagonal or lower triangular design matrix. In comparison, calculating a Jacobian is significantly harder; especially at high-level conceptual design synthesis whereat the mathematical form of the design equations is not well known.

Nevertheless, there are cases where the SISO criterion is inadequate as there are coupled, non-SISO designs with $|J| \neq 0$. Such cases are functionally independent thus admit design solutions but will be rejected per AD's SISO criterion. For example in robotics, the robot Jacobian that relates joint velocities to end-effector velocities is used routinely to plan and execute robot paths and transform forces and torques from the end effector to the manipulator. The robot Jacobian is in fact a design matrix that relates output (end-effector velocities) to input (joint velocities). In most cases, it is not SISO so that most robot designs would have been rejected per the SISO criterion.

To recap, the Jacobian matrix $[J]$ relates **ΔFR** to **ΔDP** of a conceived solution. Its determinant $|J|$ is a test for functional independence of **FR**: yes iff $|J| \neq 0$; no iff $|J| = 0$. Furthermore, iff $|J| \neq 0$, then the conceived solution in term of **DP** can satisfy the **FR**. Otherwise, it cannot. In short, $|J|$ acts as a qualifier: accept a design solution iff $|J| \neq 0$.

We end this section with Table 2.2 which provides a contrast between SISO criterion and $|J|$ criterion derived per formal definition of functional dependency.

Table 2.2 Contrasting AD SISO criterion with $|J|$ criterion

	$\begin{bmatrix} x & o & x \\ x & x & o \\ o & x & x \end{bmatrix}$	$\begin{bmatrix} x & o & x \\ x & x & o \\ o & x & x \end{bmatrix}$	$\begin{bmatrix} x & o & o \\ x & x & o \\ x & x & x \end{bmatrix}$	$\begin{bmatrix} x & o & o \\ o & x & o \\ o & o & x \end{bmatrix}$
SISO criterion:	Bad	Bad	Good	Better
–	Reject	Reject	Accept	Accept
$\|J\|$ criterion:	Bad	Good	Better	Best
–	Reject	Accept	Accept	Accept
–	if $\|J\| = 0$	if $\|J\| \neq 0$	since $\|J\| \neq 0$	since $\|J\| \neq 0$

2.2.2.3 |*J*| for Various Categories of Design

Water Faucet

Both designs **a** and **b** have the same FRs

FR_1 = control flow rate Q;
FR_2 = control water temperature T.

Both designs have the same governing physics:

$$\text{Mass conservation:} \quad Q = Q_h + Q_c; \tag{2.6a}$$

$$\text{Energy conservation:} \quad QT = Q_h T_h + Q_c T_c.$$
$$T = \frac{Q_h T_h + Q_c T_c}{Q_h + Q_c} \tag{2.6b}$$

$$= \frac{(Q_h/Q_c)T_h + T_c}{(Q_h/Q_c) + 1} \tag{2.6c}$$

For Faucet **a**, we choose the left knob controlling Q_h as DP_1; and the right knob controlling Q_c as DP_2 (see Fig. 2.1a). Substituting into Eqs. (2.6a) and (2.6b):

$$FR_1 = DP_1 + DP_2; \qquad FR_2 = \frac{DP_1 T_h + DP_2 T_c}{DP_1 + DP_2}.$$
$$\frac{\partial FR_1}{\partial DP_1} = 1; \qquad \frac{\partial FR_1}{\partial DP_2} = 1.$$
$$\frac{\partial FR_2}{\partial DP_1} = \frac{(T_h - T_c)DP_2}{(DP_1 + DP_2)^2}; \quad \frac{\partial FR_2}{\partial DP_2} = -\frac{(T_h - T_c)DP_1}{(DP_1 + DP_2)^2}$$

$$|J| = \begin{vmatrix} 1 & 1 \\ \frac{(T_h - T_c)DP_2}{(DP_1 + DP_2)^2} & -\frac{(T_h - T_c)DP_1}{(DP_1 + DP_2)^2} \end{vmatrix} = -\frac{T_h - T_c}{DP_1 + DP_2} \tag{2.7}$$

Per AD's SISO criterion, FRs in Faucet **a** are coupled and the design should be rejected. However, according to formal definition of functional dependence, FRs of the design are functionally independent since $|J| \neq 0$. It is therefore acceptable.

For Faucet **b**, we choose as DP_1, the up/down lever controlling total flow rate ($Q_h + Q_c$), and as DP_2, the left/right lever controlling ratio of the hot/cold water flow rate (Q_h/Q_c) (see Fig. 2.1b). Substituting into Eqs. (2.6a) and (2.6c), we have an uncoupled design as indicated by the design equation:

$$FR_1 = DP_1; \quad FR_2 = \frac{DP_2 T_h + T_c}{DP_2 + 1}.$$
$$\frac{\partial FR_1}{\partial DP_1} = 1; \qquad \frac{\partial FR_1}{\partial DP_2} = 0;$$
$$\frac{\partial FR_2}{\partial DP_1} = 0; \qquad \frac{\partial FR_2}{\partial DP_2} = \frac{(T_h - T_c)}{(DP_2 + 1)^2}.$$

$$|J| = \begin{vmatrix} 1 & 0 \\ 0 & \frac{(T_h - T_c)}{(DP_2 + 1)^2} \end{vmatrix} = \frac{(T_h - T_c)}{(DP_2 + 1)^2} \tag{2.8}$$

Note that as the water heater temperature T_h is set closer to outside water temperature T_c, the $|J|$ value gets closer to zero so that the faucet becomes less capable of providing the two independent functions. In short, the $|J|$ criterion provides a quantitative measure of independence which the AD SISO criterion cannot.

Note further that the physics governing both faucets are the same. Yet their $[J]$ matrices in Eqs. (2.7) and (2.8) are different: one is coupled and the other is uncoupled. This conveys a fundamental message in AD. Namely, it is the choice of design solutions DPs, not the physics that determine the goodness of a design.

Projector Design

The two FRs of a projector are as follows:

$$\begin{aligned} FR_1 &= \text{magnify the image} = \frac{D}{d}; \\ FR_2 &= \text{focus the image} = \frac{1}{D} + \frac{1}{d} + \frac{1}{f} = 0. \end{aligned} \tag{2.9}$$

In the above, $D = DP_1$ is the distance of the lens from the screen aka the throw of the projector; $d = DP_2$ is the distance of the lens from the object and $f = DP_3$ is the focal length of the lens. Thus, we have a redundant design: 2 FRs and 3 DPs. If we were to fix the extra DP to get an equal number of FRs and DPs, we would have three (= ${}_3C_2$) possible $|J|$ solutions as follows:

$$\begin{aligned} &\frac{\partial FR_1}{\partial D} = \frac{1}{d}; \quad &&\frac{\partial FR_1}{\partial d} = -\frac{D}{d^2}; \quad &&\frac{\partial FR_1}{\partial f} = 0. \\ &\frac{\partial FR_2}{\partial D} = -\frac{1}{D^2}; \quad &&\frac{\partial FR_2}{\partial d} = -\frac{1}{d^2}; \quad &&\frac{\partial FR_2}{\partial f} = -\frac{1}{f^2}. \end{aligned}$$

$$(D, d) \text{ as DPs}: \ |J| = \begin{vmatrix} \frac{1}{d} & -\frac{D}{d^2} \\ -\frac{1}{D^2} & -\frac{1}{d^2} \end{vmatrix} = -\frac{1}{d^2}\left(\frac{1}{D} + \frac{1}{d}\right); \tag{2.10a}$$

$$(D, f) \text{ as DPs}: \ |J| = \begin{vmatrix} \frac{1}{d} & 0 \\ -\frac{1}{D^2} & -\frac{1}{f^2} \end{vmatrix} = -\frac{1}{df^2}. \tag{2.10b}$$

$$(d, f) \text{ as DPs}: \ |J| = \begin{vmatrix} -\frac{D}{d^2} & 0 \\ -\frac{1}{d^2} & -\frac{1}{f^2} \end{vmatrix} = \frac{D}{d^2 f^2}. \tag{2.10c}$$

All three candidate sets of DPs for the FRs do ensure functional independence of FRs because their $|J| \neq 0$. However, AD SISO criterion will reject solution (2.10a). Nonetheless solution (2.10b), which is acceptable to both AD SISO and $|J|$ criteria, would be the preferred choice since it permits wider latitude of throw D for the projector to accommodate various room sizes.

Disbursement Algorithm

The $|J|$ values for the four possible DM solutions, Eqs. (2.1a) thru (2.1d), in Sect. 2.2.1.1 are as follows:

$$|J| = \begin{vmatrix} \$20 & \$10 & \$5 \\ 1 & 1 & 1 \\ 2 & 0 & -1 \end{vmatrix} = \$0; \quad (2.11a)$$

$$|J| = \begin{vmatrix} \$20 & \$10 & \$1 \\ 1 & 1 & 1 \\ 2 & 0 & -1 \end{vmatrix} = \$8; \quad (2.11b)$$

$$|J| = \begin{vmatrix} \$20 & \$5 & \$1 \\ 1 & 1 & 1 \\ 2 & 0 & -1 \end{vmatrix} = -\$7; \quad (2.11c)$$

$$|J| = \begin{vmatrix} \$10 & \$5 & \$1 \\ 1 & 1 & 1 \\ 2 & 0 & -1 \end{vmatrix} = \$3. \quad (2.11d)$$

We showed in Sect. 2.2.1.1 that for design solution (2.1a), FR_3 is dependent on FR_1 and FR_2. This is confirmed in Eq. (2.11a) above which shows $|J| = 0$. This example demonstrates that an improper choice of DPs can destroy functional independency as originally intended. This is why we need to continually check for it.

Design with Insufficient DPs

To show that $|J|$ of a design with insufficient DPs is zero, i.e., the design FRs are functionally dependent, consider a design with two (FR_1, FR_2) and one DP_1 whose effect on (FR_1, FR_2) are (A_{11}, A_{21}):

$$\begin{bmatrix} \Delta FR_1 \\ \Delta FR_2 \end{bmatrix} = \begin{bmatrix} A_{11} \\ A_{21} \end{bmatrix} [\Delta DP_1].$$

We conjure up a second DP_2 identical to DP_1 to make up for the insufficiency in DP. This second DP_2 has identical effects of (A_{11}, A_{21}) on (FR_1, FR_2):

$$\begin{bmatrix} \Delta FR_1 \\ \Delta FR_2 \end{bmatrix} = \begin{bmatrix} A_{11} & A_{11} \\ A_{21} & A_{21} \end{bmatrix} \begin{bmatrix} \Delta DP_1 \\ \Delta DP_2 \end{bmatrix}$$

The $|J|$ of the above design equals $A_{11}A_{21} - A_{11}A_{21} = 0$. This confirms that FR_1 and FR_2 are functionally dependent when there are insufficient DPs.

2.3 Mathematical Exposition of Information Axiom

The information axiom states that a good design solution must minimize its information contents *I*, or equivalently maximize its probability of success, P_s. Figure 2.10 illustrates the evaluation of P_s. In the presence of variability, an FR_i will exhibit a range of values called the system range or the spread. Cognizant of its variability, a designer would accept the FR_i if it falls within a specified range called the design range. The overlap of the two ranges shown shaded in Fig. 2.10 is P_s, the probability of success of the design. To consider P_s, we need to identify sources of variation that are generating the variability in FRs; how the variability is magnified by the design; and what are the countermeasures for them. We consider these in the next several sections.

2.3.1 *Recognition of Noise Variables*

Since variability in FRs is a consequence of variation, we need to recognize and identify the sources that are generating the variation. We denote these sources as the noise variables, NVs. For example in the faucet design, the temperature T_c of the cold water in Eqs. (2.6b) and (2.6c), Sect. 2.2.2.3, entering the faucet from the outside is a NV since it fluctuates with the uncontrollable temperature outside. It is a NV induced by the environment. If the water heater in a building does not have sufficient capacity to meet the demand of multiple faucets, hot water pressure will fluctuate with the number of faucets turned on or off at a given time. This will affect hot water flow Q_h in Eqs. (2.6a), (2.6b), and (2.6c). This is a NV induced by customer usage. If a projector is used for a variety of room size, it will need a variety of throws to magnify the image. Hence, the throw *D* in Eq. (2.9), Sect. 2.2.2.3 is a NV induced by customer usage. In the

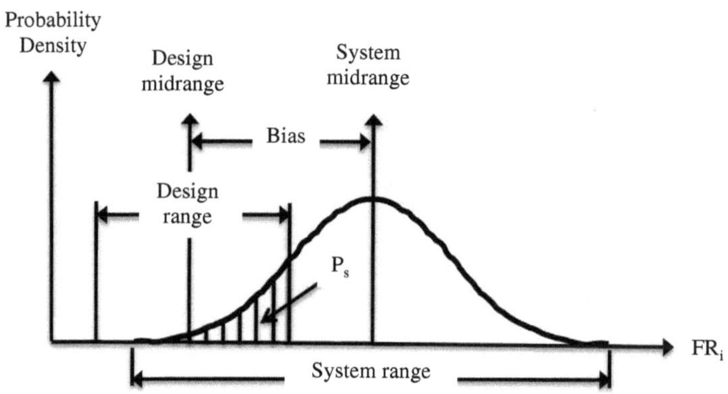

Fig. 2.10 Evaluating the probability of success

hubcap design, the NV is the interference caused by the manufacturing variation in D_{rim} and D_{clip}.

Let **NV** denotes the noise variables that cause variation in **FR**. A **NV** triggers a random deviation in **FR** from its current value **FR*** given by the amount:

$$\mathbf{FR} - \mathbf{FR}^* = [J^{NV}](\mathbf{NV} - \mathbf{NV}^*). \qquad (2.12)$$

In Eq. (2.12), **NV*** is a reference value, e.g., the midrange of **NV**; $[J^{NV}]$ is the Jacobian matrix of $\partial FR_i/\partial NV_j$ given by

$$[J^{NV}] = \begin{bmatrix} \frac{\partial FR_1}{\partial NV_1} & \cdots & \frac{\partial FR_1}{\partial NV_m} \\ \vdots & \ddots & \vdots \\ \frac{\partial FR_n}{\partial NV_1} & \cdots & \frac{\partial FR_n}{\partial NV_m} \end{bmatrix}_{n \times m}$$

The superscript NV is used to distinguish $[J^{NV}]$ from $[J]$, the Jacobian matrix of $\partial FR_i/\partial DP_j$ which hereafter will be superscripted with "DP." While $[J^{DP}]$ relates to the functional independence of **FR** in a design, $[J^{NV}]$ relates to the sensitivity of **FR** to noise **NV**. To illustrate, the $[J^{NV}]$ for some earlier designs are as follows.

For the faucet designs **a** and **b**:

$$FR_1 = Q_h + Q_c; \qquad FR_2 = \frac{Q_h T_h + Q_c T_c}{Q_h + Q_c}.$$

$$NV_1 = Q_h; \qquad NV_2 = T_c.$$

$$\frac{\partial FR_1}{\partial NV_1} = 1; \qquad \frac{\partial FR_1}{\partial NV_2} = 0.$$

$$\frac{\partial FR_2}{\partial NV_1} = \frac{Q_c(T_h - T_c)}{(Q_h + Q_c)^2} \qquad \frac{\partial FR_2}{\partial NV_2} = \frac{Q_c}{Q_h + Q_c}.$$

$$[J^{NV}] = \begin{bmatrix} 1 & 0 \\ \frac{Q_c(T_h - T_c)}{(Q_h + Q_c)^2} & \frac{Q_c}{Q_h + Q_c} \end{bmatrix}$$

For the projector with fixed d equals to a constant, Eq. (2.9) gives:

$$FR_1 = \frac{D}{d}; \qquad FR_2 = \frac{1}{D} + \frac{1}{d} + \frac{1}{f}.$$

$$NV = D.$$

$$\frac{\partial FR_1}{\partial NV} = \frac{1}{d}; \qquad \frac{\partial FR_2}{\partial NV} = -\frac{1}{D^2}.$$

$$[J^{NV}] = \begin{bmatrix} \frac{1}{d} \\ -\frac{1}{D^2} \end{bmatrix}$$

For the hubcap design,

$$\mathrm{FR} = k\delta;$$
$$\mathrm{NV} = \delta;$$
$$\frac{\partial \mathrm{FR}}{\partial \mathrm{NV}} = k.$$
$$\left[J^{\mathrm{NV}}\right] = \left[\frac{\partial \mathrm{FR}}{\partial \mathrm{NV}}\right] = k$$

2.3.2 *Countermeasures to Noise Variables*

The countermeasures to noise sources are as follows: (1) to reduce if not eliminate them, (2) to compensate for them, and (3) to desensitize the design against them. Action (1) refers to the reduction if not elimination of ($\mathbf{NV} - \mathbf{NV}^*$) in Eq. (2.12). This involves tightening the design tolerances, identifying and eliminating process variables that cause variation, and a host of other activities associated with fighting **NV** head-on.

2.3.2.1 Compensation as a Countermeasure

Compensation avoids fighting noise head-on. Instead, it provides a mechanism that further adjust **DP** to nullify $[J^{\mathrm{NV}}](\mathbf{NV} - \mathbf{NV}^*)$:

$$\mathbf{FR} - \mathbf{FR}^* = \left[J^{\mathrm{NV}}\right](\mathbf{NV} - \mathbf{NV}^*) - \left[J^{\mathrm{DP}}\right](\mathbf{DP} - \mathbf{DP}^*) = 0. \tag{2.13}$$

An example people most familiar with is tire balancing in which correction weights ($\mathbf{DP} - \mathbf{DP}^*$) are added to counteract the combined effect of the tire and wheel unbalance. Other examples are water faucet and projector designs discussed earlier. Per Eq. (2.13), the amount of compensation needed is as follows:

$$\mathbf{DP} - \mathbf{DP}^* = \left[J^{\mathrm{DP}}\right]^{-1}\left[J^{\mathrm{NV}}\right](\mathbf{NV} - \mathbf{NV}^*).$$

As revealed in above equation, a prerequisite to compensation is that the design satisfies independent axiom, i.e., $|\, J^{\mathrm{DP}} \,| \neq 0$. Otherwise, $[J^{\mathrm{DP}}]^{-1}$ does not exist, and compensation is not possible. The uncoupled design, e.g., the single-handle faucet, is most easy to compensate since its $[J^{\mathrm{DP}}]$ provides a one-to-one relationship between **DP** and **NV**. The decoupled design, e.g., projector of Eq. (2.10b), is equally easy to compensate if we follow the forward substitution scheme dictated by $[J^{\mathrm{DP}}]$ described in Sect. 2.2.1. The coupled design while possible is difficult to compensate. In short, AD criterion for independence is most applicable in designing for compensation. Per information axiom, information content in a compensated design is zero since variation is completely nullified.

2.3.2.2 Robust Design as a Countermeasure

Another countermeasure action (3) is to move activities to the design stage. Parameters are designed into the design that reduce the sensitivity $[J^{NV}]$ in Eq. (2.13), thereby reducing if not eliminating activities in countermeasures (1) and (2) altogether. For example in hubcap design, in which retention force FR equals $k\delta$, instead of fighting variability in δ head-on, we use a less stiff cantilever spring $k = [J^{NV}] \rightarrow$ small. So that variability in δ is not amplified and transmitted to the retention force. The strategy not to fight **NV** head-on but to reduce the sensitivity to **NV** is known as robust design. Robust design has been the centerpiece of Design for Six Sigma (DFFS).

2.3.3 *Implementing Countermeasures to Minimize Information Content*

Referring to Fig. 2.10, we minimize information content or equivalently maximize P_s in two steps:

1. reduce bias (= system mid-range − design mid-range) to zero by compensation;
2. minimize the system range to within the design range through robust design.

To begin with, we take the expected value of the random variables **FR** and **NV** on both sides of on Eq. (2.13) to arrive at the expression for bias:

$$\text{Bias} = \text{E}(\mathbf{FR}) - \mathbf{FR}^* = \left[J^{\text{NV}}\right]\{\text{E}(\mathbf{NV}) - \mathbf{NV}^*\} - \left[J^{\text{DP}}\right](\mathbf{DP} - \mathbf{DP}^*) \tag{2.14}$$

It follows that adjustment in **DP** needed to reduce bias to zero by compensation is as follows:

$$(\mathbf{DP} - \mathbf{DP}^*) = \left[J^{\text{DP}}\right]^{-1}\left[J^{\text{NV}}\right]\{\text{E}(\mathbf{NV}) - \mathbf{NV}^*\} \tag{2.15}$$

Note again that bias cannot be reduced to zero if FRs are functionally dependent since $|\,J^{DP}\,| = 0$ implies $[J^{DP}]^{-1}$ does not exist; thus, no solution is possible.

We next subtract Eq. (2.14) from Eq. (2.13) to obtain,

$$\mathbf{FR} - E(\mathbf{FR}) = \left[J^{\text{NV}}\right]\{\mathbf{NV} - E(\mathbf{NV})\}.$$

The variance–covariance of **FR** is then given as follows:

$$\underset{n \times n}{\left[V^{\text{FR}}\right]} = \underset{n \times m}{\left[J^{\text{NV}}\right]}\ \underset{m \times m}{\left[V^{\text{NV}}\right]}\ \underset{m \times n}{\left[J^{\text{NV}}\right]^T}$$

where $[V^{FR}]$ and $[V^{NV}]$ are the variance–covariance of **FR** and **NV** shown below.

$$\left[V^{\mathrm{FR}}\right] = E\{\{\mathbf{FR} - E(\mathbf{FR})\}\{\mathbf{FR} - E(\mathbf{FR})\}^{T}\}$$
$$\left[V^{\mathrm{NV}}\right] = E\{\{\mathbf{NV} - E(\mathbf{NV})\}\{\mathbf{NV} - E(\mathbf{NV})\}^{T}\}.$$

Assuming the **NVs** are probabilistically independent, the matrix $[V^{\mathrm{NV}}]$ would be diagonal. The variance of $\mathrm{FR_i}$ is then the ith diagonal element of $[V^{\mathrm{FR}}]$ given by

$$v_{ii}^{\mathrm{FR}} = \sum_{k=1}^{m} j_{ik}^{\mathrm{NV}} v_{kk}^{\mathrm{NV}} j_{ik}^{\mathrm{NV}}$$

The total variance of **FR** is the trace of $[V^{\mathrm{FR}}]$:

$$\text{Variance of } \mathbf{FR} = \sum_{i=1}^{n} \mathrm{v}_{\mathrm{ii}}^{\mathrm{FR}} = \sum_{i=1}^{n}\sum_{k=1}^{m} j_{ik}^{\mathrm{NV}} v_{kk}^{\mathrm{NV}} j_{ik}^{\mathrm{NV}} \qquad (2.16)$$

To maximize P_s, we reduce bias to zero by compensation per Eq. (2.15) and minimize system range, equals to squared root of variance of **FR,** to within the design range of **FR** by robust design per Eq. (2.16).

Summarizing, the steps in AD in mathematical terms are as follows.

1. Define **FR** in a solution neutral environment, free of functional inter-dependence among them.

2. Conceive solution **DP** that maintains the functional independence in **FR** so that $\mathbf{FR}^*$ can be achieved:

$$|J^{\mathrm{DP}}| \neq 0$$

3. Minimize the spread of **FR** with robust design. Namely, reduce $[J^{\mathrm{NV}}]$:

$$\sum_{i=1}^{\mathrm{n}}\sum_{k=1}^{\mathrm{m}} \mathrm{j}_{ik}^{\mathrm{NV}} v_{kk}^{\mathrm{NV}} j_{ik}^{\mathrm{NV}} \rightarrow \text{minimum}$$

4. Subject to constraint that the bias is zero:

$$(\mathbf{DP} - \mathbf{DP}^*) = \left[J^{\mathbf{DP}}\right]^{-1}\left[J^{\mathbf{NV}}\right]\{E(\mathbf{NV}) - \mathbf{NV}^*\}$$

Step 2 and 3 express, respectively, the independence axiom and information axiom in mathematical terms. Step 4 states in mathematical term that independent axiom takes precedence over information axiom. Namely, if **FRs** are not functionally independent, then $|\mathrm{J}^{\mathrm{DP}}| = 0$; $[J^{\mathrm{DP}}]^{-1}$ does not exist; and the constraint that bias = 0 cannot be satisfied. This point is missed in DFSS courses that do not include AD. Common sense tells us that Robust Design optimization has to be subsequent to requirement definition **FR**, and solution conception **DP** because performance of a poorly defined and ill-conceived design cannot be improved via subsequent optimization.

Appendix A1: FR Decomposition of Door-to-Body System

FR_1: fit door exterior to adjacent panels (showroom)
- $FR_{1.1}$: achieve uniform gap around perimeter
  - $FR_{1.1.1}$: achieve uniform gap on both edges
  - $FR_{1.1.2}$: balance leading and trailing edge gaps
  - $FR_{1.1.3}$: align feature lines
- $FR_{1.2}$: achieve flushness to adjacent panels
  - $FR_{1.2.1}$: achieve flushness below beltline
    - $FR_{1.2.1.1}$: achieve flushness at leading edge
    - $FR_{1.2.1.2}$: achieve flushness along both edges
    - $FR_{1.2.1.3}$: achieve flushness at trailing edge
  - $FR_{1.2.2}$: achieve flushness above beltline

FR_2: keep interior quiet and intrusion free
- $FR_{2.1}$: block water and airborne noise
  - $FR_{2.1.1}$: maintain contact pressure
    - $FR_{2.1.1.1}$: maintain adequate seal margin
    - $FR_{2.1.1.2}$: maintain adequate seal height
  - $FR_{2.1.2}$: maintain seal footprint
- $FR_{2.2}$: divert away water
- $FR_{2.3}$: reduce noise transmission through seal
  - $FR_{2.3.1}$: detune seal from noise transmission
  - $FR_{2.3.2}$: dissipate noise energy
- $FR_{2.4}$: eliminate noise sources from mating surfaces
  - $FR_{2.4.1}$: eliminate seal itch
  - $FR_{2.4.2}$: prevent gap-induced turbulence
  - $FR_{2.4.3}$: stop flushness-induced turbulence
- $FR_{2.5}$: prevent CO emission intrusion
  - $FR_{2.5.1}$: control leakage across seals
  - $FR_{2.5.2}$: maintain mass flow rate of inlet air

FR_3: ensure proper opening/closing of the door
- $FR_{3.1}$: hold open at distinct positions on incline
  - $FR_{3.1.1}$: ensure reaction force > gravity
  - $FR_{3.1.2}$: bar opening door swing thru stops
- $FR_{3.2}$: reduce effort to swing door
  - $FR_{3.2.1}$: let closing door swing thru stops
  - $FR_{3.2.2}$: eliminate resistance to swing
- $FR_{3.3}$: latch / unlatch easily
  - $FR_{3.3.1}$: reduce effort to latch
    - $FR_{3.3.1.1}$: decrease energy resisting latch
      - $FR_{3.3.1.1.1}$: lower KE to surmount latch misalign
      - $FR_{3.3.1.1.2}$: lower KE to compress seal
      - $FR_{3.3.1.1.3}$: lower KE to deflect header
      - $FR_{3.3.1.1.4}$: lower KE to overcome airbind
    - $FR_{3.3.1.2}$: store spring energy from opening
  - $FR_{3.3.2}$: reduce effort to unlatch

Appendix A2: DP Decomposition of Door-to-Body System

DP_1: door fitting scheme
- $DP_{1.1}$: hinge / latch system
  - + $DP_{1.1.1}$: hinge tip in x-z plane
  - + $DP_{1.1.2}$: fore/aft position of hinge axis
  - + $DP_{1.1.3}$: vertical position of hinge datum
- $DP_{1.2}$: system for achieving flushness
  - $DP_{1.2.1}$: hinge / latch system
    - + $DP_{1.2.1.1}$: in/out position of hinge axis
    - + $DP_{1.2.1.2}$: hinge tip in y-z plane
    - + $DP_{1.2.1.3}$: in/out position of striker
  - + $DP_{1.2.2}$: header over bent

DP_2: system for a quiet & intrusion-free interior
- $DP_{2.1}$: sealing energy as barrier
  - $DP_{2.1.1}$: seal indentation by header
    - + $DP_{2.1.1.1}$: position of door interior surface
    - + $DP_{2.1.1.2}$: a system to maintain uniform seal height
  - + $DP_{2.1.2}$: contour of contacting surfaces
- + $DP_{2.2}$: channel slope
- $DP_{2.3}$: noise transmission management
  - + $DP_{2.3.1}$: modal property of seal section
  - + $DP_{2.3.2}$: seal damping characteristic
- $DP_{2.4}$: noise elimination system
  - + $DP_{2.4.1}$: lubricant, substrate loss modulus
  - + $DP_{2.4.2}$: gap filler
  - + $DP_{2.4.3}$: header stiffness
- $DP_{2.5}$: positive cabin pressure
  - + $DP_{2.5.1}$: sealing energy as barrier to intrusion
  - + $DP_{2.5.2}$: fan

DP_3: system for opening/closing door
- $DP_{3.1}$: detent mechanism
  - + $DP_{3.1.1}$: stiffness & preloads of check link spring
  - + $DP_{3.1.2}$: site, depth & climb of check link valleys
- $DP_{3.2}$: system to reduce occupant effort
  - + $DP_{3.2.1}$: site, depth & descent of check link valleys
  - + $DP_{3.2.2}$: hinge axes aligned with axis of rotation
- $DP_{3.3}$: system for counterbalancing opening/closing door
  - $DP_{3.3.1}$: system to reduce effort to latch
    - $DP_{3.3.1.1}$: KE threshold reducer
      - + $DP_{3.3.1.1.1}$: up-down adjustable striker
      - + $DP_{3.3.1.1.2}$: areas under weather-strip CLD
      - + $DP_{3.3.1.1.3}$: area under header load-deflection curve
      - + $DP_{3.3.1.1.4}$: pressure relief valve
    - + $DP_{3.3.1.2}$: pre-loaded check link torsional spring
  - + $DP_{3.3.2}$: mechanism to relieve reaction at latch

References

1. N.P. Suh, *The Principles of Design* (Oxford University Press, New York, NY, 1990)
2. G. Strang, *Linear Algebra and Its Applications*, 2nd edn. (Academic Press Inc., New York, NY, 1980)
3. H.L. Oh, Modeling variation to enhance quality in manufacturing. Presented at the Conference on *Uncertainty in Engineering Design*, NBS, 10–11 May 1988
4. H.L. Oh, T. Lee, R. Lipowski, A graph theory based method for functional decoupling of a design with complex interaction structure, ASME 2010 International Design Engineering Technical Conferences, DETC2010-28609

Part II
Large Complex Products

Chapter 3
New Developments to Guide Strategic Product Design and Systematic Innovation

Efren M. Benavides and Joan B. Rodriguez

Abstract Whereas the laws of physics establish the relationships between functional requirements, constraints, and design parameters, Axiomatic Design establishes the conceptual principles that drive the engineer's decisions toward a leader product. Hence, Axiomatic Design appears as a natural kernel of any strategic product design activity. According to Axiomatic Design principles, the fulfillment of the Independence and the Information Axioms defines the "best design," which requires selecting one and only one design parameter per functional requirement. Nonetheless, due to the inherent complexity of the physical laws and social models found in most of the industry challenges, the number of design parameters which are available for the designer tends to be much larger than the number of functional requirements to satisfy. Without any other tool helping the designer to select the correct design parameters, the number of degrees of freedom is high enough to avoid a complete rational strategy in the design. In order to fix this problem, mainly during the conceptual definition of a product, new developments in the Axiomatic Design theory are necessary. This chapter explains how the Linearity Theorem refines the definition of "best design," with the aim of guiding product designers. The theorem is applied to case studies that show how to select the design parameters and how to drive the innovation procedures. The key point for leading innovation is to solve contradictions between current ideas and the concept of "best design," which must be in accordance with the design axioms. Finally, some conclusions derived from the authors' experience during technological advising are presented.

Keywords Linearity Theorem · Systematic innovation · Product design · Design parameter selection

E.M. Benavides (✉)
Department of Fluid Mechanics and Aerospace Propulsion,
Universidad Politécnica de Madrid, Madrid, Spain
e-mail: efren.moreno@upm.es

J.B. Rodriguez
Consultant Engineer—Sides, Madrid, Spain
e-mail: joanbruno@coiae.com

© Springer International Publishing Switzerland 2016
A.M. Farid and N.P. Suh (eds.), *Axiomatic Design in Large Systems*,
DOI 10.1007/978-3-319-32388-6_3

Nomenclature

A_i	Cross-sectional area in section i
c_d	Drag coefficient
d_1	Tube 1 diameter
d_d	Diameter of dust particles
d_{pores}	Diameter of filter pores
δ_0, m_0	Design parameter tolerances
DP	Design parameter
f	Transfer function
FR	Functional requirement
l	Functional requirement value
m	Design parameter value
m_a	Mass of the air
$\dot{m}_a$	Flow mass of air
$\dot{m}_f$	Flow mass of fuel
μ	Mean
n	Number of dust particles per volume unit
N	Number of filter pores
N_c	Number of cyclone turns
p_i	Air pressure in zone i
ΔP_f	Fuel pressure drop
ρ_0	Air density
ρ_d	Density of dust particles
ρ_f	Fuel density
$\ddot{r}$	Radial acceleration inside cyclone
$\dot{r}$	Radial speed inside cyclone
r	Radial position inside cyclone
R	Radius of curvature of cyclone
σ^2	Variance
u_i	Air speed in zone i
V_{23}	Dust container capacity
$\dot{W}$	Fan power
x	Dimensionless variable in physical domain
y	Dimensionless variable in functional domain

3.1 Introduction

One of the main objectives of Design Science Theories is to provide a rationale to justify design decisions. Because of the amount of uncertainty and informal information that is present during the early stages of product design, the more these theories are early applied in the design process, the higher the value they may provide [1–3]. In this context, Design Science Theories [4, 5] understand the term design both as a process and as a physical solution. Hence, the design theories focus on two complementary aspects: firstly, on defining the appropriate life cycle assessments and accurately accomplishing the different steps through which the design passes from the voice of customer to the detailed solution; secondly, on defining unequivocally and universally what the best design means. If the main purpose of Design Science is to set a rationale that justifies design decisions in order to find the best solution to a particular set of needs, the symbiosis of both aspects is necessary and a correct balance between process and solution must be assured. Although process-based and solution-based approaches are aware of this circumstance and both suggest decision-making and process criteria, in general, process-based theories lack universal decision-making criteria, and solution-based theories lack precision in the establishment of the design process. The design process definition should define the adequate steps that optimize the information flow throughout the process, and the decision criteria should ensure that in each step, the best decision has been selected according to the available information. The combination of both approaches results in the consecution of the best design with the minimum consumption of resources [2]. In other words, the best design solution should be conducted by an adequate process where decision making is based on rule-based criteria.

In general, in engineering design problems the income of parameters derived from the laws of physics is higher than the number of requirements and hence, the number of parameters to be selected is higher than the available equations. In this context, where the designer success passes through the selection of the best DP for each FR from a set of parameters with the minimum resources invested, it is necessary to formulate a definition of best design according to the inner nature of DPs by preserving Axiomatic Design principles.

Under this framework, Axiomatic Design defines the best design by identifying the best combination of DPs to satisfy a set of FRs by means of two axioms (Independence and Information Axioms). From these principles, the best design is defined as the one that maintains functional independence (FRs which are independent in a neutral solution environment preserve independence after DPs definition) and has minimum information content (transfer functions and DPs variation ensure that FRs' acceptance interval has the maximum probability of being satisfied). Naturally derived from these principles, in an Axiomatic Design each FR is satisfied by one and only one DP [3]. This definition helps the designer not only to identify the best design solution but to enrich the design process with an objective criterion in each level of the design hierarchy. Thus, the target is to find a solution whose design matrix can be written as a diagonal one such as it is represented by Eq. (3.1).

$$\overrightarrow{\mathrm{FR}} = \begin{bmatrix} A & 0 & 0 \\ 0 & B & 0 \\ 0 & 0 & C \end{bmatrix} \overrightarrow{\mathrm{DP}} \Rightarrow \text{Ideal Design Matrix}: \begin{bmatrix} X & & \\ & X & \\ & & X \end{bmatrix} \tag{3.1}$$

Axiomatic Design theory states that the ideal design matrix has to be diagonal, but nothing is said about the nature of the elements *A*, *B*, and *C* placed over the diagonal. For this reason, Axiomatic Design defines the ideal design matrix as any diagonal matrix no matter what their elements are. This fact leads to use a simplified nomenclature where the nonzero terms are all marked with an *X* just as it is represented on the right-hand side of Eq. (3.1) (a formal mathematical calculation of the *X*s and the blanks in the design matrix requires to define a dimensionless design matrix, for a deeper explanation please refer to Benavides [1]). At this point, without additional investigation, nothing avoids these elements to be nonlinear terms, i.e., the nonzero terms in the Ideal Design Matrix defined by Eq. (3.1) could be functions which depend on the value of the design parameters [3, 6]. Therefore, two independent designers could select two different ideal designs, one with a constant design matrix and the other with a non-constant design matrix. Hence, an arbitrary option still remains in the definition of the best design. The following question arises: What is the best qualitative behavior for the function $\mathrm{FR} = X \cdot \mathrm{DP}$, the linear or the nonlinear one?

Because of the universal formulation of the design principles, the Axiomatic Design definition of best design should be valid for both quantitative and qualitative approaches [1, 3]. This characteristic should allow Axiomatic Design principles to solve the previous question. This work shows that the answer to this question is the Linearity Theorem [1].

The Linearity Theorem refines the concept of the best design by stating that A, B, and C must be constant elements. This is represented in Eq. (3.2). This refined definition of the ideal design increases the consistency and applicability of the theory and allows the designer to address to a huger variety of design problems, from strategic decisions to detailed engineering solutions. This peculiarity confers to the theory an accurateness to guide strategic product design and innovation [7–9].

$$\overrightarrow{\mathrm{FR}} = \begin{bmatrix} A & 0 & 0 \\ 0 & B & 0 \\ 0 & 0 & C \end{bmatrix} \overrightarrow{\mathrm{DP}} \Rightarrow \text{Ideal Design Matrix:} \begin{bmatrix} \mathrm{cte} & & \\ & \mathrm{cte} & \\ & & \mathrm{cte} \end{bmatrix} \tag{3.2}$$

Note that, even though qualitative and quantitative approaches require a similar definition of functional requirements in a neutral solution environment, they can diverge in the way of selecting the appropriate DPs. Indeed, it can be said that in a qualitative problem, the designer creates embodied solutions that will be checked by means of the design axioms, whereas in a quantitative problem, often the designer faces how to identify the best DPs from a set of parameters derived from the laws of physics and transfer functions. An interesting approach to a similar concern founded on constraint optimization can be found in Oh [10].

In this chapter, the approach is based on the Linearity Theorem [1], which states that the best DPs are the ones that lead to a diagonal design matrix whose elements are constant as written in Eq. (3.2). As it will be shown, the derivation of the theorem is based on Suh's axioms, corollaries, and theorems [3, 6].

The obligation of fulfilling the Linearity Theorem is a new tool suitable to refine the definition of best design in terms of the inner nature of the available DPs, and hence suitable to select (or invent) the best design parameters, those that maximize the probability of success and conduct the design team to the best solution. An immediate consequence of this result is that the nature of the relations between the FRs and the DPs selected by a designer is very important for leading valuable innovation.

To present this theorem, this chapter is structured in two main sections. The first one is dedicated to the theoretical framework of the Linearity Theorem, where it will be enounced and proved. Secondly, a more practical-oriented section is presented, where basic methodological steps to apply the theorem will be suggested. For this purpose, illustrative case studies will be solved. Finally, main conclusions will be deepened by authors experience in solving industry problems by the use of Axiomatic Design.

3.2 The Linearity Theorem: Theoretical Framework

Naturally, once the design problem has been formulated in the functional domain, the designer's decisions take place in the physical domain. However, as it has been exposed, the number of available DPs is in general larger than the number of FRs which results in redundant designs. In this context, the Linearity Theorem [1] constitutes a new criterion that, derived from the axioms, helps the designer to select the best DPs.

3.2.1 Problem Description

Corollary 2 and Theorem 4 of Axiomatic Design [3] establish that the best design is the one where a single FR is related only to a single DP. Corollary 6 states that the best design must allocate the largest allowable tolerances [3]. The theory also establishes that the axioms (and hence, the theorems and the corollaries derived from them) must be applied to all of the levels of the design hierarchy [3].

The new question addressed in this chapter is, if based on the Axiomatic Design theory, the Linearity Theorem holds. The following lines are focused to prove it and, as a direct consequence, to show that the best design is the one that forced the DPs to be linear [1].

Let $[\underline{l}, \overline{l}]$ be the acceptance interval for the FR which in general derives from customer needs exigencies. Let $l = f(m)$ be the transfer function relating the FR with the DP, whose shape and formulation generally derive from the laws of physics and by the designer's decisions. This function is assumed to be continuous and differentiable. Finally, let $[\underline{m}, \overline{m}]$ be the variation interval for the DP, where the

Table 3.1 Design range, system range, common range, and information content defined in terms of the intervals of acceptance for both the customer and the designer

Design range	System range	Common range	Information content
$DR = [\underline{l}, \bar{l}]$	$SR = [\min f([\underline{m}, \bar{m}]), \max f([\underline{m}, \bar{m}])] = [l_{\min}, l_{\max}]$	$CR = [l_{\min}, l_{\max}] \cap [\underline{l}, \bar{l}]$	$I = \log \frac{[l_{\min}, l_{\max}]}{[l_{\min}, l_{\max}] \cap [\underline{l}, \bar{l}]}$

minimum and maximum values are selected by the designer to ensure that customer needs are satisfied. Note that $(\bar{m} - \underline{m})/2$ can be interpreted as the design tolerance, which is related to the customer tolerance $(\bar{l} - \underline{l})/2$ through the transfer function $l = f(m)$ and the fulfillment of the Information Axiom. Hereafter, design range, system range, common range, and information content can be defined as shown in Table 3.1.

In general, in nonlinear responses it may happen that the system range (i.e., the image of the DP variation interval) does not lie within the design range. This may happen, for example, if the system has a maximum (or a minimum) response for a point inside the design range. For instance, if $\max f([\underline{m}, \bar{m}]) = l_{\max} > \bar{l}$, Fig. 3.1 shows only the case $l_{\max} > \bar{l}$, but note that the argumentation would be the same for a minimum with $\min f([\underline{m}, \bar{m}]) = l_{\min} < \underline{l}$). Under this circumstance, $SR = [l_{\min}, l_{\max}] > \bar{l} - \underline{l}$ and the Information Axiom is not satisfied: $I = \log \frac{[l_{\min}, l_{\max}]}{[\underline{l}, \bar{l}]} = \log \frac{l_{\max} - l_{\min}}{\bar{l} - \underline{l}} > 0$ (Fig. 3.1a). As Fig. 3.1a depicts, in order to accomplish the Information Axiom, the designer should either extend the design range (see the vertical arrow in Fig. 3.1a), which may be not accepted by the customer, or reduce the variation interval (see the horizontal arrow in Fig. 3.1a), which contradicts Corollary 6 and may result in the definition of critical tolerances and new center of intervals (Fig. 3.1b). This generally adds complexity to the manufacturing process and increases the final cost of the product [1, 11].

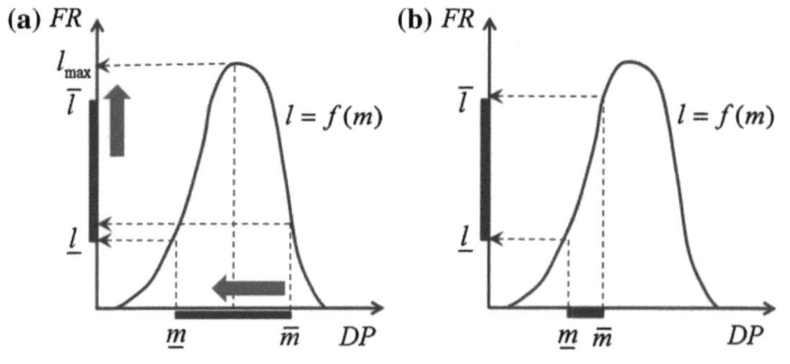

Fig. 3.1 Acceptance interval (design range), variation interval, and transfer function in nonlinear responses. **a** Represents a case where the Information Axiom is not satisfied (the information content is larger than zero), whereas Fig. **b** Represents a case where the Information Axiom is satisfied

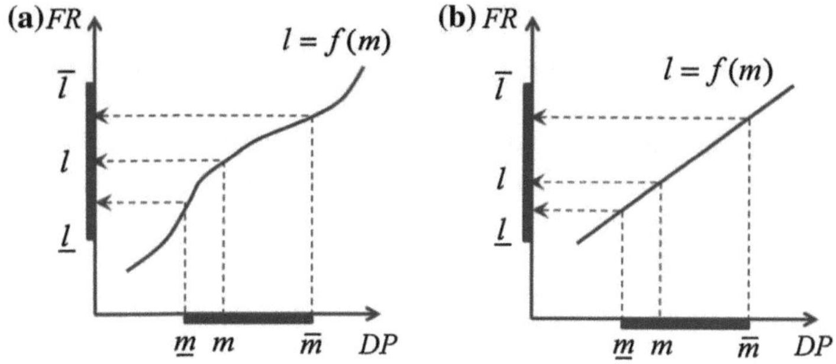

Fig. 3.2 Acceptance interval (design range), variation interval, and transfer function in nonlinear (**a**) and linear responses (**b**), both with a zero information content ($I = 0$)

Once the designer's decisions have achieved SR = CR, the following question (illustrated in Fig. 3.2) arises: Which is the shape of the best DP to control the FR? Figure 3.2a shows a nonlinear relation, whereas Fig. 3.2b shows a linear one. Which one facilitates the accomplishment of Axiomatic Design Corollary 6 (or the Information Axiom) that guides to the specification of the largest allowable tolerance in stating FRs?

To answer these questions, the Linearity Theorem will be stated and proved [1].

Statement: Linear designs are better than nonlinear designs.

Alternate statement: Linear design parameters lead to better designs than nonlinear parameters. In other words, when there is more than one DP affecting a particular FR, the one that produces the most linear variation of the functional requirement should be chosen.

Proof The proof of the theorem is based on the systematic application of Axiom 1 and Axiom 2 throughout the design hierarchy, where the axioms should be satisfied for each level. This structure imposes (1) to select the system response, i.e., the transfer function, and (2) to select the adequate DPs and their tolerances.

Without any loss of generality, when $l \in [\underline{l}, \bar{l}]$ and $m \in [\underline{m}, \bar{m}]$, the following two dimensionless variables x and y can be defined as

$$\text{FR: } l = \frac{\bar{l} + \underline{l}}{2} + y\frac{\bar{l} - \underline{l}}{2} \tag{3.3}$$

$$\text{DP: } m = \frac{\bar{m} + \underline{m}}{2} + x\frac{\bar{m} - \underline{m}}{2} \tag{3.4}$$

Note that $x \in [-1, +1]$ and $y \in [-1, +1]$ when $l \in [\underline{l}, \bar{l}]$ holds for the FR and $m \in [\underline{m}, \bar{m}]$ holds for the DP. Using these new variables, the transfer function $l = f(m)$ can be rewritten as:

$$y(x) = \frac{2}{\bar{l} - \underline{l}} \left[f\left(\frac{\bar{m} + \underline{m}}{2} + x \frac{\bar{m} - \underline{m}}{2} \right) - \frac{\bar{l} + \underline{l}}{2} \right] \tag{3.5}$$

Under this formulation, $\mathrm{SR} = [y_{\min}, y_{\max}] = [\inf[y([-1, +1])], \sup[y([-1, +1])]]$ and the Information Axiom will be satisfied if $[y_{\min}, y_{\max}] \subseteq [-1, +1]$. Equation (3.5) can be rewritten as:

$$y = \frac{2}{\bar{l} - \underline{l}} f\left(\frac{\bar{m} + \underline{m}}{2} \left(1 + x \frac{\bar{m} - \underline{m}}{\bar{m} + \underline{m}} \right) \right) - \frac{\bar{l} + \underline{l}}{\bar{l} - \underline{l}} \tag{3.6}$$

Under the hypothesis $\bar{m} - \underline{m} \ll \bar{m} + \underline{m}$ (i.e., $\bar{m} \to \underline{m}$ which means that we will restrict ourselves to study small nonlinearities), Eq. (3.6) can be expanded as a Taylor series (which implies the assumption of the regularity of the transfer function):

$$\begin{aligned} \frac{\bar{l} - \underline{l}}{2} \left[y(x) + \frac{\bar{l} + \underline{l}}{\bar{l} - \underline{l}} \right] &= f\left(\frac{\bar{m} + \underline{m}}{2} \right) + f'\left(\frac{\bar{m} + \underline{m}}{2} \right) x \frac{\bar{m} - \underline{m}}{2} \\ &\quad + \frac{1}{2} f''\left(\frac{\bar{m} + \underline{m}}{2} \right) \left(x \frac{\bar{m} - \underline{m}}{2} \right)^2 + 0\left(x \frac{\bar{m} - \underline{m}}{\bar{m} + \underline{m}} \right)^3 \end{aligned} \tag{3.7}$$

Without any loss of generality we can assume that m is a random variable with an unknown density function in the interval $[\underline{m}, \bar{m}]$, which implies that x is also randomly distributed in the interval $[-1, +1]$. The mean and the variance of $y(x)$ are as follows:

$$\begin{aligned} \mu &= E[y(x)] \\ &= \frac{2}{\bar{l} - \underline{l}} f\left(\frac{\bar{m} + \underline{m}}{2} \right) - \frac{\bar{l} + \underline{l}}{\bar{l} - \underline{l}} \\ &\quad + \frac{\bar{m} - \underline{m}}{\bar{l} - \underline{l}} \left[E[x] f'\left(\frac{\bar{m} + \underline{m}}{2} \right) + E[x^2] \frac{\bar{m} - \underline{m}}{4} f''\left(\frac{\bar{m} + \underline{m}}{2} \right) + \cdots \right] \end{aligned} \tag{3.8}$$

$$\begin{aligned} \sigma^2 &= E\left[(y(x) - E[y(x)])^2 \right] \\ &= \left(\frac{\bar{m} - \underline{m}}{\bar{l} - \underline{l}} \right)^2 E\left[x^2 - E[x]^2 \right] \left[f'\left(\frac{\bar{m} + \underline{m}}{2} \right) \right]^2 \\ &\quad + \left(\frac{\bar{m} - \underline{m}}{\bar{l} - \underline{l}} \right)^2 2(\bar{m} - \underline{m}) E\left[x^3 - x^2 E[x] \right] f'\left(\frac{\bar{m} + \underline{m}}{2} \right) f''\left(\frac{\bar{m} + \underline{m}}{2} \right) \\ &\quad + \left(\frac{\bar{m} - \underline{m}}{\bar{l} - \underline{l}} \right)^2 (\bar{m} - \underline{m})^2 E\left[\left(x^2 - E[x]^2 \right)^2 \right] \left[f''\left(\frac{\bar{m} + \underline{m}}{2} \right) \right]^2 + \cdots \end{aligned} \tag{3.9}$$

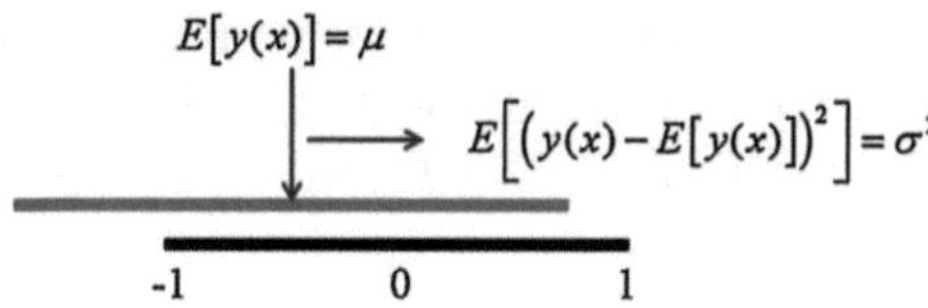

Fig. 3.3 Variation of the DP (*dark gray line*) and its influence on the system range interval (*light gray line*) when the Information Axiom is not satisfied. The extremes of the system range are the points $y(-1)$ and $y(+1)$

As it can be observed, the terms resulting from nonlinear designs, $f^{(n)}$ with $n \geq 2$, appear both in Eqs. (3.8) and (3.9). This means that nonlinear terms modify the following: (1) the center of the response interval and (2) the width of the response interval. In terms of design specifications, the nonlinearities modify the center of the system range and the width of the system range. Consequently, if nonlinear terms are large enough, $[y_{\min}, y_{\max}] \subseteq [-1, +1]$ may not be valid, and in terms of the information content, $I = \log(\mathrm{SR}/\mathrm{CR}) > 1$, and the second axiom would not be satisfied. Figure 3.3 illustrates this situation, where due to a change of the mean and the variance, the SR does not match the design specification [−1, 1].

As a result, in order to comply with Information Axiom and Corollary 6, two new conditions should be imposed, $\mu = E[y(x)] = 0$ and $\sigma^2 = E[(y(x) - E[y(x)])^2] < \sigma^2_{\max}$, which are illustrated in Fig. 3.4.

The reader can observe that in Eqs. (3.8) and (3.9), the mean μ and the variance σ^2 constitute a set of independent FR where $m_0 = \frac{\bar{m}+\underline{m}}{2}$ and $\delta_0 = \frac{\bar{m}-\underline{m}}{2}$ are the new DPs. In this context, it is derived that, in general, $\begin{bmatrix} \frac{\partial \mu}{\partial m_0} & \frac{\partial \mu}{\partial \delta_0} \\ \frac{\partial \sigma^2}{\partial m_0} & \frac{\partial \sigma^2}{\partial \delta_0} \end{bmatrix} \neq \begin{bmatrix} X & 0 \\ 0 & X \end{bmatrix}$ so the problem is coupled and does not satisfy the Independence Axiom.

Looking at the structure of Eqs. (3.8) and (3.9), it is possible to infer that the condition $\delta_0 = 0$ leads to a diagonal matrix; however, this condition does not satisfy Corollary 6 because it leads to $\bar{m} = \underline{m}$ which guides to the reduction of the design tolerance. In addition, in general, the designer can only null the mean of the variable x but not the rest of the statistical moments because he has to assume that the DP is a random variable that fluctuates inside its interval of definition. This

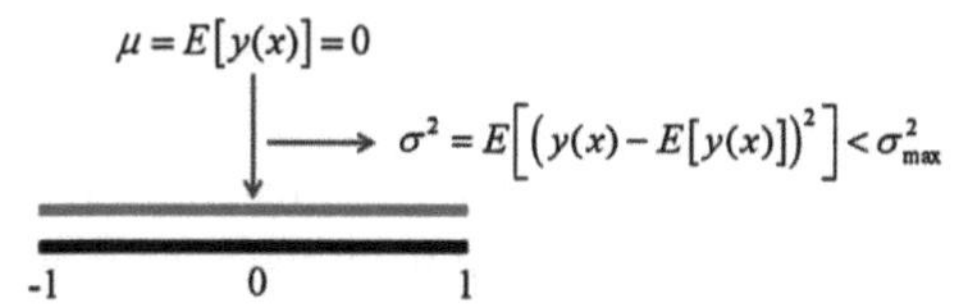

Fig. 3.4 Variation of the DP (*dark gray line*) and its influence on the system range interval (*light gray line*) when the Information Axiom and Corollary 6 are both satisfied. The extremes of the system range are the points $y(-1) = -1$ and $y(+1) = 1$. These new conditions allow to satisfy the Information Axiom

proves that the only way to obtain a diagonal matrix is to remove all the derivatives of order larger than one in Eqs. (3.8) and (3.9). Thus, the diagonal matrix is obtained when $f''(m_0) = f'''(m_0) = \cdots = f^{(n}(m_0) = 0$, which reduces Eqs. (3.8) and (3.9) to:

$$\mu = \frac{2}{\bar{l} - \underline{l}} f(m_0) - \frac{\bar{l} + \underline{l}}{\bar{l} - \underline{l}} + \frac{2\delta_0}{\bar{l} - \underline{l}} E[x] f'(m_0) \tag{3.10}$$

$$\sigma^2 = E\left[x^2 - E[x]^2\right] \left[\frac{2\delta_0}{\bar{l} - \underline{l}} f'(m_0)\right]^2 \tag{3.11}$$

Note that $f'(m_0)$ must be constant in order to null the higher order derivatives. Let us now analyze the obtained Eqs. (3.10) and (3.11) which describe a decoupled design. Designers seek to have $E[x] \to 0$ because in other case, they should reduce the standard deviation of x in order to accommodate its variation inside the fixed interval $[-1, 1]$, which would contradict Corollary 6. Thus, Corollary 6 states $E[x] = 0$, and hence, the design matrix in this level of the design hierarchy is as follows:

$$\begin{bmatrix} \frac{\partial \mu}{\partial m_0} & \frac{\partial \mu}{\partial \delta_0} \\ \frac{\partial \sigma^2}{\partial m_0} & \frac{\partial \sigma^2}{\partial \delta_0} \end{bmatrix} = \begin{bmatrix} \frac{2f'(m_0)}{\bar{l} - \underline{l}} & 0 \\ 0 & \left[\frac{2f'(m_0)}{\bar{l} - \underline{l}}\right]^2 E[x^2] \delta_0 \end{bmatrix} \tag{3.12}$$

which guides to a diagonal form as Axiomatic Design requires $\begin{bmatrix} X & 0 \\ 0 & X \end{bmatrix}$.

Therefore, canceling all derivatives higher than first order (and satisfying the Information Axiom which includes the satisfaction of Corollary 6) implies to state that the best transfer function is linear. This also proves that linear DPs are better than nonlinear ones. The Linearity Theorem is hence proved (for an extensive explanation of the proof of the Linearity Theorem please refer to Benavides [1]).

3.3 The Linearity Theorem: Practical Application

3.3.1 Basic Methodological Steps to Apply the Linearity Theorem

As the application of the Linearity Theorem requires knowledge of the quantitative relation between FR and DP, the following main steps are required. These steps are the basic ones that the authors found in common when facing different design problems and are based on the main methodological structure suggested by Suh in The Principles of Design [3].

1. **Quantitative formulation of the design problem.**
    - Definition of the challenge.
    - Selection of the minimum number of independent FR in a solution-neutral environment and, if available, definition of the acceptance intervals (design ranges).
    - Formulation of the input and system constraints and laws of physics.
    - Obtaining design equations, identification of DPs, and formulation of the design matrix.
2. **Axiomatic design analysis.**
    - Compliance with the Independence Axiom.
    - In case of redundant designs, application of the Linearity Theorem to select the most accurate DPs.
    - Compliance with the Information Axiom.
    - Compliance with the constraints.
3. **Description of the physical solution in terms of DPs and FRs.**

3.3.2 Case Study 1: Filtering System in Vacuum Cleaners

In this section, the basic methodological steps will be sequentially applied to illustrate the use of the Linearity Theorem for analyzing two different filtering systems in vacuum cleaners. For a detailed description of functional uncoupling and laws of physics for this system, please refer to Rodriguez [12].

1. **Quantitative formulation of the design problem.**
    - *Definition of the challenge*:
      Analyze two different technologies (porous filter and centrifugal separation) for filtering dust particles when vacuum cleaning. Identify their main dependences and select the best solution according to Axiomatic Design.
    - *Selection of the minimum number of independent FR in a neutral solution environment*:

      FR1: **Cleanup dust particles**, which might be represented by the speed of the air, u_1, that must remove the dust particles from the floor;
      FR2: **Retain dust particles**, which represents the functionality of separating all the particles that have a size bigger than $d_{d\min}$;
      FR3: Operate a long time, which represents the maximum period of time, $t_{\max}$, that the customer expects to operate the device.

    - *Formulation of the input and system constraints and laws of physics*:
      Figure 3.5 illustrates two simplified models for both filtering technologies:

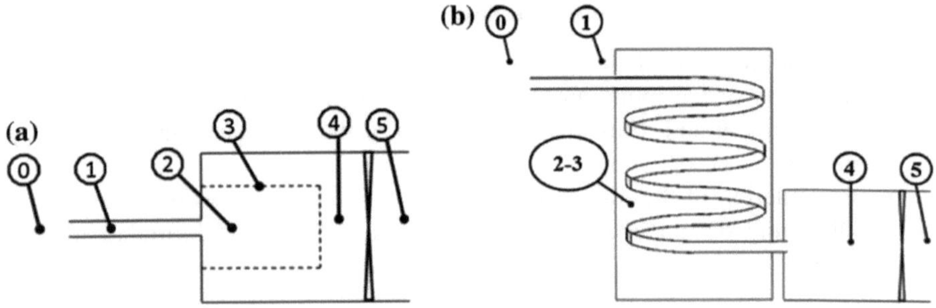

Fig. 3.5 Porous filter (**a**) and centrifuge separation (**b**) simplified models

– **Porous filter solution**:

First, fluid dynamics equations are formulated assuming that the air can be modeled as an incompressible ideal fluid. Under this assumption, the mass-flow rate is derived as a function of the geometry of the problem and the electric power of the motor (it is assumed that there is an electrical motor that transforms electrical energy in kinetic one):

$$\dot{m}_a = \sqrt[3]{\frac{2\dot{W}(\rho_0 A_1)^2}{1+\left(\frac{A_1}{A_3}\right)^2}} \tag{3.13}$$

Which combined with the mass-flow rate equation $\dot{m}_a = \rho_0 u_1 A_1$ gives an expression for u_1, where ρ_0, u_1, A_1 and A_3, and $\dot{W}$ are the air density at room conditions, the air speed at the inlet pipe, the cross-sectional area of the inlet pipe and the porous filter, and the electric power of the motor, respectively.

Secondly, the filter will be effective if the diameter of filter pores is smaller than the diameter of dust particles, which guides to the condition:

$$d_{d\min} \geq d_{\text{pores}} \tag{3.14}$$

Finally, the operational time will be a function of the number of filter pores per the velocity of clogging them, which can be formulated as:

$$t_{\max} = \frac{N}{n\frac{\dot{m}_a}{\rho_0}} \tag{3.15}$$

where N and n represent the number of filter pores and the number of dust particles per unit of volume, respectively.

– **Centrifuge separation solution**:
Considering again the air as an ideal fluid, the mass-flow rate can be derived as a function of the geometry of the problem and the electric power of the motor:

$$\rho_0 u_1 A_1 = \sqrt[3]{2\dot{W}(\rho_0 A_1)^2} \tag{3.16}$$

If it is considered that the cyclone comprises a stream tube which follows a helical stream line with a characteristic radius R and N_c turns and that d_d, ρ_d and c_d represent the dust particle's diameter, density, and drag coefficient, the differential equation that describes the radial displacement, x, of a dust particle inside the cyclone is as follows:

$$\frac{4}{3}\pi\left(\frac{d_d}{2}\right)^3 \rho_d \ddot{r} = \frac{4}{3}\pi\left(\frac{d_d}{2}\right)^3 \rho_d \frac{u_1^2}{R} - \frac{1}{2}\rho_0 \dot{r}^2 c_d \left(\frac{\pi d_d^2}{4}\right) \tag{3.17}$$

It can be shown [12] that the relevant case is the one with the smaller dust particles where aerodynamic forces turn dominant. In this circumstance, the radial velocity will become constant as stated by:

$$\dot{r} = \sqrt{\frac{3}{4c_d}\frac{\rho_d}{\rho_0} d_{\min} \frac{u_1^2}{R}} \tag{3.18}$$

Equation (3.18) can be directly integrated to obtain the radial position of a dust particle as a function of time (3.19):

$$r = \sqrt{\frac{3}{4c_d}\frac{\rho_d}{\rho_0} d_{\min} \frac{u_1^2}{R} t} \tag{3.19}$$

Taking into account that the particles retained into the container satisfy the condition $r(t) = d_1 = \frac{4}{\pi}\sqrt{A_1}$ and that the time inside the cyclone is $t = \frac{2\pi R N_c}{u_1}$, an immediate condition for the diameter of dust particles that can be separated is obtained as a function of the geometry of the cyclone. Finally, the operational time will depend on the volume of the dust container V_{23}, the velocity to fill it, and the density of dust particles per unit of volume n, which can be formulated as:

$$t_{\max} = \frac{V_{23}}{\frac{4}{3}\pi\left(\frac{d_d}{2}\right)^3 n \frac{\dot{m}_a}{\rho_0}} \tag{3.20}$$

- *Obtaining the design equations, identification of DPs, and formulation of the design matrix.*

Table 3.2 Design equations FR = *f*(DP*s*) for porous filter and centrifuge separation

Porous filter separation	Centrifuge separation
$FR1 = u_1 = \sqrt[3]{\frac{2\dot{W}/(\rho_0 A_1)}{1+\left(\frac{4A_1}{N\pi d_{pores}^2}\right)^2}}$ (3.21)	$FR1 = u_1 = \sqrt[3]{\frac{2\dot{W}}{\rho_0 A_1}}$ (3.24)
$FR2 = d_{min} = d_{pores}$ (3.22)	$FR2 = d_{min} = \frac{3c_d}{4\pi^3}\frac{\rho_0}{\rho_d}\frac{A_1}{RN_c^2}$ (3.25)
$FR3 = t_{max} = \frac{N}{n\sqrt[3]{\frac{2\dot{W}A_1^2}{1+\left(\frac{4A_1}{N\pi d_{pores}^2}\right)^2}}}$ (3.23)	$FR3 = t_{max} = \frac{\frac{3V_{23}}{2\pi d_d^3}}{n\sqrt[3]{2\dot{W}A_1^2}}$ (3.26)

Based on the previous equations, the design equations ***FR*** = *f*(***DP***) can be immediately formulated. Table 3.2 collects the results in the adequate form for identifying the FRs.

From the design Eqs. (3.21)–(3.23), the available DPs for porous filter solution are the motor power $\dot{W}$, the inlet area A_1, the diameter of filter pores d_{pores}, and the number of pores N.

From the design Eqs. (3.24)–(3.26), the available DPs for centrifuge separation are the motor power $\dot{W}$, the inlet area A_1, the volume of the dust container V_{23}, the number of cyclone turns N_c, and the radius of curvature of the cyclone R.

Once the DPs have been identified, the design equations for both solutions can be written in terms of the design matrix as $\{\Delta FR\} = [DM]\{\Delta DP\}$, where $DM_{ij} = \partial FR_i / \partial PD_j$ and where X represents nonzero elements $DM_{ij} \neq 0$.

– Porous filter separation:

$$\begin{pmatrix} \Delta u_1 \\ \Delta d_{dmin} \\ \Delta t \end{pmatrix} = \begin{pmatrix} X & X & X & X \\ 0 & 0 & X & 0 \\ X & X & X & X \end{pmatrix} \begin{pmatrix} \Delta \dot{W} \\ \Delta A_1 \\ \Delta d_{pores} \\ \Delta N \end{pmatrix} \quad (3.27)$$

– Centrifuge separation

$$\begin{pmatrix} \Delta u_1 \\ \Delta d_{dmin} \\ \Delta t_{max} \end{pmatrix} = \begin{pmatrix} X & X & 0 & 0 & 0 \\ 0 & X & X & X & 0 \\ X & X & 0 & 0 & X \end{pmatrix} \begin{pmatrix} \Delta \dot{W} \\ \Delta A_1 \\ \Delta R \\ \Delta N_c \\ \Delta V_{23} \end{pmatrix} \quad (3.28)$$

2. **Axiomatic design analysis.**
   - *Compliance with the Independence Axiom.*
   As design matrices (3.27) and (3.28) show, the filter-based solution is a coupled design, where the functionality of vacuuming $(FR1 : u_1)$ directly depends on the diameter and number of filter pores. As a consequence, two FR that where independent in a neutral-based solution (cleanup dust particles and retain dust particles) become coupled in the physical solution: The more particles are retained, the more filter pores clog; thus, the power for vacuuming and cleaning-up particles decreases. On the other hand, (3.28) shows a decoupled design matrix, where as a first approach, the functionality of vacuuming is not compromised by the physical solution for retaining the particles. Hence, based on the Independence Axiom, centrifuge separation is selected as a better solution.
   - *In case of redundant designs, application of the Linearity Theorem to select the most accurate DPs.*
   According to Axiomatic Design, (3.28) should be a better design, although the number of DPs is higher than the number of FRs, which results in a redundant design. The quest of the ideal design is generally based on DPs rearrangement, either giving constant values to some of them, or creating new DPs as a combination of them [3, 6]. This strategy, however, may require of a deeper analysis of DPs nature in order to avoid arbitrary DPs selection. As it is shown in Table 3.3, different design matrices can be obtained according to the DPs choice. (It is assumed $\partial n / \partial (RN_c^2) \approx 0$ so it impacts only the particles with minimum volume.)
   The reader can observe that the three different design matrices can be generated according to the DPs selection. According to the Independence Axiom, designs (3.30) and (3.31) would be better than (3.29), and the final decision could be solved by the Information Axiom. However, why should the inlet area A_1 be removed as a main DP?
   To help the designer to select the DP that automatically complies with Axioms 1 and 2, the Linearity Theorem can be applied. As stated, according to that theorem, parameters that generate a linear relation between DP and

Table 3.3 Three possible strategies to diminish redundancy in centrifuge filter by DPs rearrangement

Strategy 1: DP combination	$\begin{pmatrix} \Delta u_1 \\ d_{d\min} \\ t_{\max} \end{pmatrix} = \begin{pmatrix} X & 0 & 0 \\ X & X & 0 \\ X & 0 & X \end{pmatrix} \begin{pmatrix} \dot{W}/A_1 \\ RN_c^2 \\ V_{23} \end{pmatrix}$ (3.29)
Strategy 2: A_1, N_c removed	$\begin{pmatrix} \Delta u_1 \\ \Delta d_{d\min} \\ \Delta t_{\max} \end{pmatrix} = \begin{pmatrix} X & 0 & 0 \\ 0 & X & 0 \\ X & 0 & X \end{pmatrix} \begin{pmatrix} \Delta \dot{W} \\ \Delta R \\ \Delta V_{23} \end{pmatrix}$ (3.30)
Strategy 3: A_1, R removed	$\begin{pmatrix} \Delta u_1 \\ \Delta d_{d\min} \\ \Delta t_{\max} \end{pmatrix} = \begin{pmatrix} X & 0 & 0 \\ 0 & X & 0 \\ X & 0 & X \end{pmatrix} \begin{pmatrix} \Delta \dot{W} \\ \Delta N_c \\ \Delta V_{23} \end{pmatrix}$ (3.31)

FR are better than nonlinear ones. In case of FR2, Eq. (3.25) shows that the diameter of the dust particle that can be separated $d_{d\min}$ has a linear dependency with A_1, neither with R or N_c. As a consequence, A_1 should be selected as a main DP. However, this decision produces a contradiction because it leads to a coupled design that does not satisfy the Independence Axiom. The removal of this contradiction represents a new challenge for the designer, who is guided to conceive a new physical solution where the term $\partial d_{d\min}/\partial A_1$ could be removed.
This is a clear result that shows that the analysis and solving of contradictions between the results surfaced by the application of the Axioms, the Linearity Theorem, and the definition of the best design guide the innovative procedure toward a new solution.

3. **Description of the physical solution in terms of DPs and FRs.**
Figure 3.6 shows a new physical solution where the inlet area $A_{0'}$ has been added as a new DP. As a consequence, the cyclone diameter is different from the inlet diameter.
In this solution, the new DP decouples the inlet speed from the cyclone speed. Hence, the design equations and design matrix results (Table 3.4).
As it can be observed, the design matrix obtained in (3.35) is formally similar to the ones in (3.30) or (3.31); however, the physical solutions are not. The solution derived from the application of the Linearity Theorem conducted to an almost uncoupled solution, where the functionality of cleaning up dust particles is completed decoupled from the geometry of the cyclone, whose main parameters define the diameter of the minimum dust particle that can be separated. Finally, the volume of the container, also decoupled of FR2 in a first approach, will be set to obtain an operational time for a particular value of specific electrical power $\dot{W}/A_{0'}$.

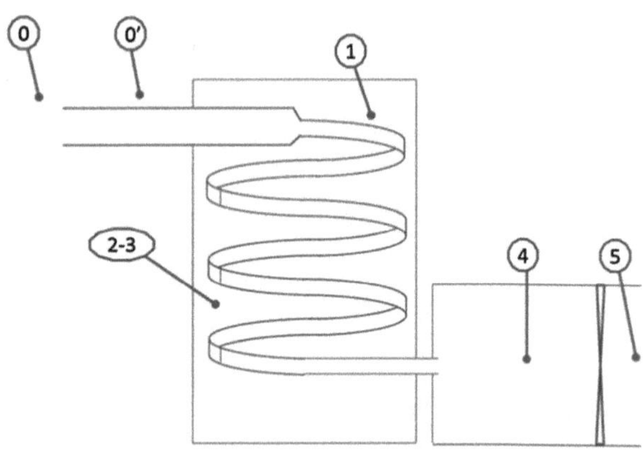

Fig. 3.6 Cyclone-based solution with decoupled air speeds

Table 3.4 Design equations and design matrix after applying the Linearity Theorem

$FR1 = \text{clean-up dust particles} = u_{0'} = \sqrt[3]{\frac{2\dot{W}}{\rho_0 A_{0'}}}$	(3.32)
$FR2 = \text{Reatain dust particles} = d_{d\min} = \frac{3c_d}{16\pi^2}\frac{\rho_0}{\rho_d}\frac{A_1}{RN_c^2}$	(3.33)
$FR3 = \text{operate long time} = t_{\max} = \frac{3V_{23}}{2\pi d_d^3 n \sqrt[3]{2\dot{W}A_{0'}^2}}$	(3.34)
$\text{Design matrix} = \begin{pmatrix} \Delta u_1 \\ d_{d\min} \\ t_{\max} \end{pmatrix} = \begin{pmatrix} cte & 0 & 0 \\ 0 & cte & 0 \\ X & 0 & X \end{pmatrix} \begin{pmatrix} (\dot{W}/A_{0'})^{1/3} \\ A_1/RN_c^2 \\ V_{23} \end{pmatrix}$	(3.35)

3.3.3 *Case Study 2: Conceptual Design of a Fuel Supply System for Gasoline Engines*

In this section, the Linearity Theorem will be used for analyzing different types of fuel metering systems for petrol engines.

1. **Quantitative formulation of the design problem.**
   - *Definition of the challenge*:
     Select the best fuel metering system for a petrol engine. This means to meter the correct fuel quantity as a function of the mass-flow rate of air, temperatures, type of operation, etc.
   - *Selection of the minimum number of independent FR in a neutral solution environment*:

     FR1: Deliver fuel at a desired rate, which is represented by the mass per unit of time metered into the engine, $\dot{m}_f$

   - *Formulation of the input and system constraints and laws of physics.*
     The fuel in the tank must be introduced into the engine by crossing a wall of the engine, and hence, the conservation of mass and energy assures that the mass-flow rate of a liquid that passes through a hole in a wall is given by $\dot{m}_f = A\sqrt{2\rho_f \Delta P_f}$, where A is a characteristic discharge area, ρ_f is the density of the fuel, and ΔP_f is the pressure drop across the hole.
   - *Obtaining the design equations, identification of DPs, and formulation of the design matrix.*
     In this problem, there is only one FR and two possible DPs, the area and the pressure drop. The direct application of the Linearity Theorem advises to select the area instead of the pressure drop because the mass-flow rate depends linearly with the area (note that the pressure drop is inside the square root). However, a designer could select the pressure drop as the design parameter. If such is the case, the solution would be a carburetor.

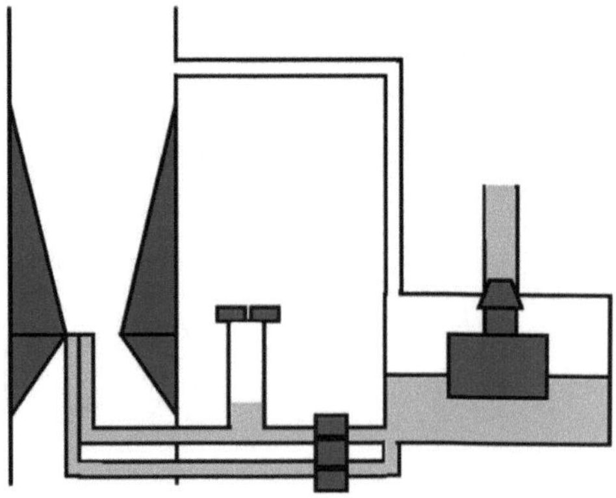

Fig. 3.7 Basic scheme of a venturi-based carburetor. The air which passes through the reduction of area is used to produce a pressure drop that aspires the fuel into the air stream

2. **Axiomatic design analysis**

- **Carburetor**:
  Figure 3.7 represents the basic scheme of a carburetor. As it is well-known, a carburetor is based on the Venturi effect, which uses the stream of air in order to produce a pressure drop which is a nonlinear function of the mass-flow rate of air. The pressure drop in the air is used to produce a pressure drop in the fuel. As it can be stated, the behavior is highly nonlinear and, based on the Linearity Theorem, we can conclude that the venturi-based carburetor is not the best design for metering fuel inside a petrol engine.
- **K-Jetronic**:
  As long as the Linear Theorem produces a contradiction between the concept "carburetor" and the concept "best design," a window for innovation remains open. At this point, Axiomatic Design can be used to establish that a linear system could be better than a carburetor. For that reason, it is convenient to invent a system which controls the mass-flow rate of fuel with the area and not with the pressure drop. This system was yet introduced in the market with the commercial name of k-Jetronic. The basic scheme of this system is shown in Fig. 3.8.

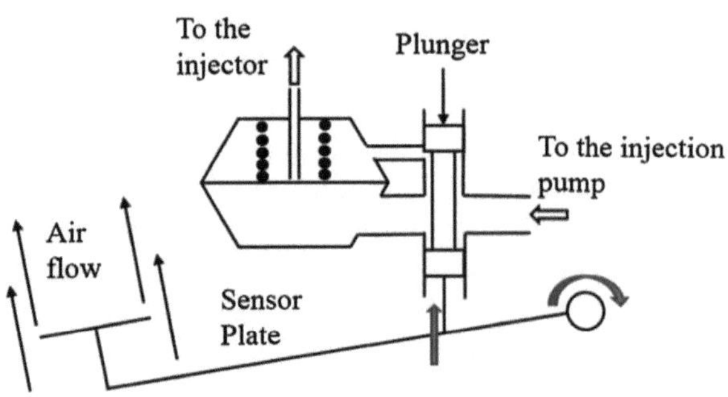

Fig. 3.8 Basic scheme of a k-Jetronic fuel metering system. The intensity of the air flow changes the position of a sensor plate which controls the area of the fuel port by means of control plunger

As the automotive industry has shown, this system was considered better than carburetors. However, although it is linear in the first level of the hierarchy of design, it is not linear in the second level of the hierarchy of design due to the nonlinear behavior of the position of the main lever with the air mass-flow rate. Therefore, there is still margin for innovation. However, how can the system be improved if we have tried the two DPs involved in the physical law? The answer is making a disruptive innovation.

3. **Description of the physical solution in terms of DPs and FRs.**

- **Discontinuous metering**:

  In the previous physical law for the fuel mass-flow rate, there is an implicit assumption: The mass-flow rate $\dot{m}_f = A\sqrt{2\rho\Delta P}$ is continuous or quasi-steady. However, there is not any reason for ensuring that steady systems are better than non-steady ones. In addition, there is a contradiction in the previous system: The linear system in the first level of the hierarchy is not the best system in the second level, and hence, we could conclude that something is missing at the first level. This contradiction can act as a motor for an innovative session and eventually can lead the designers toward a disruptive innovation. In this case, the innovation lies in conceiving a way to introduce more physical parameters in the physical law.

  - *Obtaining new design equations and formulation of the design matrix.* Let us think on a discontinuous system whose temporal-averaged discharge depends on the duty cycle of the system. If such system is used, the mass of fuel delivered is $\dot{m}_f = \frac{1}{T}\int_0^{\chi T} A(t)\sqrt{2\rho\Delta P(t)}\mathrm{d}t$ where T is the period of the control signal and χ is the fraction of time that the area is open during that period. As long as the area and the pressure drop cannot be selected as the DPs (because they were selected in the previous systems without leading to the best design), the new selected DP must be χ. However, this DP is located in the upper limit of a very nonlinear integral and hence inherits its nonlinear behavior against the advice of the Linear Theorem. Again, there is place for a disruptive innovation. A new challenge arises: Is it possible to obtain a linear system based on a discontinuous metering? This question is represented in Fig. 3.9 where the main requirement is to obtain a linear system.

    Note that $\dot{m}_f = \frac{1}{T}\int_0^{\chi T} A(t)\sqrt{2\rho\Delta P(t)}\mathrm{d}t$ can be reduced to $\dot{m}_f = A\sqrt{2\rho\Delta P}\chi$ if ΔP and A are constant. Thus, the designer must introduce into the system the required elements for doing that. For the condition $\Delta P = \text{cte}$, he must introduce a pressure regulator; for the condition $A = \text{cte}$, he must introduce a rapid response of the system in order to reduce the fly time of the needle. This last requirement can be achieved, for example, by reducing the size of the needle or by introducing hydraulic assistants. This solution is represented in Fig. 3.10. Note that in this case, the design matrix is diagonal (there is only one element) and

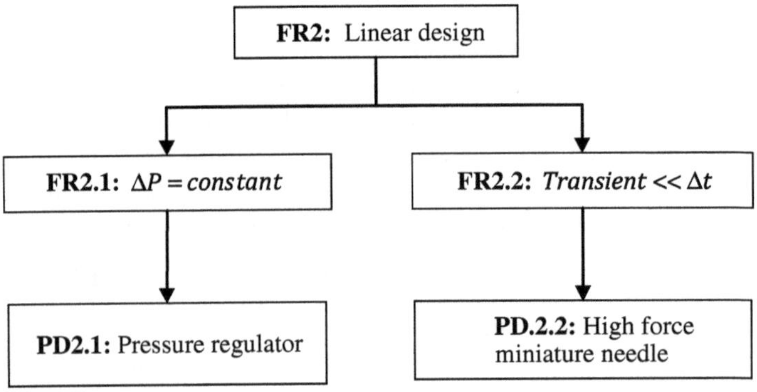

Fig. 3.9 Structure of FRs. The first level is to produce the adequate fuel mass-flow rate. The second level is to select a linear DP for it. The third level is the one that assures the subsystems required for achieved it

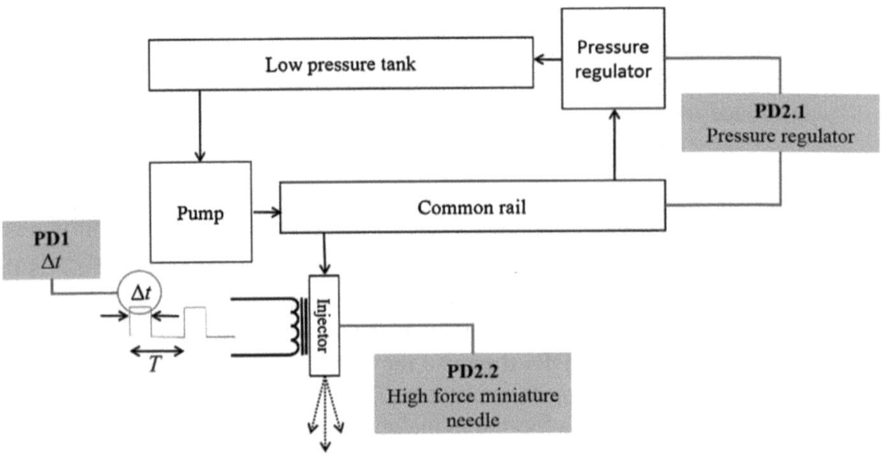

Fig. 3.10 Basic scheme of a discontinuous fuel metering system. A pump is used to feed a pressure regulator, and a high force miniaturized system is used to control a needle valve

constant because the designer has achieved the following physical law: $\dot{m}_f = \text{cte}\,\chi$ being χ the duty cycle. Therefore, the system fulfills the Linearity Theorem and the axioms, which leads to the best system and positions it as the winner system in the market. As an interesting exercise, the authors suggest to show that in the following levels of the design hierarchy, there are no additional contradictions between the chosen system and the concept of ideal design. For example, the pump can be designed with total independence of the rest of elements. As a result, the obtained solution is the best one. The solution is mature, which reduces the margin for additional disruptive innovations in this challenge.

3.3.4 Compliance with the Information Axiom and Constraints

The calculation of the information content constitutes one of the most difficult tasks in the early stages of design, basically, because the lack of precise information. However, the application of the Linearity Theorem permits to adopt some qualitative considerations, in terms of probability of success. The qualitative evaluation of information is one of the most important aspects of the use of corollaries and theorems when applying Axiomatic Design to guide strategic design. Particularly, the use of the Linearity Theorem with Corollary 6 ensures that the selected DPs allow the designer to define the widest intervals that conduct to the broadest tolerances and to the easiest transfer functions that guarantee the design specifications during the life cycle of the product. Therefore, the use of the aforementioned theorems and corollaries guide to increase the probabilities of success of the product and, consequently, to facilitate the compliance with Axiom 2.

Constraints, in general, condition the definition of FRs and DPs intervals. In those circumstances, the previous argumentation is immediately applicable. Finally, for those constraints which determine whether a solution is valid or not (as happens with product cost), the experience shows that the easier to control the DPs and their response through the design equations, the easier (1) to redefine the physical solution, or (2) to move it to the operational point that satisfies constraints. As shown in the case studies, if the constraints allow it, the designer might convert a redundant design into an uncouple one. Hence, both scenarios justify the adoption of the Linearity Theorem to guide the early phases of design.

An interesting remark surfaces when in the development of the Linearity Theorem (see, e.g., the Alternate Statement), all the DPs are considered as sources of variation causing the design to deviate from its target values. However, this is not a main concern or a drawback because that is the normal situation in almost all real designs. Let us think, for example, in the transmittal of torque from a driving shaft to a driven shaft made of a different material through press fitting of the two shafts, and assume DP1 as the amount of interference set to achieve the target value of the maximum torque transmittable and DP2 as the temperature at which torque transmission operates. It could be thought that DP1 enables a design to achieve its target values (and hence, DP1 should satisfy only Independence Axiom because it is not a source of variation), while DP2 is a source of variation (and hence, DP2 should satisfy only Information Axiom). If such was the situation, a question arises: Does the Linearity Theorem, which requires Information Axiom for its validity, apply to DP1? The real scenario is that both DP1 and DP2 are sources of variation, and hence, the Linearity Theorem applies to both: DP1 because of dimensional tolerances, geometric tolerances, position tolerances, mounting tolerances, surface finish, vibrations, operational loads, corrosion, etc., make always DP1 a source of variation and DP2 because the presence of a random variation in the operating temperature will also cause a differential expansion of the two shaft diameters and hence a deviation of the transmittable torque from the target value.

The reader should note that the selection of a linear DP does not eliminate the existence of the rest of parameters (in the fuel metering system, selecting the duty cycle does not eliminate the existence of the pressure drop and the area). It is more like selecting a DP and making it linear, and constraint the rest of variables imposed by the physics (in the same example, the area and the pressure drop must be constant to assure the linear behavior of the duty cycle). Even in the case of having a DP which is explicitly linear, it could be better to select another DP whose behavior, at a first glance, appears as nonlinear (this is the case of the area against the duty cycle). Designers must deal with the parameters allowed by the physics, which could be all of them nonlinear ones, and with the constraints imposed by their own decisions, which could over-constraint the imagination but also could boost the innovation trying to find out how to convert the system into the best linear one. Nevertheless, finding a point which is exempt of contradictions is a creative process which could yield an empty set. Certainly, Axiomatic Design and consequently the presented theorem are not properly boosters of the creative process; however, they represent a valuable tool for the designers in which they can rely to guide their creativity into the right direction.

3.4 Conclusions: Applicability and Value to Industry

The adequate selection of the design parameters that better accomplish the acceptance intervals imposed by customers' attributes is one of the most important activities in the early stages of design. This aspect turns even more critical when the number of design parameters is high and complex. An inadequate selection of them generally conducts to increase the number of iterations to find an adequate solution and the complexity of it. As a consequence, a direct impact in the product development costs occurs.

In this context, this chapter presents the Linearity Theorem as a consistent rationale which, when applicable, gives the designer the opportunity to select the most accurate design parameters to achieve the best design according to Axiomatic Design principles. As a resulting conclusion derived from the design axioms, the theorem states that linear designs are better than nonlinear ones. Additionally to the theorem proof, two case studies derived from the industry and adapted by the authors for pedagogical purposes have been presented. Through them, the reader and practitioner can experience how the systematic application of the Linearity Theorem and the observing of the design parameters nature constitutes a valuable tool in order to accomplish systematic innovation. As a result, the case studies here presented have not only the pedagogical purpose of introducing advanced concepts of Axiomatic Design, but also the aim of proving how functional independence and minimum information content, applied through the Linearity Theorem, conducts to products with high probabilities of acquiring a leader position in the market.

Certainly, when speaking about applying Axiomatic Design to the industry, it is crucial to communicate a consistent methodology. According to the authors' experience, design processes solely do not guarantee the success in terms of valuable innovations if they do not incorporate rule-based tools that help the decision making to unequivocally identify the best solution.

First of all, one of the most inspiring challenges the practitioner encounters when applying Axiomatic Design in the industry is to properly formulate the design problem. Very often, seeking for innovation in strategic product design obliges the designer to formulate the law of physics and social models that constitute the framework of the design problem. Naturally, these laws and models will establish strong relations between the functional requirements and the design parameters that, added to the customer's attributes, will consistently formulate the design problem in terms of functional requirements, acceptance intervals, constraints, transfer functions, and tolerances. Secondly, the authors encountered that laws and models could not be seen as the final point of conceptual design activities after which problems evolve to optimization, but that the appropriate use of them is at the base of accomplishing innovation. As it is derived from the proof of the theorem, the linearity of a final design is not only conditioned by the function transfer but also by the proper selection of the adequate design parameters from the available ones.

In this context, based on Axiomatic Design principles and the use of the Linearity Theorem, the authors have achieved a methodical approach to design which facilitates systematic innovation. One of the most important aspects of this perspective is that in order to solve redundancy or contradictions, the definition of the best design is not abandoned, so the Linearity Theorem naturally derives from the two axioms. As a consequence, not only it establishes a rationale to select the adequate design parameters, but if the constraints allow it, a redundant design matrix or a diagonal matrix with non-constant elements could be transformed into a diagonal matrix with constant elements. Additionally, because the availability of precise information is not high during conceptual design, the use of Linearity Theorem facilitates the ulterior satisfaction of Information Axiom, which means to increase the probabilities of success of the product.

References

1. E.M. Benavides, *Advanced Engineering Design: An Integrated Approach* (Woodhead Publishing, 2012)
2. O. Lara-Rapp, E.M. Benavides, Ideal design process characterization. The impact of preliminary decision-making tools in the consumption of resources and its uncertainty. Res. Eng. Design **02**, 97–119 (2015)
3. N.P. Suh, *The Principles of Design* (Oxford University Press, 1990)
4. N. Cross, *Engineering Design Methods: Strategies for Product Design*, 4th edn (Wiley, 2008)
5. G. Pahl, W. Beitz, J. Feldhusen, K.H. Grote, *Engineering Design. A Systematic Approach*, 3rd edn. (Springer, 2007)
6. N.P. Suh, *Axiomatic Design: Advances and Applications* (Oxford University Press, 2001)
7. E.M. Benavides, J.B. Rodriguez, Methodological contradictions solved by the linearity theorem: the vacuum cleaner as a case study, in *Proceedings of the 8th International Conference on Axiomatic Design*. Lisbon, Portugal, 2014

8. J.B. Rodriguez, E.M. Benavides, Axiomatic design as a consultancy tool in product design, in *Proceedings of the 8th International Conference on Axiomatic Design*. Lisbon, Portugal, 2014
9. J.B. Rodriguez, E.M. Benavides, Axiomatic design of an airport passenger terminal, in *Proceedings of the 6th International Conference on Axiomatic Design*. Daejon, Korea 2011
10. H.L. Oh, A constraint optimization perspective on axiomatic design, in *Proceedings of the 7th International Conference on Axiomatic Design*. Worcester, Massachusetts 2013
11. G. Taguchi, et al. *Quality Engineering in Production Systems* (McGraw-Hill, 1989)
12. J.B. Rodríguez, E.M. Benavides, The vacuum cleaner as a case study for teaching axiomatic design, in *Proceedings of the* 7th *International Conference on Axiomatic Design*. Worcester, Massachusetts 2013

Chapter 4
A Different Consideration on Information and Complexity in Axiomatic Design

Erik Puik and Darek Ceglarek

Abstract To gain competitive power, product designs and their production means have become more and more complex over the past decennia. Product designers are faced with the increasingly difficult task to guarantee steady behavior of the systems they produce. This requires thorough understanding of the complex principles that determine the behavior of these products. It starts with notion how the many parts, of which the product design consists, are cross-linked with each other and their surroundings. If the design relations act predictable then the product design behaves predictable, and the functional requirements have high certainty of being satisfied. Axiomatic Design offers a number of ways to model the relations in a product design in order to improve its predictability. The 'information content' or 'entropy' of the design is indicative for the behavior of a system. The information content in Axiomatic Design is in the jurisdiction of the Information Axiom. This chapter investigates if information could be applied in a broader context; to bring the whole of methods in AD under a single heading. According to the definition of information by Shannon and Weaver, a broader application may be applied for Axiomatic Design. Along this path, an alternative framework of different kinds of information is decomposed that can be used to analyze progression in a product design. 'Useful information,' proportional to the 'ignorance of the designer after application of all his knowledge,' is decomposed into three kinds of information that are applied to graphically monitor the design process as it evolves.

Keywords Axiomatic Design · Complexity · Information · Probability · Product design · Independence · Entropy

---

E. Puik (✉)
HU University of Applied Sciences Utrecht, 3500 AN Utrecht, The Netherlands
e-mail: erik.puik@hu.nl

D. Ceglarek
International Digital Laboratory, WMG, University of Warwick, Coventry CV4 7AL, UK
e-mail: D.J.Ceglarek@warwick.ac.uk

© Springer International Publishing Switzerland 2016
A.M. Farid and N.P. Suh (eds.), *Axiomatic Design in Large Systems*,
DOI 10.1007/978-3-319-32388-6_4

4.1 Introduction

The design of products that are increasingly complex requires the designer to define many relationships between even so many parts to determine the 'collective behaviors' of a system. These collective behaviors do not only relate internally, but also interact with their environment. Of these many relations, a large portion needs to be restricted to impose the correct behavior of the system. The designer applies structure to the product design in order to achieve this situation. The structure can be of any kind: structure to enhance the image of the envisioned system, structure to model the system in the physical domain, or structure to enable manufacturing of the system according to modern standards. Axiomatic Design (AD) was developed to find the right regularities to prevent a system from performing in a non-regular or chaotic way by restricting it to the positive functional behavior. To achieve this, AD declares 'axioms' that cannot be proven nor derived from physical phenomena. A number of seven conceptual axioms were defined back in 1978 when the first paper about AD was presented [25]. Only two of those seven axioms stood the test of time, now known as the 'Independence Axiom' and the 'Information Axiom.' In 1999, a third axiom was added, the 'Complexity Axiom,' addressing four different types of complexity in design [19]. The general guideline of AD starts with functional requirements (FRs) that are satisfied by a sensible selection of Design Parameters (DPs), and the probability of this happening is a measure of success for the system design. The probability on positive functional behavior is maximized by the right choice of regularities, in the case of AD, the 'Design Matrix,' and the overlap between 'design range' and 'system range.' The probability of DPs satisfying FRs is a measure for 'Information in Design.' Information in design or 'entropy' is a state of chaos and in good accordance with the nature of the axioms it should be reduced. Obviously, it is related to the Information Axiom, but recent investigations indicate that the Independence Axiom and the Complexity Axiom also have a relation to information in design [11]. In this chapter, an attempt is made to organize the different kinds of information, understand them, and evaluate the consequences of their application. The goal was to structure the foundation of AD by bringing the 'concept of information' and the 'concept of complexity' under a single heading, being able to unequivocally model the design process.

This chapter consists of two parts. Part A, 'A Review on Information in Axiomatic Design,' evaluates the concept of information or 'Boltzmann's entropy' according to Shannon and Weaver [18]. It investigates information in a broader sense than the general directive within AD. It decomposes information and brings information in relation to the complexity theory of AD. Part B, 'Application of Information to Monitor Development Processes,' evaluates the outcome of Part A in relation to the design process. It uses a graphical representation of the design process to visualize typical design situations and explains them from the perspective of information in design. Further, the chapter is organized as follows: In Sect. 4.2, an evaluation is made on the background of information and complexity in AD. Section 4.3 expands the application of information in AD based on the definition of Shannon and Weaver. Section 4.4 decomposes information in design into three different kinds of sub-information. Section 4.5 discusses the findings of Part A. Section 4.6 investigates the three kinds of sub-information and how they evolve during the design process. Section 4.7 discusses these findings, and Sect. 4.8 gives a general conclusion of parts A and B.

Part A: A Review on Information in Axiomatic Design

4.2 Background

Information is mostly interpreted as transferred knowledge concerning a particular state or circumstance. Information in engineering is mostly related to the notion of complexity. If a system is complex, then a lot of information is required to describe the system. This section investigates the background of information in Axiomatic Design. Section 4.2.1 starts with the kinds of information that were defined by Suh. Section 4.2.2 inventories the role of the design matrix in information. Section 4.2.3 zooms in on the definition of complexity in AD and how it is anchored to information in AD.

4.2.1 Background on Information in Axiomatic Design

Information in Axiomatic Design is derived from the information theory using a measure of Boltzmann's entropy according to Shannon and Weaver [18], Brillouin [2], and Suh [20]. It uses the logarithmic representation as introduced by Hartley to make information additive instead of multiplicative [9]. According to the information theory, information is inversely related to the probability of success of a goal being met. In Axiomatic Design, a goal is met when DPs are causing FRs and constraints to be within tolerances.

The total amount of information in a design is called 'total information.' *Total information* was split into two parts, 'Useful' and 'Superfluous' information [20, p. 148]. *Useful information* relates solely to the satisfaction of particular tasks. These tasks are specified in terms of the FRs and constraints. *Superfluous information* does not affect the relation between DPs and FRs. Where probabilities are multiplicative, information is additive due to the logarithmic function. This leads to the breakdown of *total information* as shown in Fig. 4.1.

Every product design in progress will have 'information content.' The *information content* is a measure of the probability of success of achieving the specified FRs [20, p. 156]. The probability of success is obtained by considering all FRs to be satisfied in their mapping to DPs. Then, the joint information content is determined by taking the sum of all individual 'Information.' The result gives the information content of the design. The Information Axiom dictates that the information content

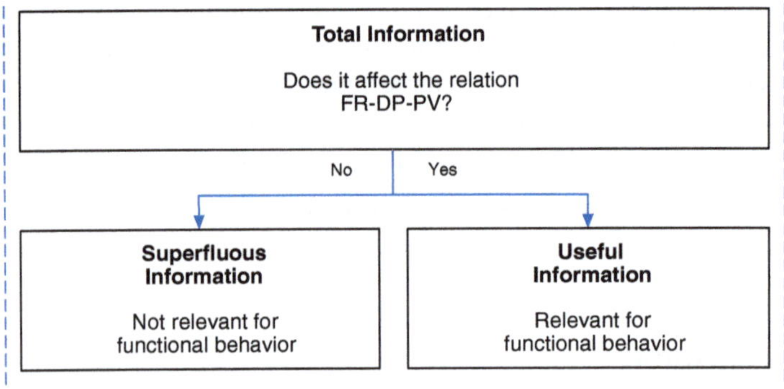

Fig. 4.1 Breakdown of *total information*

of a system should be minimized and thus, maximizing the probability of FRs to be satisfied.

This means for the information breakdown of Fig. 4.1 that *superfluous information* is no information from the axiomatic perspective and may be ignored. On the other hand, *useful information* must be properly eliminated from the design to guarantee complete satisfaction of all FRs, and therefore, the main focus in the design process should be on *useful information*.

4.2.2 Investigations for the Dependency of the Axioms

The dependency of the Axioms has been investigated a number of times. The first book about AD [20, p. 67] includes a paragraph about the relationship between axioms 1 and 2. Suh addresses the misunderstanding that the Independence Axiom is a consequence of the Information Axiom, by explaining that a coupled design could have lower information content than an uncoupled design. Without the Independence Axiom, it is not possible to choose the uncoupled design, which, from the design perspective, is more preferred than the coupled design. The second book [21, p. 175] contains some mathematical proof of the independence, based on the Boltzmann's entropy of the FR array as was published by El-Haik and Yang [6]. If the design matrix is square and non-singular with constant entries, and DPs are normally distributed random variables, then the entropy h of the FRs is given by

$$h(\{\mathrm{FR}\}) = h(\{\mathrm{DP}\}) + \ln|[A]| \quad (4.1)$$

where $|[A]|$ is the determinant of the design matrix $[A]$. Investigation of the determinant leads to the understanding that a coupled matrix can indeed have lower information content than an uncoupled matrix, which was reflected by the substantiation of corollary 7. In 2005, the book of El-Haik confirms Eq. (4.1) [5, p. 75].

Based on these investigations, it may be concluded that both axioms serve a particular goal and should be maintained.

4.2.3 Background on Axiomatic Complexity

Complexity is defined as 'A measure of uncertainty in achieving the specified FRs' [22, pp. 4, 58 and 65]. The Complexity Axiom advises to 'reduce the complexity of a system.' The theory defines two kinds of complexity, 'time-independent' and 'time-dependent.' In the case of *time-independent complexity*, the behavior is governed by the given set of FR and DP relationships. *Time-dependent complexity* depends upon the initial condition with FR and DP relationships, but unless the system goes back to the same set of initial conditions periodically, the distant future behavior is totally unpredictable as the system tends to escalate [19]. *Time-dependent complexity* is not further investigated in this chapter.

Time-independent complexity consists of two components: 'Real' and 'Imaginary' time-independent complexity, further to be referred to as *real complexity* and *imaginary complexity* (C_{R} and C_{Im}). *Real complexity* is inversely related to the probability of success that the associated FRs are satisfied according to one of the following relations

$$C_{\mathrm{R}} = -\sum_{i=1}^{m} \log_b P \tag{4.2}$$

$$C_{\mathrm{R}} = -\sum_{i=1}^{m} \log_b P_{i|\{j\}} \quad \text{for } \{j\} = \{1, 2, \ldots, i-1\} \tag{4.3}$$

depending if the system is uncoupled (4.2) or decoupled (4.3). Relation (4.2) is under the reservation that the total probability P_i is the 'joint probability of processes that are statistically independent.' Relation (4.3), for decoupled systems, is modified to correct for dependencies in the probabilistic function [22, p. 57]. 'b' is in both cases the base of the logarithm, usually in bits or nats depending of the preferred definition. Given (4.2) and (4.3), real complexity can be related to the information content in AD, which was defined in terms of the probability of success of achieving the desired set of FRs [20, p. 59], as

$$C_{\mathrm{R}} = I \tag{4.4}$$

in which C_{R} is the *real complexity* and I is the information as defined in AD [22].

Imaginary complexity is defined as complexity that exists due to 'a lack of understanding about the system design, system architecture, or system behavior' [19 p. 120]. It is caused by the absence of essential knowledge of the system. The designer cannot solve the problems in a structured manner and therefore is forced to apply trial-and-error. *Imaginary complexity* exists due to a lack of understanding of the designer.

4.2.4 Breakdown of Complexity in the Context of AD

Like most definitions in AD, complexity is also defined in the functional domain. This implies that 'Total Complexity' can be decomposed in a functional and a non-functional part analog to information in Fig. 4.1. Figure 4.2 shows the breakdown of *total complexity*.

Total complexity is broken down in a functional part 'Complexity according to the Complexity Axiom' and a 'Superfluous' part. *Superfluous complexity* has no effect on the FRs of the system and therefore is not relevant for AD. It is further ignored. *Real complexity* is by definition equal to the information of Axiom 2; their direct relation was given by Eq. (4.4). *Imaginary complexity* is harder to understand. It is defined as 'uncertainty that is no real uncertainty' and 'it arises because of the designer's lack of knowledge and understanding' [23, p. 65]. The book states further, 'when a design is uncoupled or decoupled, the imaginary component of complexity is equal to zero' [23, p. 71]. For a decoupled design, this is only guaranteed if the optimization order of the design relations is known. *Imaginary complexity* is inversely related to the satisfaction of the Independence Axiom.

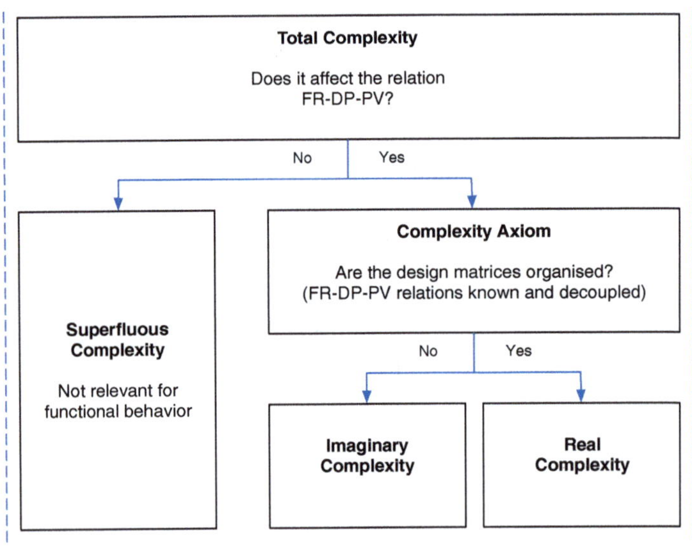

Fig. 4.2 Breakdown of *total complexity*

4.3 A Broader Application of Boltzmann's Entropy in Useful Information

Figure 4.1 shows the current decomposition of information in Axiomatic Design; *total information* of the system was divided into *superfluous information* and *useful information*. Since *superfluous information* does not affect the FRs, this means that elimination of *useful information* from the design is a prerequisite to satisfy all FRs.

4.3.1 Elimination of Useful Information from a Design

As elimination of *useful information* leads to satisfaction of the FRs, the question is how this can be achieved. A first and most straightforward hypothesis would be to assume that useful information could be eliminated by satisfaction of the Information Axiom. However, this is not the case.

Statement Elimination of *useful information* cannot be guaranteed by satisfaction of the Information Axiom alone.

Proof According to good AD practice, the information content of a design can be calculated with Suh [20, p. 157].

$$I = \log\left(\frac{\text{system range}}{\text{common range}}\right) \tag{4.5}$$

If it concerns multiple FRs, the different information contents should be summarized. Satisfaction of the Information Axiom can only take place if all system ranges are within the common ranges [7, 20, 21]. However, this does not satisfy the Independence Axiom; the design needs to be independent too. With only addressing the Information Axiom, the design could therefore still be a coupled design, and some FRs may not be satisfied. If there are unsatisfied FRs, *useful information* is not completely eliminated; the statement is true. Information as addressed by the Information Axiom will cause a design to be robust by guaranteeing overlap between system and common ranges, but it does not guarantee independence of the design.

This investigation leads to the understanding that *useful information* cannot be eliminated by satisfaction of the Information Axiom alone. This implies that a certain part of *useful information* is addressed by the Independence Axiom and that this axiom indeed is related to information. The question arises what kind of information this is.

In the books of Suh, the Independence Axiom was never associated with information according to Boltzmann's entropy, neither was *imaginary complexity*. However, *imaginary complexity* was considered to have a stochastic nature for some problems [19, p. 121, 22, p. 66]. Further, the book about complexity shows a number of examples that are clearly explaining how knowledge of the designer is related to the quality of his design outcomes. One example is a case where the designer does not realize that the design is a good design with a decoupled matrix. The designer uses trial-and-error to test many different sequences of DPs to satisfy the FRs, needing to test n! sequences, thinking that the design is quite complex. This situation describes exactly the characteristic behavior of Boltzmann's entropy in a design.

4.3.2 Information Related to the Independence Axiom

The question is if the Independence Axiom is related to Boltzmann's entropy and if this is the case, how it is embedded.

Statement 2 The Independence Axiom is related to Boltzmann's entropy.

Proof statement 2 The information theory of Shannon and Weaver states that information is 'related to the number of alternatives that remain possible to a physical system' [18]. The 'number of alternatives' indicates that the current design is not fully restricted within its delineated boundaries. Further Weaver explains, 'information does not relate to what the design is as much as what the design could be.' In an incomplete design, many alternatives in which the design can manifest itself are still open. Only a certain amount of these alternatives lead to satisfaction of FRs. The other alternatives lead to unsatisfied FRs. For an ignorant designer, this process has a stochastic nature; it increases the Boltzmann's entropy and as a result also information in design. Therefore, not only a lack of robustness causes information in design, but also every lack of boundaries that are needed to restrict the system to operate correctly.

Example 1 In a fully robust system, the Information Axiom is satisfied, because the system ranges match the common ranges. If the designer lacks understanding of the design, and hereby the design matrix is coupled, he will be surprised of the inexplicable system behavior when he tries to set up the system. To the designer, the system seems to operate randomly until he gains knowledge of the system. What first appeared random shows to behave in a structured manner, but only after acquisition of the appropriate knowledge.

Example 2 A designer overlooks a DP during the design process and as a result he assumes that the design matrix is decoupled conforming good AD practice. In a later stage of the lifecycle, this DP, which should have been properly 'fixed,' appears to drift away from its initial value. The drifting DP may cause coupling of the design matrix and randomly deprives satisfaction of the FRs.

Explanation The statement claims that information is not solely restricted to the Information Axiom. Example 1 explains that the dissatisfaction of the Independence Axiom may introduce features with a stochastic nature, and therefore, it also deals with information in design. This information is related to missing structure of the design that is a requisite to make a design independent. Gell-Mann and Lloyd call this missing structure a 'lack of regularities in the system.' The lack of regularities increases entropy in the system and 'the smaller the entropy, the less spread there will be among the entities that follow these regularities' [8, p. 50]. A lack of regularities in the design will increase its chaotic behavior and thus increases information. The definition of well-chosen FRs, the process of selecting matching DPs, decoupling the relations between FRs and DPs, making sure that all DPs are relevant, and ensuring that all relevant DPs are known, are all regularities that contribute to a more predictable behavior of the design, and hence, they eliminate information from the design.

Examples 1 and 2 can be clarified further by experiments that were described by Shannon [17] and Brillouin [2]. This experiment studies the transfer of a message in the English language over a telegraph line. The total character set exists of 26 characters of the alphabet and a space between words. Initially, the transfer per character is studied when no a priori knowledge of the English alphabet is present. The information content for all characters is the same and is calculated at 4.76 bits. This number can be roughly confirmed when realizing that five bits of information give a total of $2^5 = 32$ combinations; so a total of 27 combinations are expected to come just under five bits. For the second experiment, knowledge is made available that the a priori probability of occurrence of the characters in the English language is not equally distributed; for example, the space and the character '*E*' appear more frequently than others. Availability of this knowledge reduces the *total information* needed to transfer characters. Reconstruction of a corrupted message can be performed on a basis of statistical knowledge of the English character distribution instead of mere coincidence, thus increasing the chance on a successful outcome. The information per character indeed appears to be lower and is determined to be 4.03 bits. For the third experiment, the knowledge of the English words and grammar is made available to the receiving end. This knowledge helps rejecting unsuccessfully reconstructed messages and in this way further increasing the chances on a successful reconstruction of the message. Depending on the situation, the actual amount of information is estimated to be between 1 and 2 bits. This example clarifies that every type of knowledge-based condition, imposed on the possible freedom of choice, immediately results in a decrease of information. The same applies to the synthesis of a product design where every good definition of an FR and its DP limits the possible variation of the behavior of the system and thus reduces information or entropy in the design. Adding regularities in a design decreases information; it quantifies the extent to which an entity is taken to be regular, nonrandom, and hence predictable. For AD, this is not only limited to the Information Axiom since the description of rule-based features for the 'Structure of the Design' also adds-up to the predictability of the product design and therefore also reduces information. Finally, decoupling of the design matrix is a process that eliminates wrong outcomes in a structured manner. The remaining stochastic process has no other options than to operate within the remaining boundaries of the system. In a good design, all remaining boundaries lead to a successful outcome and thus satisfaction of the FRs.

4.3.3 Disruptive Character of the Independence Axiom

The recommended way to address design problems advises to address the Independence Axiom first, followed by the Information Axiom. The reason is that the Independence Axiom initially determines the design matrix and by doing this it sets the design relations to be optimized by the Information Axiom. This causes the Independence Axiom to be disruptive for the Information Axiom. Therefore, according to good AD practice, the satisfaction of the Independence Axiom precedes the satisfaction of the Information Axiom. From this perspective, the more general concept arises that the Independence Axiom is about 'doing the right things' and the Information Axiom is about 'doing things right' [11]. Though these statements are not meant to be inexhaustible, they well contribute to general understanding of how these kinds of information address the product design.

4.4 Total Decomposition of Information

This section decomposes information in AD according to the situation as proposed in the former paragraph. The decomposition starts with *useful information* since this is the highest kind of information that is relevant for the functional behavior of systems.

4.4.1 Decomposition of Useful Information

The claim that information is in principle related to both the Independence Axiom and the Information Axiom makes *useful information* the aggregate of these kinds of information conforming

$$I_{\text{Useful}} = I_{\text{Related to Axiom 1}} + I_{\text{Related to Axiom 2}} \tag{4.6}$$

where both kinds of information are the result of irregularities in the design; Axiom 1 dealing with the structure of the design, and Axiom 2 dealing with robustness in the design. The information related to Axiom 1 disappears when the design matrix is decoupled and the information related to Axiom 2 disappears when a design becomes robust. As a result, *useful information* measures the lack of total regularities and therefore the 'Ignorance of the designer'; this is exactly according to the conclusion of Gell-Mann and Lloyd, which leads to the following equation

$$I_{\text{Useful}} = \text{IGN}_{\text{Designer}} \tag{4.7}$$

where $\text{IGN}_{\text{Designer}}$ is the total ignorance of the designer under proviso that there was enough time to apply the designer's knowledge to the design. As indicated $I_{\text{Related to Axiom 1}}$ is a different kind of information and not the same as defined for the Information Axiom in AD. Consequently, a new definition is needed to differentiate these two kinds of information.

Definition 1 The information caused by the irregularities in the *structure of the design* is called 'Unorganized Information' since it only exists when the design matrix has not yet been organized. *Organized information* is information that resides in the system because not all FRs, DPs, and PVs are known and/or the design matrix is not uncoupled or decoupled.

Definition 2 The information that concerns robustness of the design, which is traditionally indicated by the Information Axiom in AD, is further called 'axiomatic information.'

The breakdown of *total information* as shown in Fig. 4.1 can be expanded by applying this definition and is shown in Fig. 4.3.

Unorganized information is determined by the organization of FRs and DPs in the design matrix and their decoupling but it has no impact on the common range of the system; they are situated at the same hierarchical level.

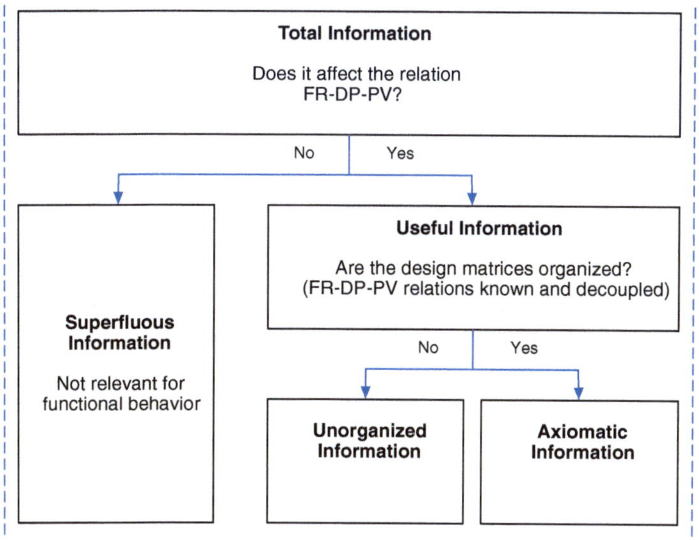

Fig. 4.3 Expanded breakdown of *total information*

4.4.2 Decomposition of Unorganized Information

If a design matrix is properly developed, then all FRs and DPs are known, and if the design matrix is decoupled, only *axiomatic information* is left in the system. *Axiomatic information* typically gives feedback to the designer about his lacking knowledge. If a system range does not satisfy the design range, the designer will notice that a particular FR is unsatisfied. The designer will also know what DPs are responsible for the problem because of his understanding of the design matrix. This is not the case for *unorganized information*; lacking knowledge does not automatically come to the surface and information may remain hidden. The first example of Sect. 4.3.3 shows a situation in which the designer does know that he is missing knowledge to set up the system. In this case, a design shows inexplicable system behavior to the designer, which warns the designer that he does not yet fully understand the design. The second example shows a different case. The designer misses a DP, but is not aware of this problem. His lacking knowledge is essential to prevent malfunction in the future, when changing circumstances that are not clear to the designer, enable the DP to cause problems. Missing knowledge hinders the designer to make the right choices for the essential regularities in a design and therefore, *unorganized information* may manifest itself in at two different appearances: 'unrecognized' and 'recognized.'

Definition 3 'Unrecognized Information' is a part of *unorganized information* that is not recognized by the designer and therefore remains hidden in the system. It is addressed by finding the right FRs, DPs, and PVs.

Definition 4 'Recognized Information' is the part of *unorganized information* that is recognized by the designer but as the knowledge to address the problem is lacking, it cannot yet be eliminated from the design. It is addressed by preparation of the design matrix and decoupling it.

The next paragraph will give an overview of all kinds of information that are covered in this chapter.

4.4.3 Overview of Information in Design

This chapter has explained the seven kinds of information. An overview of the different kinds of information as is shown in Fig. 4.4:

- *Total information:* the total information content of the design (full entropy of the design);
- *Superfluous information:* information that does not affect the relation between FRs and DPs;
- *Useful information*: the part of total information that affects the relation between FRs and DPs;

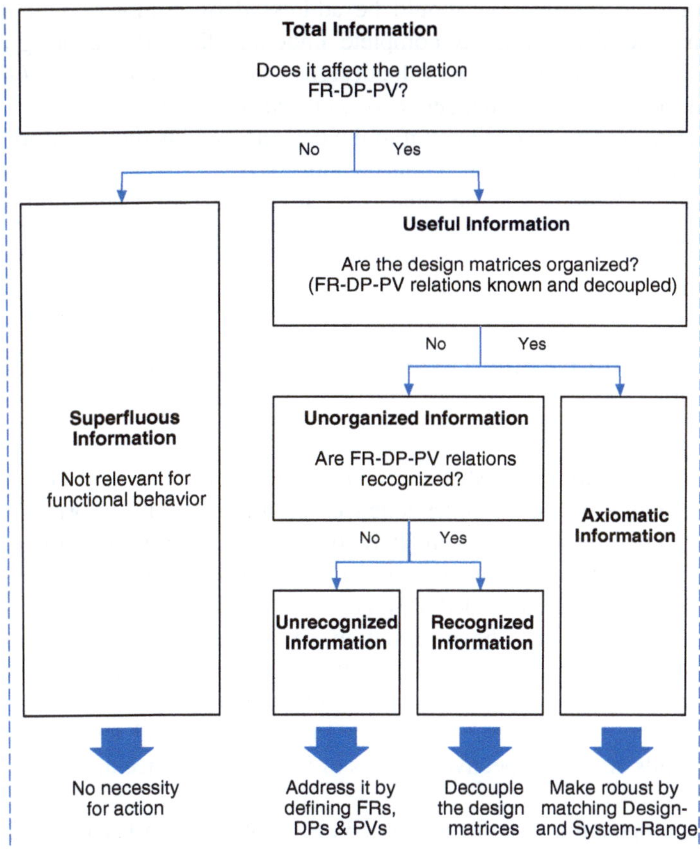

Fig. 4.4 Final breakdown of *total information*

- *Axiomatic information*: a specific kind of *useful information* due to a discrepancy in design ranges and system ranges according to Axiom 2;
- *Unorganized information:* a specific kind of *useful information* that is caused by insufficient relational regularities of the design (FRs and DPs);
- *Unrecognized information:* a specific kind of *unorganized information* that is not recognized by the designer and therefore remains unaddressed;
- *Recognized information*: a specific kind of *unorganized information* that is recognized by the designer but the knowledge to address the problem in an appropriate manner is lacking.

Figure 4.4 completes the breakdown of information in AD. A number of three kinds of information should be addressed to ensure a good design:

- *Unrecognized information* should be addressed by completion of the design relations and decoupling as complete understanding of the design relations leaves no room for ignorance of the designer. Once *unrecognized information* is recognized, it instantly changes to *recognized information*;
- *Recognized information* is known to the designer, and it should be addressed by investigation of its cause so it can be addressed;
- *Axiomatic information* is eliminated by matching the system and the design ranges.

4.5 Discussion on the Review of Information in Axiomatic Design

Based on the definitions of Shannon and Weaver, Brillouin, and Gell-Mann and Lloyd, the statement that the Independence Axiom deals with information may be considered to be true. It leads to an alternative definition of information within AD. Accepting the definition means that useful information is the basis for AD since it covers every aspect that is needed to satisfy the FRs. But it also means that both the Independence Axiom and the Information Axiom are addressing information in design. However, both axioms address different kinds of regularities and therefore deal with different kinds of information; regularities in the product design that deal with its structure are different from the regularities that deal with robustness and these two kinds are the respective habitats of the Independence and the Information Axioms.

This decomposition of information gives a new view to the discussion of the interdependency of the axioms, especially the discussion in which the Independence Axiom was considered to be a subset of the Information Axiom. The term 'subset' is clearly not true since the axioms are found on the same hierarchical level. The fact that the axioms address their stand-alone conceptual weaknesses [6, 20, 21] is further reinforced by this research. Yet, the axioms are not entirely independent. The regularities that deal with the structure of the design will set the arena for operation of the regularities that deal with robustness. Changes in the structure of the product design may therefore be disruptive to regularities for robustness. This makes the Information Axiom submissive to the Independence Axiom though their fields of operation are fully separated.

The decomposition of complexity, as shown in Fig. 4.2, is comparable to the decomposition of information in Fig. 4.3. In this alternative definition of information, the Complexity Axiom and *useful information* are the same concept. This is in line with the observation that: (1) *real complexity* and information content in a design according to the Information Axiom were defined being equal, and (2) *imaginary complexity* and the Independence Axiom are both aiming for equal goals. Both are derived in the functional domain and in absence of either one of them all FRs will be satisfied. As a result, the absence of *useful information* from a product design satisfies the Complexity Axiom and vice versa. Satisfaction of the Complexity Axiom leads in its turn not only to elimination of both kinds of complexity, but it leads also to satisfaction of both axioms. This makes the Independence Axiom and the Information Axiom a subset of the Complexity Axiom. This observation is not affected whether the Independence Axiom is considered to be information or not. The operation and practical application of AD are not much affected by this last observation for at least three reasons. First, the possibility to develop the axioms 1 and 2 from the Complexity Axiom has already been reported in Suh [23, p. 83]. It mentions the Complexity Axiom being 'less explicit than particularly the Independence Axiom' and advises to apply the axioms as they are, because both axioms serve their specific goals. Secondly, the Complexity Axiom will still be axiomatic in a sense that it cannot be derived from a higher truth. Third, the Complexity Axiom alone is not easily applied in the design process. It does not structure the order in which the information content of the design should be addressed. The Independence Axiom and the Information Axiom are best maintained as starting points of the design process and should be addressed in that particular order. In its application, it leaves AD intact as is.

A limitation of this consideration on information in Axiomatic Design is that it does not expand AD. Although the definition of *unorganized information* and its two derivatives *unrecognized* and *recognized information* is new, these concepts could have been defined without an analysis based on information. As such, the approach does not change AD and the way it is applied. However, it does structure AD being a method that defines FRs that are met by suitable DPs and PVs and if their relation is not crystal clear, there will be information in design. Adding regularities in the design may reduce that information. It approaches the design process other way round; it forces good behavior of the system by elimination of potential bad conducts, instead of focusing on what the design does. It is not said that this approach prevails over the other, but it might help to locate *unrecognized information* more effectively. There is room for enhancement of this view on the design process. The number of methods to determine information in the conceptual stage is limited and their application is laborious.

Time-dependent complexity was not yet investigated because it deals with a lack of knowledge to the designer and simultaneously with a changing system range. Its escalating character requires profound investigations. It remains for future research.

Part B: Application of Information to Monitor Development Processes

Part B of this chapter will show how complex development processes can be analyzed by application of the three kinds of information that were defined in Part A of the paper: *unrecognized information, recognized information,* and *axiomatic information.* Typical problems that may occur during the design process will be visualized in an 'Axiomatic Maturity Diagram.' The Axiomatic Maturity Diagram was first introduced in Puik and Ceglarek [12] but is now applied for the three kinds of information as mentioned above.

4.6 Monitoring Process Progression

Monitoring of the development process gives feedback to the designer about the course of the project. Ideally, the actual state of the project would be known on a real-time basis. As accurate determination of the state of the project is effortful, and because of that may deal with latencies, it can also be done in retrospect. In this case, the project will be monitored using a graphical representation. It not only shows errors due to typical patterns representing the different kinds of information, but it is also possible to monitor the desired dynamics in the process of design, e.g.,:

- Most efficient development path in terms of investment (SMEs);
- Optimized development path for project lead time (semiconductor industry); and
- Lowest chance for development errors (safety systems, medical).

4.6.1 *The Axiomatic Maturity Diagram*

The Axiomatic Maturity Diagram is based on the information content in a product design as represented by the Independence Axiom and the Information Axiom [12]. The diagram, shown in Fig. 4.5, uses two axes, one for each axiom, plotting the degree, in which the axioms are satisfied. The diagram in itself has no axiomatic properties but it takes its name from the fact that it applies the axioms as a premise.

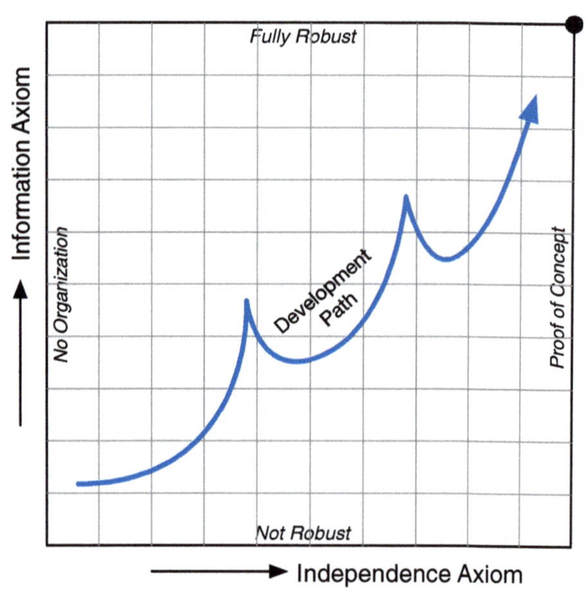

Fig. 4.5 The Axiomatic Maturity Diagram. The *horizontal axis* plots the Independence Axiom and the *vertical axis* plots the Information Axiom. The development path is arbitrary

The horizontal axis is the 'axis of organization' starting at 'No Organization' and ending with 'Proof of Concept.' Proof of concept indicates that the design matrix is decoupled, and therefore, *unorganized information* has become equal to zero. As shown in Fig. 4.4, this implies that both kinds of sub-related information have been eliminated: *unrecognized information* and *recognized information*. The vertical axis represents robustness of the design from 'Not Robust' to 'Fully Robust.' As was explained in Part A of the paper, a fully robust system implies that the *axiomatic information* has become equal to zero (the traditional information in AD coming forth from an inadequate common range). The lower left corner indicates a high level of ignorance; the designer has little knowledge how to satisfy FRs with his DPs and therefore, the 'Axiomatic Maturity' is low. The upper right corner shows low information content and maximum probability of FRs being satisfied. This is the area of high *axiomatic maturity*. Development of products starts in the lower left and moves to the upper right. Products are fully mature when they reach the upper right corner of the Axiomatic Maturity Diagram, as marked with a dot.

The diagram is plotted in Fig. 4.5. The shown development path is arbitrary. The axes of the Axiomatic Maturity Diagram are swapped in comparison with the real and imaginary axes in the complexity diagram of Suh [22, p. 71]. Two reasons apply to deviate from that definition; firstly, because the Independence Axiom and the Information Axiom are simply plotted in that order, and secondly, because the level of independence, as set by the Independence Axiom, never moves backwards (as long as no knowledge of the designer is lost, it will typically increase). By choosing this way of plotting, the maturity development path takes the form of a mathematical function. This makes reading the Axiomatic Maturity Diagram more natural.

4.6.2 Presumed and Legitimate Position in the Axiomatic Maturity Diagram

At any moment of development, the designer may presume an actual position in the diagram according to the current status of the design, but this position may differ from the real and legitimate position of the design; the presumed and legitimate positions may have discrepancies. The discrepancy is caused by a lack of knowledge of the designer because he has missed some essential design artifacts. As a result, the designer rates the level of engineering of the current product design higher than it actually is good for. When he finds the design error that causes the discrepancy, the problem can be addressed. However, if it is not discovered, then the discrepancy will present itself at some point in the remaining part of the development process as a surprise to the designer. The presumed position in the diagram needs to be corrected and that may lead to a project delay. Discrepancies between the presumed and legitimate position in the Axiomatic Maturity Diagram are the result of *unrecognized information* and due to its disruptive character, it may have large impact on the remaining product development process. Therefore, the goal is to discover discrepancies between presumed and legitimate position as early as possible.

4.6.3 Determination of the Legitimate Position

Finding *unrecognized information* is the key challenge for product designers and there is no method that comprehensively enables this. However, it is possible to apply methodologies that objectively determine the position of a design in the Axiomatic Maturity Diagram. This forces the presumed position to be based on facts instead of gut feeling. It will contribute to a higher degree of realism of the designer. A number of methods that focus on the conceptual design have been described in literature. These methods could be applied to characterize *unrecognized information*. Li et al. present a method based on AD with alternative domains that defines the conceptual design process as an integrated system with five stages and four mappings to apply mathematical descriptions as input for an expert system [28]. Similarly, Tay and Gu [27] apply the hierarchical topology of the design from the functional and physical domains into a relational data model. Chen et al. expand this method with a production framework [3] and the architecture framework for manufacturing system design of Benkamoun et al. also uses the axiomatic domains and the hierarchical structure [1]. This framework applies IDEF0 to define relations between the domains. Zhang and Chu [29] have developed an approach for the design of product and maintenance by combining AD, QFD, and FMEA. Suh has also reported a sequence of steps to follow that are based on FMEA [24]. More recent work was done by Puik et al. [13]. It defines a framework of seven steps to follow the Independence Axiom during design progression, starting with decomposition of the design, finding the DPs, decoupling the matrix, and testing the system to make sure that all DPs have been found. By performing regular checks, based on these methods, risks of discrepancies between presumed and legitimate positions can be optimized.

Unrecognized information only exits in its hidden state. It instantly changes to *recognized information* when it is discovered. In the form of *recognized information*, the designer can address it by completing and decoupling the design matrix. Quantification of *recognized information* may be done with the 'Independence Measure' as described by Do [4].

Axiomatic information is easier to quantify. It does not blur the perception of the designer with discrepancies between perceived and legitimate positions. The common ranges of the system can be quantified with the known statistical methods such as methods for six sigma Yang [29] and Taguchi [26]. Remaining risks could be quantified by FMEA [14, 24] or qualitative analysis [15].

4.6.4 Ideal Development Path in Product Design

Product development, as indicated above, will start somewhere at the lower left and will move diagonally upwards. The exact starting point will depend on the complexity of the project definition. A high-tech project that is new to the world might start with high amount of ignorance in the deep lower left corner. A project that aims to develop according to the first-time-right philosophy should start without *unrecognized information* and starts further to the lower right side of the diagram. Also the chosen path may be dependent on the amount of risk that is acceptable to the company, e.g., the most efficient development path in terms of investment (SME), a path that reduces lead time (semiconductor industry), or a path that minimizes development errors (medical or avionics). As explained in Sect. 4.3.3, it is preferred to start with the Independence Axiom followed by the Information Axiom due to the disruptive character of unorganized information, thus:

- Define FRs and find all relevant DPs to address *unrecognized information*;
- Decouple the design matrix to address *recognized information*;
- Match the design ranges and system ranges to guarantee an adequate common range to address *axiomatic information.*

Completion of these three steps leads to a preferred path that first moves to the right and then angles upwards. It is plotted in the left graph in Fig. 4.6. Depending on the preferred project strategy, a more or less risky path could be followed. In case of the rather conservative and slow but safe path of the 'Waterfall Method,' [16] the procedure of following Independence and Information Axioms in that order would be persistent (Fig. 4.6, right graph). A slightly more risky path that in practice enhances the development speed of projects is the path of 'Concurrent Engineering' [10, 21]. This gives the designer more room to start early work on robustness, process technology, and other life cycle elements. This merges the work on Independence and Information Axioms and possibly shortens the project lead time.

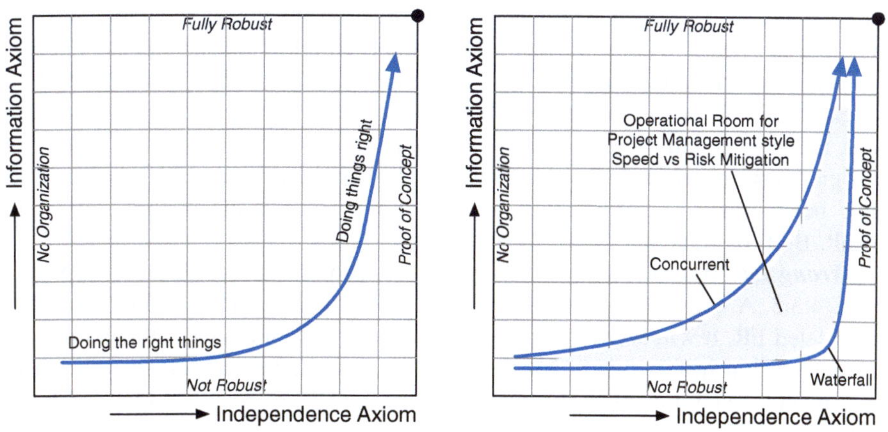

Fig. 4.6 *Left* Preferred development path through the Axiomatic Maturity Diagram, as indicated in literature, first moves to the right to satisfy Axiom 1. After this, Axiom 2 is satisfied in an upward direction. *Right* Depending on the nature of the project, a different strategy may be followed. The *right lower curve* would represent a waterfall management approach, while the *upper* would represent the path in the case of a concurrent engineering strategy

4.6.5 Consequences of Typical Errors

Unexpected errors in the development process are mostly related to the discovery of *unrecognized information*. This reveals discrepancies between the presumed and the legitimate position. It will divert the development path in the Axiomatic Maturity Diagram. Depending on the kind of error, a certain discontinuity will appear. This discontinuity is the result of the conversion of *unrecognized information* to *recognized information*. It may show as a kink in the development path or a jump to a different position in the diagram, depending on the following:

- Availability of a solution to address the problem; and
- Robustness of the current design being affected or not.

Availability of a solution will cause a jump in horizontal direction because *unrecognized* information is converted to *organized information* and that is addressed right away. If robustness of the design is affected, this means that the design matrix changes and robust DPs are replaced by non-robust DPs. This will cause a drop in vertical direction because *axiomatic information* increases. Based on some specific examples, it is shown that which effects occur in a number of typical problems.

No decoupling: The first typical problem is the example that was applied in Sect. 4.3.3, where relevant FRs and DPs are known but the design matrix is coupled. As a result, the designer will have problems setting up the system and it will show inexplicable system behavior. It is possible to optimize the design conforming Axiom 2 and have adequate common ranges, but *recognized information* remains in the system. An example is the combination lock as described by Suh [22, p. 65, 24, p. 5]. If a combination lock is to be opened without knowing the code, it is a matter of trial-and-error to open it. Even if the instruction manual is available, it is not possible to open it without further knowledge (being the code).

The designer knows he is missing essential knowledge. The result depends on whether the DPs need replacement. If this is the case, then replacing DPs will lead to a fallback in satisfaction of *axiomatic information* on the vertical axis. If the design matrix is decoupled and the known DPs can be maintained, then the effects may be minimal; all information will be eliminated and the mature state is the result. Both options are shown in Fig. 4.7.

Wrong DP: Another typical problem is the second example that was applied in Sect. 4.3.3. A wrongly chosen DP leads to the situation that the DP does not satisfy the related FR. It will seem to the designer that the design matrix is understood and decoupled, but in fact this is not the case. Time and effort are spent to match the system and design ranges of this DP, but since the DP has no effect these efforts do not succeed. The designer may deduce that something is wrong but does not know that the particular DP is causing the problem. As such, this situation leads to *unrecognized information*. To correct the problem, the designer needs to locate the wrong DP. As a result, the design matrix will need corrections and to address the

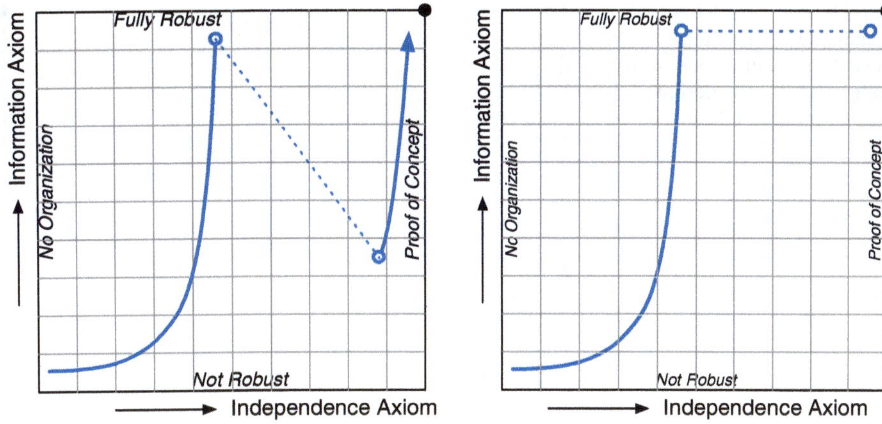

Fig. 4.7 A coupled design matrix does not conflict with satisfaction of the Information Axiom. However, if decoupling of the matrix needs replacement of DPs, then the Information Axiom is not automatically satisfied for the new DPs and efforts may be lost (*left*). The second option shows a luckier situation that the DPs can be maintained. In this case, the impact on the design is minimal (*right*)

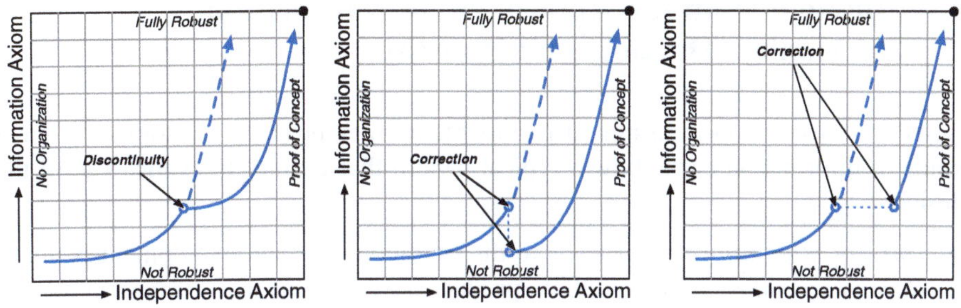

Fig. 4.8 Discovery of a wrong DP leads to a discontinuity in the development process (*left*). In the unlucky situation that an obsolete DP was already optimized, efforts are lost and the new DP again needs optimization and a correction takes place (*middle*). In a lucky situation, the problem can be solved with minor efforts. In this case, the related *unrecognized* as well the *recognized information* disappears (*right*)

related FR, a new and relevant DP will need to be installed. In Fig. 4.8, the possible discontinuities are plotted when a wrong DP in the design matrix is discovered.

Non-matching system and design ranges: A non-matching system and design ranges for one or more of the design relations between FRs and DPs lead to the situation that the Information Axiom cannot be fully satisfied. Note that *axiomatic information* is defined with joint probability (quantified product of all probabilities) or the sum of all information in the design relation [conform Eqs. (4.2)–(4.4)]. Therefore, the mature state is only reached if all system and design ranges are matched (Fig. 4.9). In this case, there is no discrepancy between presumed and legitimate positions.

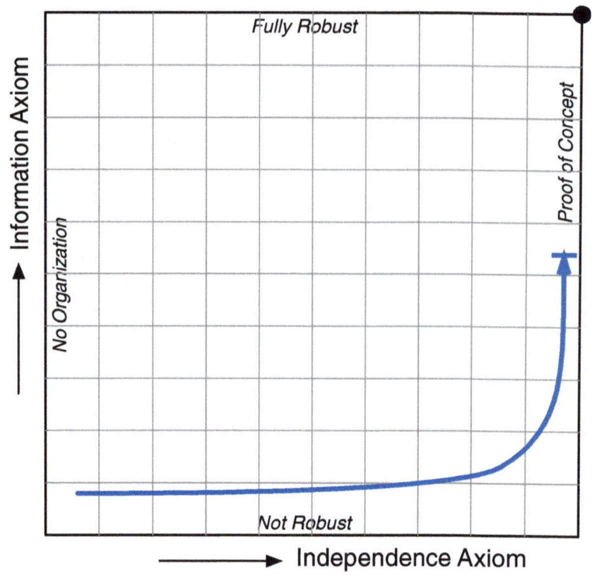

Fig. 4.9 A non-matching system and design ranges prevent the mature state from being reached. The design will not become robust

4.7 Discussion on the Application of Information to Monitor Development Processes

Based on information in design, it is possible to track product development. Three kinds of information each show a typical pattern that characterizes the state of the design:

- *Unrecognized information* leads to discontinuities in the development path;
- *Recognized information* prohibits product development path to reach proof of concept in the horizontal direction;
- *Axiomatic information* prohibits product development to reach the state of robustness in the vertical direction.

The innovative contribution in this chapter is largely carried out by the concept of *unrecognized information* that as such was not defined in AD yet. The impact of this kind of information is substantial; it can make or break the process of product design due to the discontinuities that can set the design back and might appear as a total surprise to the designer.

It also uncovers the largest weakness of this analysis; *unrecognized information*, as the name indicates, is hard to recognize and that is also the reason why it remains hidden. Good understanding of the design, e.g., by mathematical, quantitative, or qualitative modeling, increases the chance to perceive *unrecognized information*. The reason is that good understanding leads to well-chosen regularities in design and this eliminates information in general. Providing a graphical overview of the product status does not necessarily reveal missing information, but it may help to understand the stages in the development process and how to act accordingly. Faulty scenarios, eventually from the past, can be analyzed, characterized, and corrected. This learning experience might help understanding of future projects if similar patterns occur and are indeed recognized by the designer. However, in the execution of design projects, it is never completely clear if discrepancies in the Axiomatic Maturity Diagram are latent. Till now, this cannot be guaranteed.

In learning organizations, universities as well as companies, visualization of the design process can serve as a tool to explain the origin of errors made in the projects to students and novice designers. Causes and consequences become clear lessons for future design projects and it will contribute to the learning experience of the designer (design team). Graphical communication could function as a universal language to widen the scope of personnel, increasingly being capable of understanding what went wrong, for not only students and engineers, but also managers and executives.

The order in which the three kinds of information are addressed is preferably the same as in the above-mentioned bullet list. The safe way is to apply the waterfall method and address *unrecognized information* as soon as possible by functional modeling of the system. This transfers *unrecognized information* to *recognized information* so it can be addressed. Further, *recognized information* and *axiomatic information* will be addressed in that order conforming the basic practice of AD. The principle of concurrent design proposes a simultaneous approach of *recognized information* and *axiomatic information* up to some extent. This consciously trades speed of development for development risks. The right path to choose should be an executive decision.

4.8 General Conclusion

This chapter on monitoring information in complex products was divided into two parts. Part A has analyzed information in Axiomatic Design. It provides an unconventional view on information in AD, where not only robustness of a design, but also independence in design is related to information according to the Boltzmann's entropy. This makes both the Independence Axiom and the Information Axiom a subset of the Complexity Axiom. It explains that both axioms address structurally different kinds of information. It emphasizes the interdependence of the two axioms; however, the design principle of independence has disruptive dominance over robustness. *Useful information* forms the basis for AD and when eliminated from the design the functionality of the product should be guaranteed. The traditional information in AD, related to robustness was renamed to *axiomatic information* to be able to distinguish it from a new kind that was named *unorganized information. Unorganized information* was further decomposed to *unrecognized information* and *recognized information.* The former is problematic in the design because of its hidden nature; the designer does not know of its existence and will be surprised when it reveals itself, which can happen at any remaining moment in the design. The latter can be addressed by completion and decoupling of the design matrix. The answer to *axiomatic information* is to match the system and the design ranges.

Part B of this paper applies the different kinds of information for a graphical analysis of the product development process. The Axiomatic Maturity Diagram is applied for the visual representation. *Unrecognized information* may present itself with discontinuities in the diagram and causes bends or sudden jumps in the development path. There are two causes for jumps: a fallback on robustness of the design due to the fact that optimized DPs become obsolete, or the recognition of a problem for which a solution is available. In the first situation, a vertical drop is the result and the second case leads to a horizontal progression. If no jump is caused, then the information is converted to *recognized information,* which may be addressed by the decoupling of the design matrix. It results in a steady move to the right in the diagram. A preferred path is found by addressing *unrecognized information, recognized information,* and *axiomatic information* in that order, analogue to good practice in AD. Finally, the visualized analysis in the Axiomatic Maturity Diagram contributes to the understanding of imperfections during the execution of projects. Because of this, it is especially suitable for learning environments. The strengths are not particularly recognized in the prediction of imperfections in projects; this remains a challenge for future investigations.

Acknowledgements This research was funded by the Dutch Stichting Innovatie Alliantie and the HU University of Applied Sciences Utrecht.

References

1. N. Benkamoun, W. ElMaraghy, A.-L. Huyet, K. Kouiss, Architecture framework for manufacturing system design. Procedia CIRP **17**, 88–93 (2014). doi:10.1016/j.procir.2014.01.101
2. L. Brillouin, *Science and Information Theory* (Courier Corporation, 1962)
3. X. Chen, S. Gao, Y. Yang, S. Zhang, Multi-level assembly model for top-down design of mechanical products. Comput. Aided Des. **44**, 1033–1048 (2012). doi:10.1016/j.cad.2010.12.008
4. S.-H. Do, *Independence Measure (IM) Using the Independence Priority Number (IPN) in an AD Framework* (Houston, 2015)
5. B. El-Haik, *Axiomatic Quality* (Wiley-Blackwell, 2005)
6. B. El-Haik, K. Yang, The components of complexity in engineering design. IIE Trans. **31**, 925–934 (1999). doi:10.1023/A:1007650829429
7. W.H. ElMaraghy, H.A. ElMaraghy, T. Tomiyama, Complexity in engineering design and manufacturing (2012)
8. M. Gell-Mann, S. Lloyd, Information measures, effective complexity, and total information. Complexity **2**, 44–52 (1996). doi:10.1002/(SICI)1099-0526(199609/10)2:1<44:AID-CPLX10>3.0.CO;2-X
9. R.V.L. Hartley, R.V.L. Hartley, Transmission of information. Bell Syst. Tech. J. **7**, 535–563 (1928). doi:10.1002/j.1538-7305.1928.tb01236.x
10. A. Kusiak, *Concurrent Engineering* (Wiley, 1993)

11. E.C.N. Puik, D. Ceglarek, *A Review on Information in Design*, 1st edn. (Lisboa, 2014)
12. E.C.N. Puik, D. Ceglarek, *A Theory of Maturity*, 1st edn. (Lisboa, 2014)
13. E.C.N. Puik, E. Smulders, J. Gerritsen et al., in *A Method for Indexing Axiomatic Independence Applied to Reconfigurable Manufacturing Systems*, ed. by M.K. Thompson, 1st edn. (Worcester, 2013), pp. 186–194
14. E.C.N. Puik, D. Telgen, D. Ceglarek, L. van Moergestel, in *Structured Analysis of Reconfigurable Manufacturing Systems*, ed. by A. Azevedo (Porto, 2013)
15. E.C.N. Puik, D. Telgen, D. Ceglarek, L. van Moergestel, in *Qualitative Product/Process Modelling For Reconfigurable Manufacturing Systems* (IEEE, 2013)
16. W.W. Royce, Managing the development of large software systems (1970)
17. C.E. Shannon, *A Mathematical Theory of Communication* (1948), p. 125
18. C.E. Shannon, W. Weaver, *The Mathematical Theory of Communication* (The University of Illinois Press, 1949)
19. N.P. Suh, A theory of complexity, periodicity and the design axioms. Res. Eng. Des. **11**, 116–132 (1999). doi:10.1007/PL00003883
20. N.P. Suh, *The Principles of Design*. Oxford Series on Advanced Manufacturing (Oxford University Press on Demand, 1990)
21. N.P. Suh, Axiomatic design—advances and applications, in *Advances and Applications*, Chapter 5 (Oxford University Press, New York, 2001), pp. 239–298
22. N.P. Suh, *Complexity: Theory and Applications* (2005)
23. N.P. Suh, Complexity in engineering. CIRP Ann. Manuf. Technol. **54**, 46–63 (2005). doi:10.1016/S0007-8506(07)60019-5
24. N.P. Suh, *Axiomatic Design Framework for Large-Scale Complex Systems Design* (Boston, 2004)
25. N.P. Suh, A.C. Bell, D. Gossard, On an axiomatic approach to manufacturing and manufacturing systems. J. Eng. Ind. (1978)
26. G. Taguchi, S. Chowdhury, Y. Wu, *Taguchi's Quality Engineering Handbook* (Wiley-Interscience, 2005)
27. F.E.H. Tay, J. Gu, Product modeling for conceptual design support. Comput. Ind. **48**, 143–155 (2002). doi:10.1016/S0166-3615(02)00014-3
28. L. Wenqiang, W. Li, L. Yan et al., The process model to aid innovation of products conceptual design. Expert Syst. Appl. **37**, 3574–3587 (2010). doi:10.1016/j.eswa.2009.10.034
29. K. Yang, B. EI-Haik, *Design for Six Sigma* (McGraw Hill Professional, 2008)
30. Z. Zhang, X. Chu, A new approach for conceptual design of product and maintenance. Int. J. Comput. Integr. Manuf. **23**, 603–618 (2010). doi:10.1080/09511921003736766

Chapter 5
A Novel Approach for Axiomatic-Based Design for the Environment

Alessandro Giorgetti, Andrea Girgenti, Paolo Citti and Massimo Delogu

Abstract The Eco-design approach for new product development is becoming progressively more and more important for competitive and legislative reasons, especially in advanced markets (EU USA, East Asia, etc.). Its importance is increasingly growing since the decisions made in early design stages largely affect not only the cost but also the environmental impact of a product. This paper introduces a novel approach that could be used to increase the potential capability of an Eco-design approach. This aim is achieved through a better fit between the critical environmental issues and the development of new solutions using AD. The introduced approach, first, considers a meta-product point of view that uses a customized Smart Eco-design Platform and the Axiomatic Design (AD) for the improvement of the eco-sustainability of products. Then, the approach introduces the meta-system level as the reference level for detecting the system Design Matrix and developing an uncoupled design. This goal could be achieved through the use of AD and the implementation of the environmental information as a tool to reduce the space of the available design solutions. The first axiom aims to define the Design Matrix of the Functional System in order to detect its best configuration. The purpose is to avoid an optimization without appropriate knowledge in terms of interaction among meta-product and resources. Then, the Functional Requirements definition, used in AD, could represent the ideal index for the ease of sharing information and knowledge on a wide scale among different industrial sectors.

A. Giorgetti (✉) · A. Girgenti · P. Citti
Dipartimento di Ingegneria dell'Innovazione e dell'Informazione,
Università degli Studi Guglielmo Marconi, Via Plinio 44, 00193 Rome, Italy
e-mail: a.giorgetti@unimarconi.it

A. Girgenti
e-mail: a.girgenti@unimarconi.it

P. Citti
e-mail: p.citti@unimarconi.it

M. Delogu
Dipartimento di Ingegneria Industriale, Università degli Studi di Firenze,
Via di Santa Marta 3, 50139 Florence, Italy
e-mail: massimo.delogu@unifi.it

© Springer International Publishing Switzerland 2016

A.M. Farid and N.P. Suh (eds.), *Axiomatic Design in Large Systems*,
DOI 10.1007/978-3-319-32388-6_5

The development of the Smart Eco-design Platform could encourage the use of this approach in real product development. The sharing of the database enables obtaining information for reducing the field of design parameters that satisfy the Functional Requirements. In this way, it could be possible to develop a system of products with an overall higher level of eco-sustainability and a better use of resources through information derived from other fields and experiences. Typical goals that are reachable are, for instance, represented by a system that needs less consumption of energy and material during the whole product life cycle.

Keywords Eco-design · Life cycle assessment · Design for environment · Functional System · Design approach · Product development · Meta-product

5.1 Introduction

Today, environmental sustainability represents a critical driver of innovation: Scientific researchers and industrial results have been developed with a focus on the effective Eco-design approach, emerged during the 1990s [5] for the development of new products [6–9, 31, 41]. Several approaches that deal with the assessment of the environmental impact also cover an important role in Eco-design. Such approaches and their related tools have been the object of studies and investigation in recent years, by many organizations and research centers (ISO, SETAC, SPOLD, CSA, OECD, UNEP) both at national and international levels. Over the last decade, many tools and indicators for assessing and benchmarking the environmental impact of different systems have been developed [17, 30]. Examples include Life Cycle Assessment, Strategic Environmental Assessment (SEA), Environmental Impact Assessment (EIA), Environmental Risk Assessment (ERA), Cost-Benefit Analysis (CBA), Material Flow Analysis (MFA), and Ecological Footprint. Among them, one of the most commonly used technique in Design for Environment (DfE) methodologies to assess the environmental impact is Life Cycle Analysis (LCA) that is also implemented in an international standard (ISO 14040 standards, 2006). LCA aims to characterize the eco-profile of a product (and its related processes) during life cycle stages (i.e., from the raw material acquisition, via production and use phases, to waste management). The high prospect of this technique has been acknowledged, and strong development and harmonization have occurred. Other measurement methods and technological guidelines have been introduced to measure and manage the environmental impact performance of a large set of industrial products [27, 33]. All of these skills are strongly connected with a smart use of raw materials (reduction of weight, recyclability, and management of hazardous substances) and of energy (CO_2 emission reduction and energy efficiency) and the reduction of waste (scrap and hazardous materials). The proposed approach focuses on the creation of as wide as possible shared database of knowledge from different fields of experience, applications, and industries, to have a bigger reference for the product design. The shared database acts as a collector for acknowledging various fields concerning the execution of single functions or single aspects of the production in a specific context (e.g., from material science in the aircraft industry to dismantling techniques in the automotive field). The aim of this database is to drive the accomplishment of Functional Requirements (FRs) more effectively. Using the database as a reference is thus very useful to accomplish more Eco-friendly the functions of the product, through the avoidance of several solutions and the selection of the best one from an environmental point of view. The approach may have an important role in Eco-design activities. In fact, two critical issues currently limit the impact of Eco-design activities for the development of new products: the single-product point of view over the designing process and the difficulties in sharing the Eco-design knowledge and practices [10]. The single-product point of view is the most used for Eco-design analysis and considerations. In this way, traditional approaches obtain a certain level of optimization through the progressive extension of the utilization of the knowledge (transformed into useful information for designers and process managers) in the overall supply chain of each product [43]. In other words, currently it is infrequent to obtain design solutions for each product that can optimize the environmental impact of a supply chain with different levels of interactions between the customer and the supplier furthermore a huge number of processes. Therefore, the potential result of a wider eco-design action is of undoubted value because it allows to reduce harmful effects on the environment with the minimum amount of costs. On the other hand this usual approach could lead to local uncoupled solutions for achieving the Eco-design scope whereas the usual Eco-design approach often deals with the single product independently of the system synergies related to the products that belong to the same meta-product during their life cycle [10]. In particular the sharing and the use of the Eco-design knowledge and experience in a large set of the industrial sectors results in difficulties for industrial users. Consequently, only a few industrial sectors—the transport and automotive ones for instance [1, 3, 11, 14, 16, 20, 22, 25, 26, 34, 35, 38]—have successfully introduced Eco-design activities in their new product development processes with different levels of effectiveness and take into account the whole life cycle phases. One of the most important root causes of this aspect of the design is that all the information is organized in a database based on the Technical Characteristics of the products, which are often very different for each industrial sector.

In this scenario, International Standards, and local legislation become more accurate and strict year-by-year, and are diffused into a growing number of industrial fields, e.g., EU Directive 2009/125/EC (Energy Using Product—EUP); 2002/95/CE (RoHS); 2002/96/CE (WEEE); and 2006/12/EC (Waste). For this reason, Eco-design becomes more necessary year-by-year for an always larger set of industries. This trend is making even more difficult achieving the requested ecological constraints so that the use of multiproduct design approaches is a growing need [21].

On the other hand, the AD is often applied to the development of a new product or a complex system to achieve the Functional Requirements for the maximum effectiveness and efficiency in accordance with the Eco-design point of view for product development. In this paper, the AD method will be applied to a wider system of products called meta-product in order to achieve a higher efficiency in taking into account several aspects that can be harmful from an environmental point of view. The AD theory can be applied through the several steps of the development of a product, from the definition of the customer needs to the choice of the best parameters for production processes. In this paper, the AD method is going to be applied to the design process that involves a widely shared information database. The database is composed of a collection of tested practices coming from different industrial fields. In this way, it is aimed to filter the space of the available solutions from the beginning of the design process, according to environmental criteria. Consequently, the use of the database could make a better fit between the environmental critical issues and the development of new solutions.

This paper is focused on the management of the FRs and their corresponding DPs, and the other stages of the product development follow the usual AD approach [36] starting from the highest level of the zigzagging [29, 36]. The highest FRs can deal with different topics and are derived from the engineering synthesis and selection of the whole set of the stakeholder needs. The stakeholder is anybody who is involved in the production, use, and dismantling of the product (suppliers, contractors, landfills, the final users, etc.). A special case is given by those FRs that come from the manufacturing system since they establish a connection between the physical domain and the process domain, which needs to be explicitly dealt in the AD.

Three issues will be presented in this paper and could be useful to reach the sustainability of complex system:

- The definition of a Smart Eco-design Platform and its integration using AD;
- The definition of FRs with different time or dimensional scale;
- The introduction of the "meta-product" or "Functional System" approach, which means designing a product of products, or in other words, a good or a service which together provides the same set of functions for the same user(s) at the same time.

5.2 The Smart Eco-design Platform

The AD point of view takes into account all the stages of the product development, from the definition of the customer needs to that of the process variables. The proposed approach aims to extend the use of AD to a wider design platform to achieve a more ecological solution, using the Environmental Information database as a search tool for available DPs. The integration of the ecological aspects of the designing processes could be obtained through the development of a framework able to detect and classify all of the necessary information. This architecture called the "Smart Eco-design Platform" consists of all of the knowledge about the

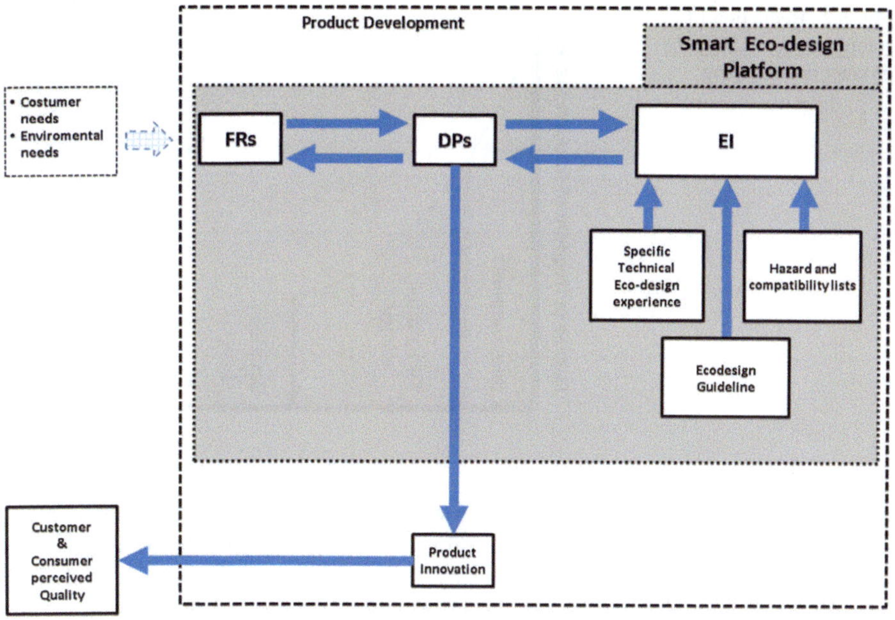

Fig. 5.1 Smart Eco-design Platform internal and external connections architecture

correlations among the FRs, the DPs of the product, the environmental information (EI) based on each particular eco-design technical experience related to the product and the sharing of the overall knowledge (see Fig. 5.1). The EI database gathers several different references that are normally available from different sources but often spread out, rather than organically collected to make easier, faster, and more efficient their use. The information of EI database could be obtained by the use of the following kind of sources of Eco-design guidelines. For example:

- Normative references (IEC 62430, ISO Guide 64, VDI 2243, UNI ISO/TR14062, IEC guide114);
- Web resources and existing database [15, 23, 32, 42];
- Research literature and industrial experience [20].

The use of the Smart Eco-design Platform fits with the AD approach since it involves the mapping between FRs and DPs. The detailed description of how the FRs are derived from the stakeholder needs (i.e., an expanded Customer Needs Domain) is not dealt in this paper; see [36] for details about the definition of the FRs [39] for useful clarifications and [4, 5] for useful FRs extensions and their classifications.

The use of the Smart Eco-design Platform deeply affects the definition of the space of the available technical solutions that is the amount of degrees of freedom for the designer. Indeed, from a theoretical point of view, there are several ways to

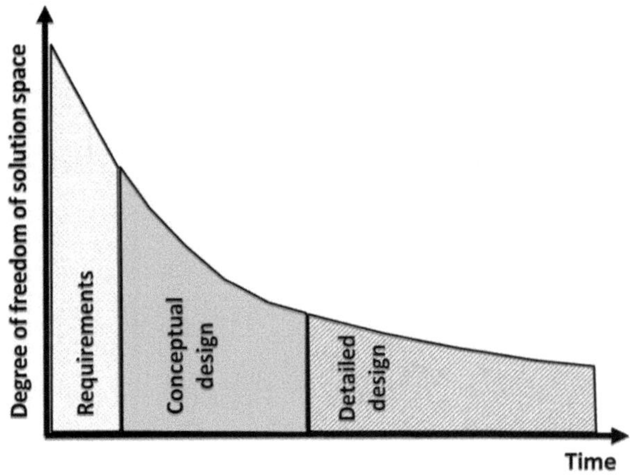

Fig. 5.2 Reduction of solution space during the product development

satisfy the customer needs and their consequential FRs. The space of the available solutions usually has the widest dimension at the beginning of the design process, and constraints naturally reduce it during the ongoing design (see Fig. 5.2) [13]. The proposed approach introduces the information from the EI database at the first design stage to identify constraints or practices that make the design better and more eco-sustainable from the beginning of the project.

The use of the EI database is useful for identifying the best manner to deploy the DPs. This means that if the information from the EI database excludes a DP, another one shall be found to satisfy the corresponding FR. For example, in a real design problem we could find out an FR which is "to stiff a polypropylene structure." The designer thinks about insert a reinforcement in the plastic to satisfy the FR and queries the EI database that suggests the boundaries to obtain an Eco-design solution satisfying the FR (see Fig. 5.3). In this particular case, the EI comes from Renault's best practice [20]. In other words, when a designer is trying to increase the stiffness of a structure and decides to use polypropylene, then he has to use a small metal sections whose the thickness is lower than 1 mm that are acceptable from a recycling point of view.

In that way, the EI drives the design toward the solutions that have the minimum environmental impact, both reducing the range of available degrees of freedom to satisfy a particular FR and identifying the best use in the service field. From both these points of view, the EI database works as a tool for spotting the most eco-sustainable DPs among available solutions both at the higher level FRs/DPs and at the lower level ones, introducing a selection or implementation criteria for DPs and making the design process more guided (see Fig. 5.4).

In this way, the use of the EI database for the addressing of the candidate DPs which best satisfy their corresponding FR can be a driver to reduce the imaginary complexity of time independent systems. Otherwise, this issue is just left to the designer experience and knowledge that are obviously partial. In fact, whereas the real complexity of a time independent system is reduced making the system range

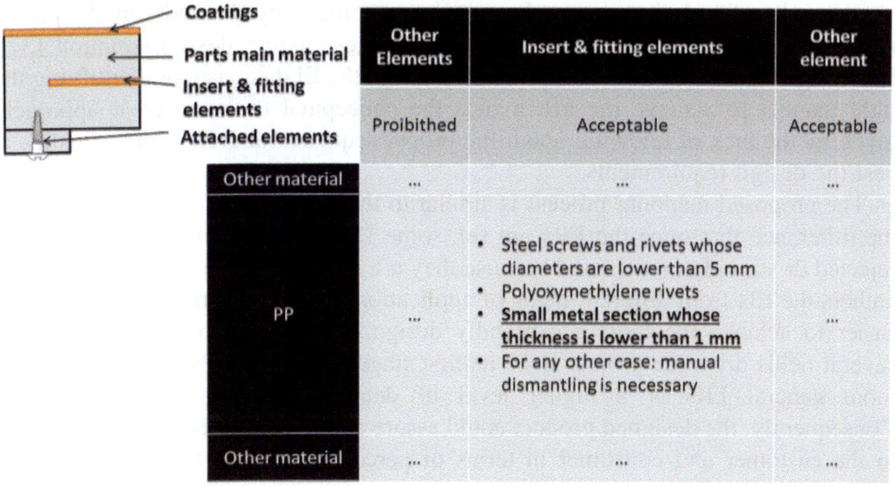

Fig. 5.3 Example of best practice for recycling specification of PP by Renault [20]

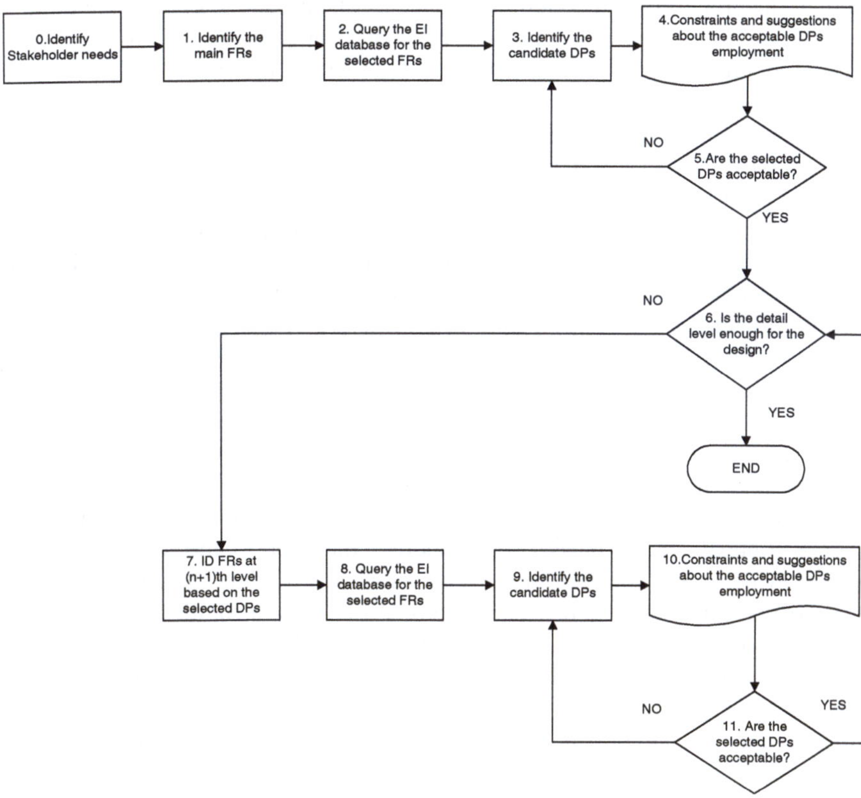

Fig. 5.4 Flowchart of the AD decomposition process through the Smart Eco-design Platform

overlapped to the design range, the imaginary complexity depends on the partial knowledge of the system and, therefore, its corresponding design equation [37]. Reducing at first the set of candidate DPs through the EI database, or identifying the right manner to achieve the FR, avoids the conceptual trial and error approach, given by the lack of knowing about the proper sequence of DPs which satisfies at best the design requirements.

The proposed mapping process is similar to the one proposed by Suh [36] with the difference that once the FRs are set, some DPs which could satisfy them are rejected or strongly constrained because they are contrary to the EI database which gathers the EIs from different fields of application. This use of the EIs makes even easier to achieve the most eco-friendly design. In fact, the best practices from several fields drive the project since its first stages, spotting DPs which satisfy the more general FRs at a higher level of detail as the more specific ones. Consequently, the designed product could assure a new higher level of performance to the customer and consumer in terms of perceived quality. Introducing the EI database as a tool for improving the environmental impact of products from the beginning of a project allows the designer to minimize costs for adapting an existing product in order to reduce its harmful effects on the environment. This feature of the proposed method aims to reduce a critical issue in Eco-design because often the product is improved with high costs in a late stage of the project or even during production in order to make it less polluting [2, 18, 40].

This passage has an important impact on the described critical issue of Eco-design implementation, i.e., the difficulties in sharing information in a wide range of industrial sectors. A platform based on FRs is clearly more general and easy to be used by users who come from different fields of application.

The EI database is composed of suggested technical solutions coming from different industrial sectors for making them available to other companies. In other terms, the database is queried through the FRs in order to obtain candidate DPs which have been successfully adopted in other sectors to achieve the same FRs or recommendations to avoid some other DPs. Such database makes available a widespread set of candidate solutions to identify the most effective for achieving a specific FR. Although collecting and sharing the technical knowledge may be not so easy in some industrial fields, developing the proposed Smart Eco-design Platform is crucial in order to identify the most effective design. Such practice is already used in several cases where this is mandatory due to laws and regulations. In some contexts, the database could address the user to the specific field or company which owns the requested information or procedure, avoiding the introduction of harmful elements or leading optimal Eco-design solutions. The creation of this shared database with suggested DPs for achieving a function will be undoubtedly a competitive factor also in cases when this is not a mandatory condition since it exploits the experiences and the technical knowledge from different sectors for identifying the most efficient and effective design. Furthermore, the development and the management of such a shared database may be a new kind of business if the information is paid.

5.3 The Definition of FRs with a Different Time or Dimensional Scale

A second topic that could be useful during the design process to develop more effective solutions through the AD is the analysis and use of the FRs in a time or dimensional scale. This is a key topic from a "design for environment" point of view since the most efficient exploitation of resources (material, space, and energy) shall take place within the entire life cycle of a product. The analysis on a timescale means splitting the FRs into short-term functions and long-term functions, while the analysis on a dimensional scale means to distinguish between functions which take place on a macrospace from those that take place in a microspace. This comes from the principle of splitting the functions and their corresponding physical variables in order to achieve solutions whose the ratio between the desired effect and the spent resources to achieve it is favorable.

Splitting FRs through a timescale makes possible to take into account various effects given by different times in the life of the system and to minimize their impact on the environment. Often, the FRs could also be split according to a spatial criterion, considering, for example, macro- and microscale.

According to this practice, the development of knowledge in the long term and short term, or in macroscale and microscale, for each product is recommended. This knowledge could be used for different products to reach longer lifetimes. In fact, the products could be optimized through a specific technical capability in the linking of technical characteristics with a different time or dimensional governance of FRs. In other words, looking at the system, with time or spatial criterion, allows one to allocate the use of specific resources or to achieve a specific FR in different times or positions. This knowledge permits the reduction of the use of materials and energy connected with the over-engineering of one or more technical characteristics during the development process (Fig. 5.5). The Design Matrix of the product is able to identify and to solve the most critical FR/DP relationships (Fig. 5.6). These design relationships are usually related to an over-consumption of energy and a higher production of scrap and hazardous materials. Then, the application of the first axiom of AD helps to evolve the product design to a higher level of recyclability and reuse.

Splitting the FRs according to a time or spatial criterion allows to maximize the system efficiency and efficacy. The available resources may be required in different times or places, thus, achieving the most efficient solution means exploiting them in the right amount when and where they are needed, avoiding their over-consumption, and adjusting the demand in these two domains. For example, considering a refrigerator, a different management of the available volume is required in different periods (due to the difference in needs of food between the summer and the winter or to the stock management). Two possible FRs could be "to minimize energy consumption" and "to optimize refrigerator capacity." These

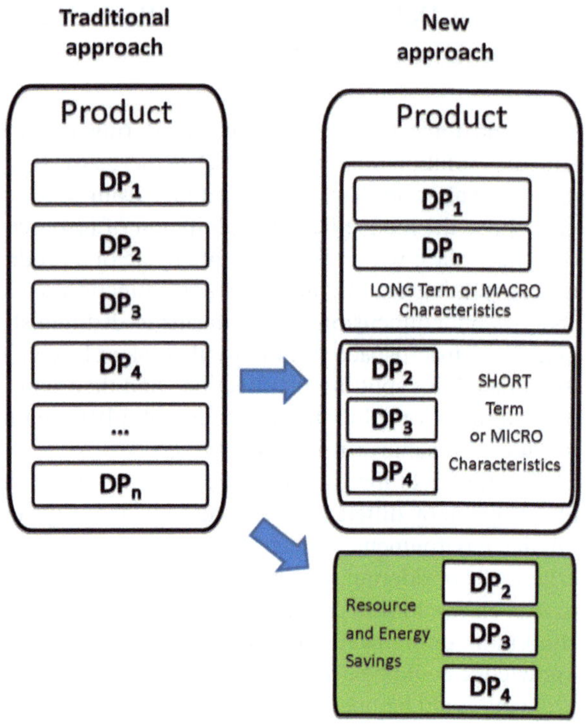

Fig. 5.5 Over-engineering detection (the dimension of the *rectangles* show the resource consumption)

elements usually produce a trade-off situation between the volume capacity and the energy consumption. The splitting of the FR in short term and long term ones: "To minimize partial load energy consumption" and "to optimize partial load refrigerator capacity" (long term) and "to minimize full load energy consumption" and "to optimize full load refrigerator capacity" (short term) could drive to a less redundant solution. For example, it could be possible to identify solutions with a variable volume of the refrigerator that could optimize the energy consumption and the load capacity both in the short and long term. The result is a reduction in redundant use of energy and material to make cold a variable amount of the food volume.

Looking at the use of resources according to these principles leads to reach the highest efficacy since the system response is never a compromise, but it can be made uncoupled in different spatial or temporal stages. These techniques are very useful in order to reduce redundancy and achieve an uncoupled design because splitting the FR into several ones allows to switch from a "one-to-many" to a "one-to-one" configuration.

Fig. 5.6 Design Matrix modified by overdesign detection

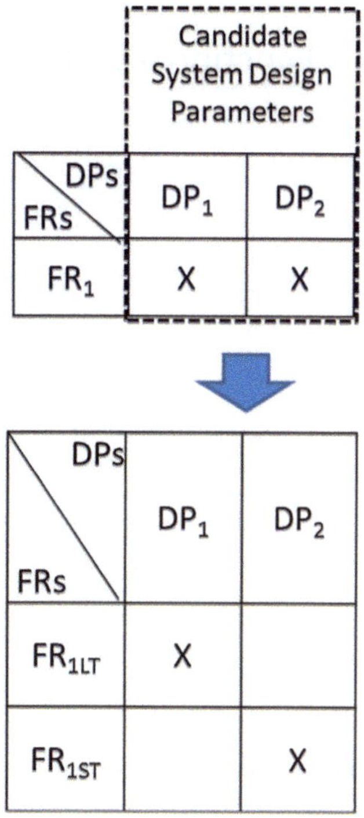

5.4 Functional System Approach

As described previously, a critical issue for the Eco-design approach is connected with the need for an expanded definition of the eco-sustainability of a product [12, 28]. In particular, a very useful consideration is that the most suitable life cycle for each product is strongly connected with both its particular features and the meta-product. We define the Functional System or the meta-Product as a system composed of many products (Fig. 5.7) that together provide the same set of functions for the same user(s) at the same time. The Functional System could be used as a basis for calculations and as a basis for the comparison among different systems that fulfill the same function. Each product could belong to a different Functional System in a different part of its life cycle, considering the production, the use, and the disposal. This means that the traditional single-product approach does not show the complete Design Matrix of a meta-product. That additional information could indicate many improvement paths to identify the uncoupled solution. In other words, without the meta-product functional scheme, it is difficult to detect and consequently to solve many FR/DP relationships linked with the Functional System perspective (products interfaces, system synergies, and system risks) that results in losing many potential better improvements.

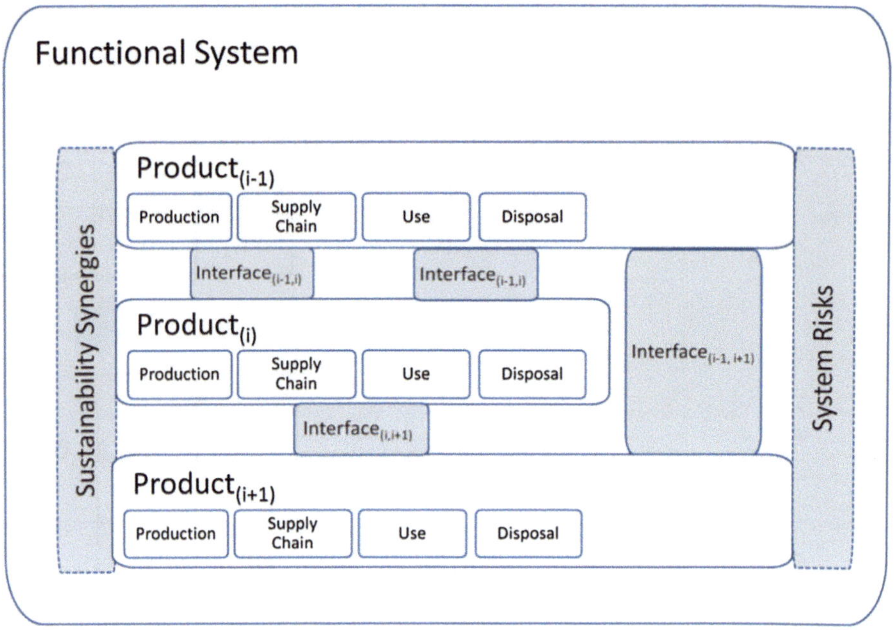

Fig. 5.7 Functional System composition: information studied in traditional approach (*white boxes*) and information added in the new approach (*gray boxes*)

Looking at an expanded system which is made by several single products is useful for having a wider point of view and identifying a bigger number of harmful features or potential interactions among components. This wider view of the system makes easier to evaluate the overall system risk rather than useful interactions with other single components and the system synergies. Looking just at a single product, like the traditional approach suggests, is a very tricky habit since it makes difficult to take into account all the features of a product and how it relates to the others.

In particular, it is important to apply the main Eco-design guidelines which are related to different stages of the life cycle of the product or the extended complex environment that is the Functional System. These instruments usually go under the name of DfX (design for X) where X stands for a specific material property or another characteristic of the meta-product. Typical examples of this approach are the design for energy consumption, the design for materials, the design for extension of life, the design for end-life, etc. [24]. As a consequence, the introduction of the Functional System concept will create the potential to achieve a higher level of optimization for Eco-designed products. This optimization through the Functional System analysis assures that more degrees of freedom will be taken into account, and more conceptual and physical resources will be introduced for the implementation of better Eco-designed solutions.

Without the Functional System analysis, it would be impossible to obtain the necessary information for an important reduction of the environmental impact of

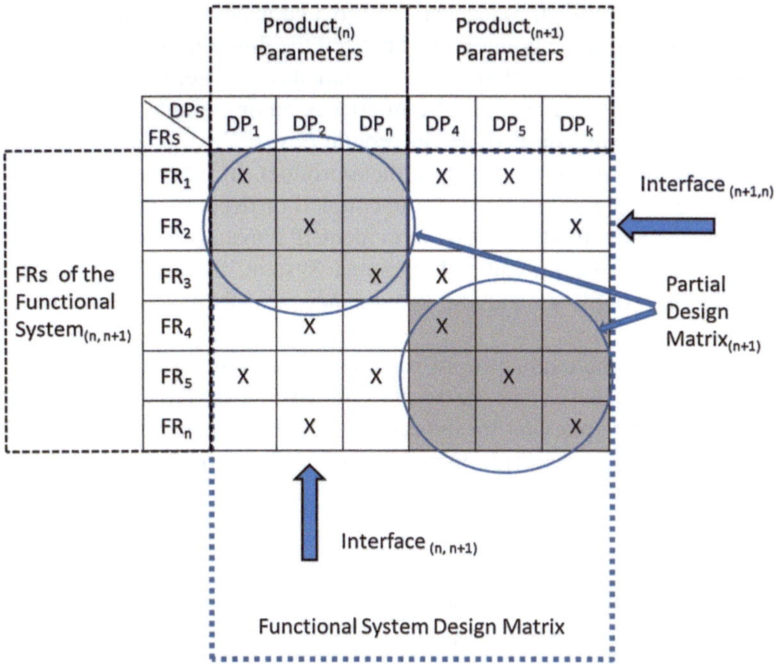

Fig. 5.8 Example of the Design Matrix for the Functional System

products since it allows to extend the Axiomatic Design method to a wider object, taking into account also those relationships among its components rather than a single stand-alone product. The Design Matrix of a single product would not be able to consider the contribution of the interface. This leads not to show all the relationships among modules and underestimate the effects of the design on the environment. We defined the interface $(n, n + 1)$ as the DPs of product *n*th that are related to the achieving the FRs of the product $(n + 1)$th. Figure 5.8 gives an example of the information that could be missed in a single-product analysis, showing a Design Matrix which contains an interface. A single-product analysis provides an insufficient level of information for the selection of the most sustainable solution. The overall design matrix, which includes the Design Matrices of each single product and the others of the interfaces, may be not really uncoupled although the matrices of single products appear uncoupled. This may happen since the interfaces may present coupling, but the designer is not able to detect them if he does not take into account the meta-product.

This aspect is particularly clear regarding the existence of the designed redundancy. Extending the usual single-product point of view to a meta-product one, i.e., from the product environmental impact to the Functional System one, is necessary in order to identify the overall Design Matrix characteristics. In this way, the focus of Eco-design actions is moved to the most critical aspects of the system sustainability. These critical aspects arise from a large amount of attention on single products and also from the boundaries (between the different products that compose the Functional System) considering both sustainability synergies of the system and system risks. Taking into account the interfaces among single products allows to find out and evaluate coupled design conditions since the interfaces represent critical sides of the overall design of a meta-product. In other words, although the design matrices of single products are uncoupled or decoupled, the corresponding interfaces may be coupled, and this is evident if a meta-product point of view is adopted. If we consider the entire Functional System, the amounts of the FR/DP correlations, the coupling, and the redundancy are displayed more clearly and completely.

In this scenario, the evaluation of the interfaces among all the single products, the management of these, and the level of holistic integration among the always new products become relevant for the eco-sustainability of each Functional System. This evaluation has to be made using a top-down approach at each level of the zigzagging.

5.5 Merge of Smart Eco-design Platform and Functional System Approach

The potential impact on the eco-sustainability due to the introduction of both the Eco-design Platform and the Functional System approach is the improvement of the eco-efficiency through different stages of the product life cycle. The Functional System and the connected meta-product can be considered as the optimal design level to develop extended eco-sustainability of products (Fig. 5.9) since it is wider than the traditional approaches.

Fig. 5.9 The path to eco-efficiency

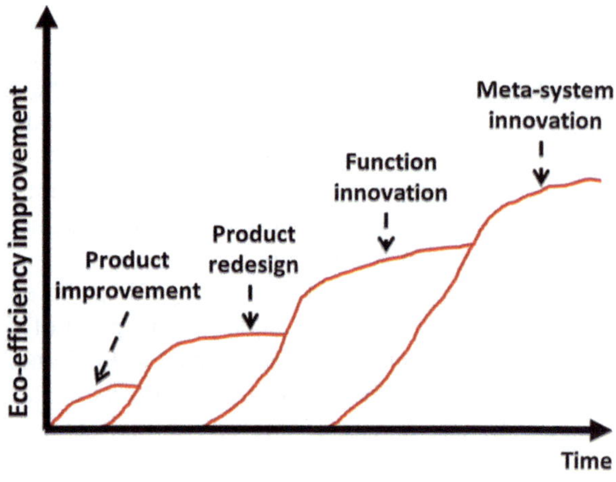

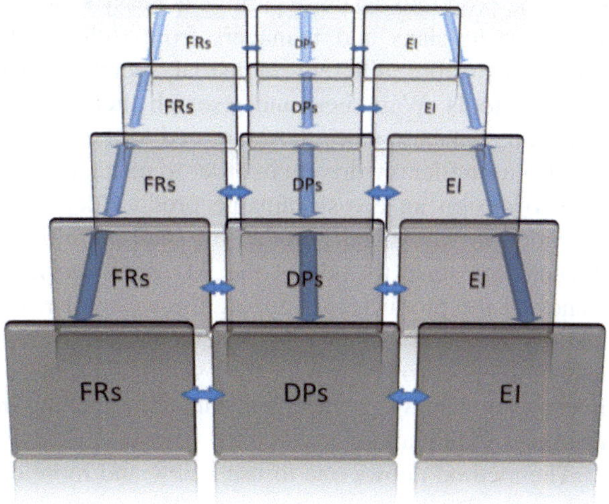

Fig. 5.10 The multilayer structure of the Smart Eco-design Platform for the Functional System

The optimal design can be obtained using the previously introduced Smart Eco-design Platform that collects all of the needed information as a filter to find out the more efficient solution in order to accomplish the FRs. The creation of this platform manages and shares the specific technical knowledge for all of the multiproduct supply chains of the Functional System. The common approach for the sharing, the collection, and the analysis of the data is based on the AD FRs definition [36]. The knowledge developed and diffused by the Smart Eco-design Platform allows the selection of a solution with a larger product flexibility and with a longer lifetime, avoiding harmful elements. In particular, it is possible to define a more intelligent use of raw materials and energies based on resources sharing inside the Functional System. The Smart Eco-design platform shall have a multilayered structure able to store and connect all FRs, EI, and DPs of the meta-product along all the step of its life cycle from the cradle to grave (Fig. 5.10).

5.6 Conclusion

The proposed approach to the Eco-design of new products drives the designer toward more eco-sustainable solutions. In particular, this approach helps the existing environmental management of products in a broader vision that takes into account a more integrated system. This path is possible through the use of a Functional System approach and permits the creation of a more integrated and holistic analysis of eco-sustainable products. Therefore, this method goes beyond the second critical issue through the introduction of a EI database which is queried through the FRs. The EI database acts suggesting constraints about the DPs employment, according the most ecological manner to fulfill their corresponding FRs. The proposed database permits the easy sharing of the Eco-design experiences among technicians and managers from different kind of businesses and fields (experience from different industrial sectors, normative references, guidelines, software tools, Web sites, and research literature). The FRs are the keywords to query this database since these are tied to the specific technical problem. The FRs are also user-friendly drivers in order to manage the Smart Eco-design Platform and help to design an eco-sustainable product depending on different contexts. The definition of where and when an FR shall be fulfilled by its DP, defining a time or dimensional scale to look at the AD decomposition, improves strongly the efficiency of the product, minimizing the waste of resources. The introduction of the meta-system approach and its implementation within the design process is useful to extend the point of view over the overall chain of goods, services, and tools which interact to achieve a specified function, making more logic and rationale the resources exploitation.

The method allows one to use a new and more ecological way to design goods and to develop innovative solutions. This paper shows how AD can be used as the core of the Smart Eco-design Platform for the detection of the potential areas for the improvement and the introduction of innovative solution with regard to the sustainability of products.

References

1. H. Andriankaja, F. Vallet, J. Le Duigou, B. Eynard, A method to ecodesign structural parts in the transport sector based on product life cycle management. J. Clean. Prod. **94**, 165–176 (2015). doi:10.1016/j.jclepro.2015.02.026
2. J. Behrisch, M. Ramirez, D. Giurco, Representation of ecodesign practice: international comparison of industrial design consultancies. Sustainability **3**(10), 1778–1791 (2011). doi:10.3390/su3101778
3. L. Berzi, M. Delogu, A. Giorgetti, M. Pierini, On-field investigation and process modelling of End-of-Life Vehicles treatment in the context of Italian craft-type Authorized Treatment Facilities. Waste Manag. **33**(4), 892–906 (2013)
4. D.M. Beude, The engineering design of systems: models and methods, 2nd edn (Wiley, Hoboken, 2009)
5. H. Brezet, C. Van Hemel, *Ecodesign: A Promising Approach to Sustainable Production and Consumption* (UNEP, 1997)
6. F. Brones, M. Monteiro de Carvalho, From 50 to 1: integrating literature toward a systemic ecodesign model. J. Clean. Prod. Integr. Cleaner Prod. Sustain. Strat. **96**, 44–57 (2015). doi:10.1016/j.jclepro.2014.07.036
7. F. Cappelli, M. Delogu, M. Pierini, F. Schiavone, Design for disassembly: a methodology for identifying the optimal disassembly sequence. J. Eng. Des. **18**(6), 563–575 (2007)
8. F. Cappelli, M. Massimo Delogu, M. Pierini, Integration of LCA and EcoDesign guideline in a virtual cad frame work, in International Conference on Life Cycle Engineering, LCE'06, Leuven, 31 May–2 June 2006
9. D. Collado-Ruiza, H. Ostad-Ahmad-Ghorabi, Fuon theory: standardizing functional units for product design. Resour. Conserv. Recycl. **54**, 683–691 (2010)

10. C. Dalhammar, E. Machacek, A. Bungaard, K. Overgaard Zacho, A. Remmen, *Addressing Resource Efficiency Through the Ecodesign Directive. A Review of Opportunities and Barriers* (2014). ISBN 978–92-893-2720-6
11. F. Del Pero, M. Delogu, M. Pierini, D. Bonaffini, LifeCycleAssessment of a heavy metro train. J. Clean. Prod. **87**(15), 787–799 (2015)
12. P. Deutz, G. Neighbour, M. McGuire, Integrating sustainable waste management into product design: sustainability as a functional requirement. Sustain. Dev. **18**, 229–239 (2010)
13. K. Dewulf, Sustainable product innovation: the importance of the front-end stage in the innovation process, in *Advances in Industrial Design Engineering*, Chap. 7, ed. by A.D. Coelho (InTech, 2013), pp. 139–166. doi:10.5772/3415
14. L. Dray, An analysis of the impact of aircraft lifecycles on aviation emissions mitigation policies. J. Air Transp. Manage. Selected papers from the 15th Air Transport Research Society Conference, Sydney, 2011(28), pp. 62–69. doi:10.1016/j.jairtraman.2012.12.012
15. EDGE, The Engineering Design Guide and Environment (2015) http://edge.rit.edu/edge/EDGE/public/Home. Accessed 21 Oct 2015
16. M. Finkbeiner, R. Hoffmann, Application of life cycle assessment for the environmental certificate of the Mercedes-Benz S-Class (7 pp). Int. J. Life Cycle Assess. **11**, 240–246 (2006). doi:10.1065/lca2006.05.248
17. G. Finnveden, A. Moberg, Environmental systems analysis tools—An overview. J. Clean. Prod. **13**, 1165–1173 (2005)
18. K. Fletcher, P. Goggin, The dominant stances on ecodesign: a critique. Des. Issues **17**, 15–25 (2001)
19. D. Froelich, E. Maris, N. Haoues, L. Chemineau, H. Renard, F. Abraham, R. Lassartesses, State of the art of plastic sorting and recycling: feedback to vehicle design. Miner. Eng. **20**, 902–912 (2007)
20. D. Froelich, N. Haoues, Y. Leroy, H. Renard, Development of a new methodology to integrate ELV treatment limits into requirements for metal automotive part design. Miner. Eng. Selected papers from Material, Minerals & Metal Ecology '06, Cape Town, South Africa, 20 Nov 2006, pp. 891–901. doi:10.1016/j.mineng.2007.04.019
21. A. Giorgetti, P. Citti, G. Arcidiacono, M. Delogu, Axiomatic design for the development of eco-sustainable metaproducts, in Proceedings of the Sixth International Conference on Axiomatic Design (ICAD2011), Daejeon, Korea, pp. 88–94. ISBN 978-89-89693-30-7
22. S. Howe, A.J. Kolios, F.P. Brennan, Environmental life cycle assessment of commercial passenger jet airliners. Transp. Res. Part D: Transp. Environ. **19**, 34–41 (2013). doi:10.1016/j.trd.2012.12.004
23. IDIS International Dismantling Information System (IDIS) (2015). http://www.idis2.com. Accessed 21 Oct 2015
24. I.S. Jawahir, K.E. Rouch, O.W. Dillon, L. Holloway, A. Hall, Design for sustainability (DFS): new challenges in developing and implementing a curriculum for next generation design and manufacturing engineers. Int. J. Eng. Edu. **23**(6) (2007)
25. G. Johansson, Success factors for integration of ecodesign in product development a review of state-of-the-art. Environ. Manage. Health **13**(1), 98–107 (2002)
26. C. Koffler, Life cycle assessment of automotive lightweighting through polymers under US boundary conditions. Int. J. Life Cycle Assess **19**, 538–545 (2013). doi:10.1007/s11367-013-0652-7
27. S. Le Pochat, G. Bertolucci, D. Froelich, Integrating ecodesign by conducting changes in SMEs. J. Clean. Prod. **15**, 671–680 (2007)
28. V. Lofthouse, Ecodesign tools for designers: defining the requirements. J. Cleaner Prod. **14** (15–16), 1386–1395 (2006)
29. J. Morrison, M. Azhar, T. Lee, H. Suh, Axiomatic design for eco-design: eAD^{+}. J. Eng. Des. **24**(10), 711–737 (2013)
30. B. Ness, E. Urbel-Piirsalu, S. Anderberg, L. Olsson, Categorising tools for sustainability assessment. Ecol. Econ. **60**(3), 498–508 (2007)

31. R. Nidumolu, C.K. Prahalad, M.R. Rangaswami, Why sustainability is now the key driver of innovation. Harv. Bus. Rev. **87**(9), 56–64 (2009)
32. PILOT Ecodesign Pilot (2015). http://www.ecodesign.at/pilot. Accessed 21 Oct 2015
33. G. Ries, R. Winkler, R. Ziist, Barriers for a successful integration of environmental aspects in product design, in Environmentally Conscious Design and Inverse Manufacturing, Tokyo, 01–03 Feb 1999
34. F. Schiavone, M. Pierini, V. Eckert, Strategy-based approach to eco-design: an innovative methodology for systematic integration of ecologic/economic considerations into product development process. Int. J. Sustain. Des. **1**(1), 29–44 (2008)
35. P. Stasinopoulos, P. Compston, B. Newell, H.M. Jones, A system dynamics approach in LCA to account for temporal effects—a consequential energy LCI of car body-in-whites. Int. J. Life Cycle Assess. **17**, 199–207 (2011). doi:10.1007/s11367-011-0344-0
36. N.P. Suh, *Axiomatic design: advances and applications* (Oxford University Press, New York, 2001)
37. N.P. Suh, Complexity: theory and applications. MIT-Pappalardo Series in Mechanical Engineering (2005)
38. N. Tchertchian, P.-A. Yvars, D. Millet, Benefits and limits of a Constraint Satisfaction Problem/Life Cycle Assessment approach for the ecodesign of complex systems: a case applied to a hybrid passenger ferry. J. Clean. Prod. **42**, 1–18 (2013). doi:10.1016/j.jclepro.2012.10.048
39. M.K. Thompson, A classification of procedural errors in the definition functional requirements in axiomatic design theory, in Proceedings of the 7th International Conference on Axiomatic Design (ICAD'13), Worchester, 2013
40. A. Tukker, U. Tischner, *New business for old Europe* (Greenleaf Publishing Ltd, Sheffield, 2006)
41. S.A. Waage, Re-considering product design: a practical "road-map" for integration of sustainability issues. J. Clean. Prod. **15**, 638–649 (2007)
42. WEEE & RoHS PILOT. http://www.ecodesign.at/pilot/eeg/ENGLISH/INDEX.HTM. Accessed 21 Oct 2015
43. Q. Zhu, Y. Geng, K. Lai, Circular economy practices among Chinese manufacturers varying in environmental-oriented supply chain cooperation and the performance implications. J. Environ. Manage. **91**(6), 1324–1331 (2010)

Part III
Buildings

Chapter 6
Application of Axiomatic Design to the Design of the Built Environment: A Literature Review

Marianna Marchesi and Dominik T. Matt

Abstract Built environment design has become increasingly complex due to environmental/energy constraints and socio-economic changes. Pivotal decisions at early phases of design impact the accomplishment of the expected functionalities and performances at competitive costs. Therefore, design theories and methods supporting the accomplishment of the foreseen project outcomes are valuable especially in the initial design phase. Dedicated design methodologies for the accomplishment of general and specific project objectives and problem solving are widely available in engineering, while a comparatively reduced number of methodologies exist for designing the built environment. The axiomatic design theory (AD), one of the available methodologies, shows features for which it may be an appropriate approach for architects and engineers involved in this design area. It provides a decision-making framework with a systematic approach and general principles to support designers on the generation and evaluation of the idea, and to select the best design from several candidate alternatives. Therefore, the present study proposes the introduction of AD for designing the built environment, and it reviews and classifies available applications of AD in this wide area. This literature review shows the effort of research to understand how AD may be used to improve this design field and consequent benefits. By this analysis, it emerges that AD can be effectively used to support design team's decision-making in the conceptual phase. In particular AD results in being mainly applied for addressing the development of effective designs with respect to specified technical requirements. Some applications show the use of AD for evaluating existing designs in order to identify associated problems with their conceptual design and for selecting the best idea from available alternatives. General results and specific outcomes regarding the various areas of application in the design of the built environment are presented, and suggestions for future research developments are recommended.

M. Marchesi (✉) · D.T. Matt
Faculty of Science and Technology, Free University of Bozen-Bolzano,
Piazza Università, 5, 39100 Bolzano, Italy
e-mail: marianna.marchesi@outlook.com

D.T. Matt
e-mail: dominik.matt@unibz.it

© Springer International Publishing Switzerland 2016

A.M. Farid and N.P. Suh (eds.), *Axiomatic Design in Large Systems*,
DOI 10.1007/978-3-319-32388-6_6

6.1 Introduction

Built environment design is now increasingly complex due to environmental and energy constraints and adjustment to evolving socio-economic conditions. Traditional decision-making models based on quality, cost and schedule now have to include and integrate additional aspects concerning sustainability [1]. Consequently, numerous and occasionally conflicting requirements and constraints have to be evaluated from the early phases of the design process. Moreover, since customers' demand is becoming varied and segmented and each market segment requires devoted solutions, architectural design has to consider and evaluate alternative designs in order to select the optimal solution according to customers' demand.

The conceptual phase is the most challenging stage of the design process in which decisions with fundamental and extensive effects on the project outcomes are made. In this phase, the opportunity to influence them is highest. On the other hand, design decisions made in the later stages have a reduced influence on the project outcomes, and the opportunity to influence those decreases over time during the several phases of the design process. In addition, poor initial decisions might be impossible to correct in the later stages [2]. Despite the decisive role of the initial design phase, this phase is not well understood. Only few dedicated tools are available to support this phase while most of the available design tools support later detail phases [3]. Moreover, traditional design tools, such as design-by-drawing, cannot always adequately solve the current design complexity in which the desired design solution is not easily found because of multi-criteria design problems, the design task is inherently complex, the cost of failure is extremely high, and several stakeholders are involved in the project [4]. Differences emerge among the design team on the direction of progression because of distinct approaches to the design process between architects and engineers [5]. Design process in architecture starts by a selection of a small set of design objectives depending on architects' subjective judgment, acquired knowledge and heuristics in order to confine the potential solutions to a manageable set. Subsequently, a solution conjecture is produced and delivered as a first proposal to the customer in order to increase the depth of information concerning the design [6]. As a consequence, an iterative process emerges: both problem and solution are reformulated multiple times and co-evolve while all relevant issues of the design task are progressively recorded [7]. On the other hand, design process in engineering consists of a systematic linear sequence of phases and activities: starts from rigorous and exhaustive analysis of the problem from the abstract to the particular decomposing it into sub-problems. Accordingly, sub-solutions are then defined and integrated into one solution. Therefore, a solution concept is synthesized only after rigorous and exhaustive analysis of the problem avoiding preconceived notions [7].

Due to the complexity of the current design tasks and the shortfall of design tools in the most challenging phase of the design process, design team requires shared approaches that integrate the strengths of both architects and engineers' models avoiding their weaknesses in order to jointly manage the process. In addition, they need design methodologies for addressing early designs towards the accomplishment of the expected outcomes and for supporting decisions and compromises among varied and conflicting parameters and complex sets of constraints. Suitable design approaches are available in engineering: the systematic approaches. These support rational and formal design thinking, provide decision-making procedures [4] and principles for the accomplishment of general and specific goals as well as for solving specific problems and linking design activities. They guide the abilities of designers in order to improve the efficiency of the design process and the effectiveness of the results, reduce workload, save time and prevent human error, without impairing the designer's creativity [8]. Design research states that an appropriate shared approach to the design process for interdisciplinary design teams should provide a succession of phases and varying cognitive procedures to perform the design process and a sequenced creation process of going to-and-from problem and solution, and sub-problems and sub-solutions [9]. In order to support design team's decision-making in the conceptual phase of the built environment design, a systematic approach is herein proposed: axiomatic design theory (AD). AD is a formal design theory differentiated from other systematic approaches by an axiomatic scientific basis that underlies the decision-making process [10, 11]. AD provides a decision-making framework with a systematic procedure and general principles to support designers during the creative and analytical phases on the generation and evaluation of the idea and on the selection of the best design from several candidate alternatives. AD principles are axioms that can be generalized. This method has shown strong potential by applications in different design areas included non-engineering fields [12]. Since AD shows features for which it may be an appropriate approach for architects and engineers involved in the built environment design and due to its effectiveness in different design areas, this study proposes the introduction of AD for designing the built environment, and inspects available applications of AD in this large area. A classification scheme on the basis of the application area, design phases, design activities, pursued goal, applied methods and axioms, evaluation type and results is performed. This analysis shows the effort to understand how AD may be used to improve the design of the built environment and consequent benefits. By this study, it results in being a suitable design approach for design team's decision-making in the conceptual phase. In practical and theoretical articles, it emerges the value of AD on the evolution of design concepts in the built environment design. On the other hand, in these applications AD is rarely used to analyses pre-existing systems in order to identify problems associated with their conceptual design. Moreover, it is observed that AD is mainly applied to improve technical functions; social and aesthetic functions are rarely addressed. In many studies AD is the solely applied method whereas in certain studies AD is combined to different design methods to manage specific activities or address particular design problems. In most articles, the first axiom is generally applied since it allows the designer to define a solution better able to satisfy the expected functionalities despite being subject to external uncertainties and disturbances [13] reducing the need for random research and the trial-and-error activity. On the other hand, the second axiom, that supports the selection of the most appropriate alternative within specific criteria, and fosters the complexity minimization, is rarely applied. In addition, each specific application area is examined with respect to the reviewed articles. Efforts regarding each of them are discussed, and recommendations for future investigations are suggested.

6.2 Literature Review

6.2.1 Conceptual Design Phase

Design of the built environment is a process of creating built solutions in order to satisfy technical, social and aesthetic functions according to existing constraints and available resources. In a built work the valuable characteristics having different degrees of importance consist of space suitability for human usage and adaptability to specific activities, capability of the construction to protect and supply resources, and aesthetics [14]. Design is constrained by context features and finite budget, time, resources and whole-life value. Therefore, a design has to satisfy various demands in a utilitarian and aesthetic form within given socio-economic constraints, and also with respect to available resources, existing context and cultural features [1]. With the goal of developing effective solutions, different and frequently conflicting requirements and constraints have to be considered from the early design phases. Consequently, the complexity of designing the built environment increases exponentially. In the conceptual design stage, design decisions have the greatest impact on the project outcomes with respect to functionality, performance, appearance, costs and sustainability. In contrast, in later design stages, decisions have a minimal influence on the project outcomes, and might not be able to correct poor decisions made initially [2]. In spite of the decisive role of the conceptual design phase on the generation of the appropriate design concepts, there is little or no guidance on what should be performed and how it should be achieved [5]. This phase is not well understood. Designers expect that all requirements are identically satisfied without considering the possible conflicts. Differences concerning direction of progression often emerge among interdisciplinary design teams because of the lack of a shared understanding of the design process due to distinct approaches between architects and engineers [5]. Traditional design tools of design such as design-by-drawing cannot always adequately support designers in the current complexity [4]. Moreover, most of the design tools are usually specialized in supporting late design development, and relatively few have been developed to support the conceptual phase [3]. Also in sustainable building design, tools are designed to assess different aspects of sustainability or the overall building sustainability on the basis of detailed design information only available at the later phases [15]. Due to the current complexity of the built environment design, the traditional ways of designing should be improved. Especially in the conceptual phase, interdisciplinary design teams should be supported by shared approaches to the design process, and designers' decision-making should be assisted by design methodologies that are able to early address the design development to the accomplishment of the expected outcomes.

6.2.2 Models of the Design Process in Engineering and Architecture

Initially from the 1960s and 1970s in both architecture and engineering, it was proposed that at beginning solution concepts should be synthesized only after rigorous and exhaustive analysis of user requirements and other basic features of the problem. Therefore, models of the design process in architecture were very similar to models in engineering design. After starting from common origins, models of the design process in architectural design have diverged from the engineering models due to criticisms from theorists and practitioners on linear and sequential processes based on an analysis-synthesis-evaluation sequence. Architectural design methodologists stressed the importance of generating solution concepts early in the design process drawing upon presuppositions. Therefore, significant differences between engineering and architectural design models emerged [7]. They are herein discussed. In architecture, design problem is considered ill-defined, and therefore an initial rigorous and exhaustive analysis appears to be superfluous. This determines that the design process does not begin with an exhaustive problem specification from which solution concept may be synthesized [7]. Instead architects identify solution concepts very early, drawing up premises. In other words, architects identify the important aspects of the problem based on previous experiences and knowledge. Then they develop a preliminary design on the basis of this analysis, and they examine it to see what else they can learn about the problem [6]. The problem is reformulated according to the client's feedback, and problem and solution co-evolve in parallel during the design process [7] until they are firmly defined. In essence, usually architectural approaches adopt solution-oriented approaches to design problems generating solution concepts early in the design process and refining them through a spiral structure of conjecture-analysis cycles [7]. In contrast in engineering design, the overall problem structure is considered well-defined as a tree, and therefore it can be split into sub-problems [7]. Engineers firstly analyse the problem from the abstract to the particular decomposing it into sub-problems for which sub-solutions are then found and synthesized into an overall solution [7]. This determines that a solution concept is synthesized only after rigorous and exhaustive analysis of the problem avoiding preconceived notions. Therefore, the design process consists of a systematic linear sequence of phases and activities starting from the general and abstract and evolving to the particular and concrete. In essence, engineers usually adopt problem-oriented approaches focused on the analysis of the problem and based on prescriptive multi-phase procedures [7].

Macmillan et al. [5] identify some common features among design process models from architecture and engineering by comparison. Most models start with a requirements statement followed by the generation of possible solutions showing progression. Most set out only what should be undertaken, not why or how it should be performed. Most models imply convergence to one solution quite early in the design process, and only a few explicitly encourage the generation of alternative concepts for evaluation. None of the models makes explicit reference to techniques for generating alternative solutions, or to formal measurement, evaluation or assessment methods [5].

Since it is proven that designers supported by a common design process framework are better able to focus on the demands of a problem than those without it [16], various authors consider that the re-integration of the two models in common models suitable for both disciplines is needed. They should integrate the strengths of architectural and engineering models avoiding their weaknesses [9]. All designers need to progress their project in a sequence of stages similar to the engineering model. Designers must also employ varying cognitive procedures during the design process similar to the architectural model. On the other hand, the emphasis placed by the engineering model on problem analysis and specification could limit innovative solution generation while the emphasis placed by the architectural model on early solution conjectures could penalize an adequate problem clarification [9].

Some authors have attempted to move towards the development of integrated procedural approaches [9] in order to improve practice across the disciplines and support transdisciplinary collaboration. In particular Roozenburg and Cross [9] assert that an appropriate model of the design process for practitioners should be able to reproduce the sequenced process of going to-and-from problem and solution, and sub-problems and sub-solutions. Accordingly, Cross has proposed a hybrid model in which the designer's thinking oscillates to-and-from problem solution and to-and-from sub-problem and sub-solution [4]. This indicates that problem definition depends on the solution concept and sub-problems identification depends on sub-solutions generation. Moreover, there is a hierarchical relationship between problem and sub-problems and between solution and sub-solutions. A given problem is clarified by its decomposition in sub-problems, and consists of identifying sub-functions and specifying performance requirements. A solution is derived from sub-solutions by generating, combining, evaluating and choosing appropriate sub-solutions. In this framework a set of design activities and correlated design methods is proposed in order to perform the design process [4].

6.2.3 Axiomatic Design Theory

The conceptual phase is the most challenging stage of the overall design process, but at the present it is not well understood. Design team lacks a common understanding of the manner in which the design process is being performed and the direction of progressing. Many designers emphasize intuition and experience combined with conventional design methods. Nowadays, these methods are not sufficient since projects are too complex. This is especially acute in the early phase when design decisions will have extensive effects on the project outcomes. Early design generation needs to be addressed in order to produce the expected outcomes and to eliminate or reduce the need for design compromise at later and more critical phases of the process. In this way, systematic approaches may be helpful to improve design in the built environment. They are design methodologies largely applied in product design, industrial engineering, and manufacturing where the focus is on the development of mass-produced products. On the other hand, design methodologies are rarely applied in the design of the built environment since it tends to produce mainly unique systems. However, because it involves large budgets, scales and development and construction time scales, it may benefit from the introduction and application of design methodologies, particularly from systematic approaches. They manage the design process and address, from an early phase, the solution towards the fulfilment of attended outcomes. These approaches, developed in engineering design, externalize design thinking, and formalize rational design procedures [4]. They provide strategies, rules and principles for the accomplishment of general and specific goals as well as methods to solve specific problems and tasks, and link design activities and phases. Systematic approaches direct the abilities of designers in order to improve the efficiency of the design process and the effectiveness of the results. They provide a rational basis for reducing workload, save time and prevent human error, but at the same time without confining designer's creativity [8]. AD theory is a systematic approach distinguished from the others by having basic principles of decision-making. AD, developed by Nam P. Suh at MIT in engineering field, establishes that there are design principles governing good design decisions. AD can be applied from synthesis to analysis of the synthesized idea, and for the selection of good ideas from plausible alternatives to all situations of problem solving in the form of products, processes or systems [10, 11] such as product design, large and small scale system design, and manufacturing process design [12]. It has been demonstrated that AD supports the development of solutions better able to guarantee the expected functionalities and performances in the presence of uncertainties and disturbances in the context (e.g. changing customers or functions or physical components) [13]. In AD, the design process is performed through the thinking interplay between what should be undertaken and how it should be developed until designers produce an acceptable result [10, 11]. During this process, by principles of functional independence and complexity minimization, designers are able to evaluate the synthesized idea before and during the analytical phase, and to select the best idea from several plausible designs within a set of criteria, even in the early design phase [10, 11].

According to Roozenburg and Cross [9], a common model of the design process appropriate for architects and engineers should provide a sequence of stages similar to the engineering model and varying cognitive procedures during the design process similar to the architectural model. It should avoid emphasis on problem analysis and specification, and at the same time emphasis on early solution conjectures [9]. It also should be able to reproduce the sequenced process of going to-and-from problem and solution, and sub-problems and sub-solutions [4]. AD shows to provide a sequence of stages to progress the project and a sequenced creation process based on going to-and-from problem and solution, and from sub-problems and sub-solutions. In AD, problem and solution are systematically and consistently specified in parallel, moving down along the hierarchy, and design decisions are made in an explicit way maintaining data. The process is supported by general principles of decision-making in order to define effective designs with respect to specified requirements, to evaluate the synthetized ideas and to select the most feasible solution among valuable alternatives. On the basis of these premises, AD may be an appropriate common approach for supporting architects and engineers in performing decision-making in the conceptual design of the built environment.

6.2.4 Applications of Axiomatic Design to the Design of the Built Environment

An analysis of published studies on applications of AD to the design of the built environment is now performed, followed by the corresponding classification. This analysis takes into account papers published between the years 2000 and 2014. Studies published in academic journals outside of databases and non-English papers are not included. The number of papers on this topic is not very high. Nevertheless, this literature review is not able to be comprehensive. The analysed articles are classified according to seven criteria (Tables 6.1, 6.2 and 6.3): application area, corresponding design phase, proposed aim, applied methods, adopted axiom, type of evaluation and finally results. The "Application area" shows the major fields of application into the design of the built environment, and consists of five subsections: civil and environmental engineering, mechanical engineering, urban design, building design, and interior product design. The "Design phase" column is created to highlight in which phase of the design process the AD approach has been applied. The proposed "Aim" section intends to show the objectives on the basis of which AD is involved in each study. This section includes applications of AD (solely and combined with other methods) and theoretical developments. The "Theoretical development" column identifies the studies that propose theoretical improvements of the design approach. The "Methods" section analyses in detail the methods adopted or proposed in each study. The "Axioms" section deals with which kind of axiom is used in the selected papers: the first axiom (the independence axiom) and the second axiom (the information axiom). The "Evaluation Type" and "Results" sections analyses the type of assessments performed between FRs and DPs and the outcomes obtained.

Table 6.1 Literature review classification: application area and design phase

	Application area					Design phase			
	Civil and environment engineering	Mechanical engineering	Urban design	Building design	Interior product design	Strategic	Conceptual	Developed	Detailed
[17]				Housing development		•	–	–	–
[18]				Housing development		•	–	–	–
[19]					Seated work place	–	•	–	–
[20]				One-family house		–	•	–	–
[21]				Housing development		•	–	–	–
[22]			Housing area development			–	•	–	–
[23]		Acoustics of auditorium				–	•	–	–
[24]		Energy consumption reduction				–	•	–	–
[25]	Dam					–	•	–	–
[26]				Emergency department		•	–	–	–
[27]	Traffic intersections					–	•	–	–
[28]	Traffic intersections					–	•	–	–

(continued)

Table 6.1 (continued)

	Application area					Design phase			
	Civil and environment engineering	Mechanical engineering	Urban design	Building design	Interior product design	Strategic	Conceptual	Developed	Detailed
[29]				Airport terminal		–	•	–	–
[30]	Traffic intersections					–	•	–	–
[31]				Temporary modular house		–	•	–	–
[32]				Architectural design		–	•	–	–
[33]	Electric power grid systems					–	•	–	–
[34]				Temporary house		–	•	–	–
[35]	Electricity, water and wastewater systems					–	•	–	–
[36]				One-family house		–	•	–	–
[37]	Transportation system					–	•	–	–
[38]	Transportation system					–	•	–	–

Table 6.2 Literature review classification: aim and methods

	Aim		Methods				
	Theoretical development	Application	Problem definition	Creative process		Analytic process	Ultimate check
			Recognition of needs	Determination of FRs and Cs	Creation of solution in terms of DPs	Analysis of solution	Check ultimate solution
[17]	•		Customer needs survey	AD	AD	–	–
[18]	•		Qualitative methods	AD	AD	–	–
[19]		•	Literature review	AD	AD	AD	AD
[20]		•	Literature review	AD	AD	AD	–
[21]	•		Market analysis, Kano model	AD, QFD, TIPS	AD, robust design, theory of flexibility	–	–
[22]	•		Literature review, POE study	AD	AD	AD	AD
[23]		•	Literature review, EMS models	AD	AD, TRIZ	AD	–
[24]		•	Literature review	AD	AD	AD	–
[25]	•	•	Literature review	–	–	AD, user-centred design, TRIZ	–
[26]		•	User interviews, user observation	–	–	AD	–
[27]	•		Literature review	AD, traffic conflict theory	AD	AD, traffic conflict theory	–
[28]		•	Literature review	AD, traffic conflict theory	AD	AD, TRIZ, traffic conflict theory	–
[29]		•	Customer needs survey	AD	AD	AD	–

(continued)

Table 6.2 (continued)

	Aim		Methods				
	Theoretical development	Application	Problem definition	Creative process		Analytic process	Ultimate check
			Recognition of needs	Determination of FRs and Cs	Creation of solution in terms of DPs	Analysis of solution	Check ultimate solution
[30]	•	•	Literature review	AD, traffic conflict theory	AD	AD	Concept selection method
[31]	•	•	Literature review	AD, product platform	AD	AD	–
[32]	•		–	–	–	–	–
[33]	•		Literature review	–	–	AD	–
[34]	•	•	Literature review	QFD, AD	AD	AD	–
[35]	•		Literature review	AD	AD	–	–
[36]		•	Literature review, user interviews	AD	AD	AD	–
[37]	•		Literature review	AD	AD	–	–
[38]	•		Literature review	AD	AD	–	–

Table 6.3 Literature review classification: axioms, evaluation type, and results

			Axioms		Evaluation type
1					Results
2			Crisp	Fuzzy	
[17]	•	–	–	•	Analysis of housing building markets and comparison
[18]	•	–	–	•	Combined design framework
[19]	•	•	•	–	Design methodology improvement in ergonomics
[20]	•	–	–	•	AD application to the design of a single-family house
[21]	–	–	–	–	Combined approach for improving early decision-making
[22]	•	•	–	•	Systematic design evaluation method
[23]	•	–	•	–	AD application combined to TRIZ
[24]	•	–	•	–	Energy-efficiency design framework
[25]	•	–	•	–	Problems associated to the dam's conceptual design
[26]	•	–	•	–	Problems in an emergency department
[27]	•	–	•	–	Conflict and coupling understanding into traffic intersection design and analysis
[28]	•	–	•	–	Functional efficiency improvement of traffic intersection design
[29]	•	–	•	–	Functional design approach
[30]	•	–	•	–	Quantification of coupling impact in traffic intersection design
[31]	•	–	•	–	Approach to address stakeholder's needs and modularity
[32]	–	–	–	–	Applicability of AD to architectural design and potential
[33]	–	–	•	–	Design principles for resilient coordination and control of electric power grid systems
[34]	•	–	•	–	Combined approach for customer needs' identification, translation into requirements and design definition
[35]	–	–	•	–	System architecture for electricity, water and wastewater systems
[36]	•	–	•	–	AD application on the conceptual design of a single high-performance house
[37]	–	–	•	–	Theory for the reconfigurable design and operation of transportation systems
[38]	–	–	•	–	Hybrid dynamic system model for the electrification of transportation systems

Hereafter the analysed papers are arranged on the basis of the application area, and their content is briefly summarized.

6.2.4.1 Civil and Environmental Engineering

Civil and environmental engineering design is a wide engineering discipline that deals with the design, construction and maintenance of the physical and built environment, including works like roads, bridges, canals, and dams. In the literature, the articles that have been identified regarding civil and environmental engineering design based on AD principles are summarized as follow.

Ibragimova et al. [25] analyse the conceptual design of Sihwa dam in Ansan (South Korea) in an attempt to understand the successes and failures of the project from the perspective of a combination of three formal design theories (axiomatic design theory, TRIZ—theory of inventive problem solving and user-centred design) and to provide guidelines for the design of similar systems in the future. In this study, the need for including functional requirements associated with environmental protection is shown. The presence and influence of coupling, conflict and compromise in the three stages of the dam are used to explain some of the problems associated with Lake Sihwa. The result is an improved understanding of the design of tidal dams and barrages which can be used to avoid repeating the mistakes of the past. It also highlights some of the similarities, differences and shortcomings of the theories used in this study [25].

Thompson et al. [27] examine the suitability of AD for traffic intersection design. They combine AD with traditional traffic conflict techniques to examine strategies for the design of a generic 4-way intersection. The conflicts for the intersection are identified, and the various types of coupling are highlighted by a hybrid design matrix (DM). Then two design strategies for the improvement of the intersection are considered: separation in space (two-dimensional separation, and three-dimensional separation) and separation in time (periodicity). The application of both techniques and their combination with AD produces a reduction in the number and severity of conflicts at intersections and the elimination of strong couplings. The most suitable techniques should be chosen on the basis of the existing Cs in the system (available space and financial resources) and by identification of the design criteria (that include minimizing the total travel time of vehicles in the system). This study proves the valuable benefits of AD combined with traditional traffic conflict techniques to the design and analysis of traffic intersections [27].

Thompson et al. [28] focus on understanding and improving the design of urban intersections to increase the traffic system efficiency through the combination of traditional traffic conflict techniques with AD and TRIZ. They are applied on two case studies. The first case study involves the redesign of a generic 4-way intersection. The second case study concerns the conceptual redesign of an existing intersection located in Daejeon, Republic of Korea. The analysis evaluates the impact of the selected design strategies on the FRs, on the traffic intersection conflicts and on the couplings in the DM. It is shown that common strategies to redesign intersection result in an unnecessary loss of FRs. The existing intersection is reviewed by holistic and modular approaches and the results are compared in terms of FRs and DM. Finally, redundancy in intersection design and symmetry in the DM are discussed [28].

Yi and Thompson [30] propose a new concept selection method to quantify coupling in hybrid DM for traffic intersections by the calculation of the coupling impact index. This value can be used to identify the level of safety and efficiency of intersection designs in a specific situation. This index is evaluated by hybrid DM that is used to identify couplings, types of conflict determined by couplings and impact of conflicts in intersections. A case study is analysed, and the most desirable solution is identified using the coupling impact index. Advantages and limitations of this coupling index were discussed [30].

Farid [33] identifies a set of multi-agent system design principles for resilient coordination and control of future electric power grid systems. In addition, the paper assesses the adherence of existing multi-agent system implementations with respect to these design principles. It concludes that while many multi-agent systems have been developed for power grids, they have been primarily intended as the decentralization of a particular decision-making/control algorithm. Therefore, they only partially contribute to power grid resilience [33].

Lubega and Farid [35] present a system architecture for electricity, water and wastewater systems in order to improve resilience and sustainability of these critical systems through integrated management. It describes how these systems interact with each other and with the environment and specifies related system parameters. The presented models can serve qualitative discussions on where and how the supply and demand of water and energy are interdependent. Second, within the operations phase, they can support the development of automated Information Technology and control solutions that integrate energy and water management. Finally, at a planning phase, they can inform quantitative decisions on how to best grow and reconfigure the water, wastewater and energy infrastructure [35].

Viswanath et al. [37] apply AD to develop a theory in transportation systems for their reconfigurable design and operation. This methodological development is demonstrated on a small subsection of the Mexico City transportation system to show its benefits on the decision-making at the planning and operation phases. In addition, comparisons of axiomatic design to traditional graph theory are made: the two approaches are complementary, but AD presents advantages that this study explains [37].

Viswanath and Farid [38] develop a hybrid dynamic system model for the electrification of transportation systems in order to successfully integrate electric vehicles on the infrastructure systems that support them. Since electric vehicles and their supporting charging infrastructure couples the transportation and electrical systems with consequent delays on the electrical grid, this study proposes a model that is capable of resolving the kinematic state of the vehicle fleet while keep track of each vehicle's state of charge. In such a way, this hybrid dynamic model manages both the transportation as well as electrical functionality in a transportation–energy nexus. The application of the model is demonstrated on an illustrative example [38].

6.2.4.2 Mechanical Engineering

Mechanical engineering is the discipline that applies the principles of physics, and materials science for the design, analysis, manufacturing and maintenance of mechanical systems and correlated aspects of the built environment. The following papers are examples of mechanical engineering design based on axiomatic design principles.

Kankey and Ogot [23] apply AD combined to TRIZ in order to solve a problem of poor acoustics in a historical auditorium. The study intends to develop an affordable permanent solution that guarantees an enjoyable listening experience for most of the audience and retains the historical aspect of the building. By the energy-material-signal (EMS) model, the problem is correctly defined decomposing it and identifying poor phenomena aspects (energy, material or signal flows). FRs are specified, and a solution is identified and expressed in terms of DPs. Then, unsought couplings are highlighted by the design matrix (DM). Since the solution is a decoupled design, TRIZ is employed to solve contradictions. Using AD combined with TRIZ, the defined solution results in an uncoupled design [23].

Cavique and Gonçalves-Coelho [24] develop a general design framework by AD on how to reduce energy consumption in buildings equipped with heat, ventilation and air conditioning (HVAC) systems. Since energy consumption of HVAC systems depends on the characteristics of the building where systems are installed, this study analyses both requirements: the reduction of energy consumption in a building and the decrease of energy consumption of HVAC systems. By the AD mapping process, FRs and DPs are decomposed in a general framework in which the reduction of the energy consumption in buildings is evaluated on the basis of the improvement of the energy building envelope behaviour, the reduction of internal loads and energy systems consumption and the production of energy on site by renewable sources [24].

6.2.4.3 Urban Design

Urban design deals with the design of urban areas up to entire cities and concerns larger scale of groups of buildings, streets and public spaces, whole neighbourhoods and districts. The following paper presents an application of axiomatic design principles on the design of urban spaces.

Kowaltowski et al. [22] suggest a systematic method for design evaluation based on AD to improve the design quality of low-income housing projects through the assessment of its environmental-life quality impact. Authors sustain that, through

AD, qualitative information can be included in the design process increasing the design quality. This method should support designers in the evaluation of numerous factors that affect the quality of user's life and environmental sustainability. A literature review is elaborated to establish architectural and urban indicators influencing environmental and life quality in low-income housing areas [22]. To include people's perception of quality into the design process, POE (post-occupation evaluations) method is proposed for directly linking design criteria to users' desires, and verifying the effectiveness of selected indicators. The selected indicators are included in the AD framework to support the decision-making design process. The evaluation and optimization of solutions should be then performed by specific analysis methods such as simulation, checklist and multi-criteria optimization [22].

6.2.4.4 Building Design

Building design refers to the architectural, engineering and technical applications to the design of buildings. In this subsection, the papers based on axiomatic design principles are taken into consideration.

Eliasson and Psilander [17] propose the application of specific methods to guarantee customer satisfaction and home building industry profit. Since entrepreneurs' goal in the home building industry is to place maximum value on the product offered providing housing for a chosen group of customers, customer preferences are carefully identified by an identification and classification of customer types and needs and then are linked to functional requirements. AD is suggested in order to reach an efficient production process by maximizing product quality and customer satisfaction with variety and minimum inputs variability. Three different home building markets are analysed and compared [17].

Sohlenius [18] suggests the application of manufacturing design methods, such as AD, to the building process since building industry shares many characteristics with the manufacturing industry. The goal is to maximize profitability in the construction industry in terms of income, cost and capital keeping a high customer value in the short and long terms. The AD decision-making framework is proposed for the development of large residential projects in which qualitative methods are combined with AD to interpret all customer needs. This study establishes the potential improvements on the definition of goals and decisions determined by the application of AD to the early stages of the design process in the building industry, specifically in real-estate development [18].

Psilander [20] proposes the application of AD to the design of housing with the goal of maximizing profits costs through the match between customers' preferences and design, limiting costs by rejection of bad designs in the initial design phase. Moreover, since by AD the design thinking is externalized, possible deviations are easily identified during the design process determining where they appear and why they are made. The consequences of deviations are evaluated. This application concerns the conceptual design of a single-family house using only qualitative information and intends to suggest an operative framework for decision-making during the design process. Standardization combined with architectural variations is suggested in order to satisfy varied customer's preferences at affordable costs [20].

Sohlenius and Johansson [21] suggest a design framework based on AD combined with quality function deployment (QFD), TRIZ, and robust design in order to improve the decision-making process in the conceptual design phase of housing developments for achieving high customer value and high productivity. Moreover, theory of flexibility and LOLA-rule are included to achieve flexibility and to define bounds on design changes. Modularity and design variety are suggested to achieve efficient production and, simultaneously, to satisfy different customer needs. The profile of the real-estate development needs to be expressed clearly through a market analysis to identify needs and Cs and meet the customer's demand. This analysis consists in assessing costumers' housing requirements, site conditions, laws and regulations. Kano model is proposed to structure customer needs and to focus on the right quality [21].

Peck et al. [26] apply AD to an emergency department analysis to identify, understand and communicate problems. The emergency department (ED) design is decomposed by AD and the decomposition is used to identify the inherent functional couplings in the design of an ED system. Many of the ED problems can be attributed to the identified couplings. They are validated and judged using a suburban community hospital case study. By the detailed functional decomposition, it is easy to fully understand the interactions that form coupling and the ways to eliminate or limit the problem [26].

Pastor and Benavides [29] apply AD to the functional design of a passenger terminal in a small tourist airport to manage its high design complexity due to many variables. Usually, in the early design phase, basic dimensions and infrastructures are estimated using formulas indicated by each national regulatory authority and international organizations to guarantee a certain level of service and safety. Subsequently, the distribution and configuration are determined according to architectural and functional criteria [30]. In this paper, the aim is to test AD as functional design approach for solving simultaneously functional layout and size according to the identified stakeholders' needs. Therefore, at the beginning, an analysis of the motion path followed by passengers is conducted. Moreover, a survey is carried out to collect the customers' needs and establish a FRs list for each functional area. Only minimum sets of FRs are defined. Each set represents the basic functions that each area should provide to guarantee customer satisfaction. The conceptual design of each functional area is then defined individually, and the derived concept for the whole system is composed linking optimally the sub-systems [29].

Lindsey et al. [31] propose an approach based on product platform and AD for the improvement of the conceptual design of modular and reusable temporary housing. This approach is used at an early design phase of a temporary house to ensure the satisfaction of the stakeholders' needs and the achievement of modularity. Since stakeholder's needs and Cs change rapidly over time and with location, AD is proposed to systematically map user needs into functions and functions into a form and to ensure the satisfaction of high-level needs. Moreover, modularity is addressed by the use of the product platform philosophy [31].

Marchesi et al. [32] review current understanding on architectural design in terms of approaches, process and tools in order to assess compatibility and potentials of applying AD to the design of architectural systems. In addition, available applications of AD in this design wide area are reviewed and classified [32].

Lindsey et al. [34] suggest a combination of two design methodologies, QFD and AD to fulfil customer's needs by their early identification, their translation to design objectives and accordingly the definition of a consistent design. The combination of these methodologies intends to provide a systematic approach for the conceptual design of construction projects, in particular temporary housing, and it is applied to a case study of a refugee temporary house unit [34].

Marchesi et al. [38] apply AD to the conceptual design development of a sustainable high-performance house. The goal is to test the AD approach in addressing early overall technical functions such as sustainability and energy efficiency. A case study is analysed: based on clients' needs, an initial set of independent FRs is defined and gradually decomposed in a systematic manner, and related Cs are identified. The corresponding design is progressively developed according to the specified requirements and existing Cs. This study presents a systematic approach to the architect's decision-making in the initial phase of the design process for the development of performing building concepts [38].

6.2.4.5 Interior Product Design

Interior product design deals with the design of furniture placed into the interior of a room or a building. An example of AD approach applied to interior product design is given in the following.

Helander and Lin [19] apply AD with the purpose of displaying advantages on the design of ergonomics. Thanks to AD, design of ergonomics benefits from a clear design framework consistent with user needs and a systematic procedure by hierarchical decomposition in order to formalize design solutions and to identify critical design parameters. Principles are provided to designs and alternative solutions are compared. Using the Independence axiom, this application results in a better unconventional solution than the conventional design solution recommended in the classical literature. The Information axiom is then applied to evaluate quantitatively alternative designs for the selection of the best solution [19].

6.3 Conclusions

In this literature review, it has shown that even though, in the context of design of the built environment, early phase decisions have the strongest effect on the design outcomes, this phase is usually not well understood. Design team has little or no guidance on what should be done and how it should be achieved. Confusion on the direction of progression appears because of the lacks of a shared understanding of the design process in the design team due to distinct approaches between architects and engineers. In this phase, architects define early a simplified problem based on previous experiences and knowledge and formulate an initial conjecture of possible solutions accordingly. Then problem and solution are evaluated with the clients and reformulated by a cyclic and spiral process until they are explicitly defined, while the opportunity to influence the design outcomes decreases over time. On the other hand, engineers assume that the overall problem can be broken down into distinct sub-problems. Therefore, by a linear and sequential process the problem is decomposed into sub-problem. Related sub-solutions are defined accordingly and then they are integrated into an overall solution.

Due to the current complexity in the design of the built environment, design team needs to be supported by shared approaches to design process in order to jointly manage it and by appropriate design methodologies to address early decision-making towards the accomplishment of the expected project outcomes. A suitable approach should be able to integrate the strengths of both architect and engineer's approaches avoiding their weaknesses. Specific approaches are available in engineering design: the systematic approaches. They provide rational design procedures, strategies, rules and principles to address the fulfilment of attended functionalities and performances without restricting designer's creativity. In this study, AD has been proposed as a suitable design approach for both engineers and architects in the design of the built environment in order to support design team in performing decision-making in the conceptual phase. AD has resulted in being adequate for architects, engineers and interdisciplinary design teams because it shows features that design research considers decisive for an appropriate design approach. It provides a framework based on sequence of stages to develop the design process. The creation phase is a sequenced process of going to-and-from problem and solution in which problem and solution are specified in parallel, in a systematic, progressive and consistent manner. During the creation and analytic phases, general principles of decision-making support designers to develop effective designs, to evaluate the synthesized idea and to select the best idea from several plausible alternatives. Since this study has proposed the introduction of AD in the design of the built environment, an analysis of available theoretic and practical applications has been performed, and a classification scheme has been introduced based on specified criteria. By this analysis, this study has shown the effort of the research to understand how AD, a formal design theory from engineering field, may be used to improve design in the built environment. Practical and theoretical articles have highlighted the value of AD on the evolution of design concepts in the built environment design. In addition, AD has been used to identify and explain problems in existing systems associated with their conceptual design; therefore, it can be also used to avoid similar problems in the design of future proposals. However, in few articles only, AD has been used to analyses pre-existing systems.

The applications of AD have covered various fields and built-in artefacts, mostly in civil and environmental engineering and building design. On the basis of these results, AD appears to be employed in both engineering and architectural disciplines as well as in interdisciplinary design areas of the built environment. In civil and environmental engineering area, AD has been applied to the design of water resources management systems and infrastructures that includes tidal dams, barrages [25], transportation (traffic intersection and transportation infrastructure) [27, 28, 30, 37, 38] and water and power networks [33, 35]. It has been mainly used to support the design concept generation, but also to analyses pre-existing systems and associated problems with their conceptual design [25]. In some of these studies, AD has been applied in a standalone form without the participation of other methods. More in detail they have employed the AD of large flexible systems [33, 35, 37, 38]. In other studies, AD has been combined with conventional techniques (e.g. traffic conflict theory) [27, 28], or it has been integrated by other design methods (e.g. TRIZ, user-centred design) [25]. The articles in the building design area concern theoretical analysis for understanding the applicability of AD in building design [32], methodological frameworks for the development of large residential areas from the project manager's viewpoint [17, 18, 21], methodological improvements [34] and practical applications in the conceptual design of prefabricated housing [20], passenger terminal [29], temporary housing [31] and high-performance housing [36]. These applications have shown benefits of applying AD to the concept generation with respect to the spatial configuration in complex built artefacts with large flow of people such as airport [29], with regard to the modularity of temporary prefabricated housing [31] or functionalities and performances of residential buildings [36]. Meanwhile, the application of AD for the analysis of the nature of pre-existing building systems and problems associated with their conceptual design has not been observed with an exception. There is only one paper included in this application area that applies AD to analyse problems: it concerns the examination of an emergency department [26]. Since it is proven that if a design theory can identify and explain the problems with an existing design, it can also be used to avoid those problems in future designs. This type of analysis may be helpful in the early phase of the design process when strategic inputs are specified. Analyses of the nature of pre-existing building systems are therefore recommended in this design area for the future. AD is applied in a standalone form [20, 26, 29, 36] and also integrated by other design methods such as QFD [34]. In mechanical engineering, the number of the identified applications is low. These applications concern conceptual proposals for building acoustic [23] and energy-efficiency improvements [34]. Since this area presents a large potential of research development, future AD applications and theoretical studies are encouraged. Also in urban and interior product design, the number of applications is currently modest with respect to other areas of application

[19, 22]. Further investigations by applications and existing systems analysis may determine significant benefits on both disciplines.

The present analysis has also found that AD has been mainly used to address technical functions; social and aesthetic functions have been rarely included [20, 25, 36]. Proposals for theoretical improvements are observed in theoretical and practical articles in which AD has been combined with conventional techniques (e.g. traffic conflict theory) to improve them, or it has been integrated by other design methods (e.g. QFD, TRIZ, user-centred design) to manage specific aspects of the design problem or to improve a particular design activity. QFD is integrated to AD in order to guarantee the fulfilment of the customer's needs by their early identification and their translation to design objectives. TRIZ is combined to AD in order to reduce coupling and solve conflicts and contradictions in the design. By the user-centred design, affordances are used to identify hidden or latent needs and absent but necessary functional requirements. In most practical articles, the first axiom has been mainly applied since it allows designers to evaluate the synthesized idea while reducing random research and minimizing repeated trial-and-error activities. The application of the second axiom has been rarely observed in the analysed applications [19]. Since it allows the evaluation of a set of criteria together, it supports the selection of the best and the most valuable alternative to solve a multi-criteria decision-making problem, but in these studies different alternative solutions have been seldom proposed and evaluated. In the future, studies directed at this aspect are recommended.

In this work, it has been shown that AD can be effectively applied to support design teams' decision-making in the conceptual phase in the design of the built environment. In future, limitations of this theory for the various design areas of the built environment, adaptations and modifications required for its use in these areas, need to be further investigated.

References

1. Commission of the European Communities, Communication from the Commission to the Council and the European Parliament on Thematic Strategy on the Urban Environment (2005)
2. American Institute of Architects, *Integrated Project Delivery: A Guide* (AIA, 2007), p. 21
3. L. Wang, W. Shen, H. Xie, J. Neelamkavil, A. Pardasani, Collaborative conceptual design—State of the art and future trends. Comput. Aided Des. **34**(13), 981–996 (2002)
4. N. Cross, *Engineering Design Methods: Strategies for Product Design* (Wiley, Chichester, 2008)
5. S. Macmillan, J. Steele, S. Austin, P. Kirby, R. Spence, Development and verification of a generic framework for conceptual design. Des. Stud. **22**(2), 169–191 (2001)
6. J. Darke, The primary generator and the design process. Des. Stud. **1**(1), 36–44 (1979)
7. N. Cross, N. Roozenburg, Modelling the design process in engineering and in architecture. J. Eng. Des. **3**(4), 325–337 (1992)
8. G. Pahl, W. Beitz, *Engineering Design: A Systematic Approach* (L. Blessing, F. Bauert, trans.) (Springer, London, 1996)

9. N.F.M. Roozenburg, N.G. Cross, Models of the design process: integrating across the disciplines. Des. Stud. **12**(4), 215–220 (1991)
10. N.P. Suh, *The Principles of Design* (Oxford University Press, New York, 1990)
11. N.P. Suh, *Axiomatic Design: Advances and Applications* (Oxford University Press, Oxford, 2001)
12. O. Kulak, S. Cebi, C. Kahraman, Applications of axiomatic design principles: a literature review. Expert Syst. Appl. **37**(9), 6705–6717 (2010)
13. A. Hatchuel, P. Le Masson, Y. Reich, B. Weil, A systematic approach of design theories using generativeness and robustness, in Proceedings of the Eighteenth International Conference on Engineering Design, Copenhagen, 15–19 Aug 2011
14. J.S. Ackerman, *Architecture* (Encyclopædia Britannica Online Academic Edition, Encyclopædia Britannica Inc., 2013)
15. G.K.C. Ding, Sustainable construction—The role of environmental assessment tools. J. Environ. Manage. **86**(3), 451–464 (2008)
16. B.L. Archer, Systematic method for designers, in *Developments in Design Methodology*, ed. by N. Cross (Wiley, Chichester, 1984)
17. G. Eliasson, K. Psilander, Entrepreneurship in customer satisfaction. The Case of Swedish, Danish and Californian Home Building, in Proceedings of the Eighth International Joseph A. Schumpeter Society Conference, Manchester, June 28–July 2 2000
18. U. Sohlenius, Can axiomatic design improve the building process? in Proceedings of the First International Conference on Axiomatic Design, Cambridge, 21–23 June 2000
19. M.G. Helander, L. Lin, Anthropometric design of workstations, in Proceedings of the First International Conference on Axiomatic Design, Cambridge, 21–23 June 2000
20. K. Psilander, Axiomatic design in the customizing home building industry. Eng. Constr. Arch. Manage. **4**(4), 318–324 (2002)
21. U. Sohlenius, C. Johansson, A framework for decision making in construction—Based on axiomatic design, in Proceedings of the Second International Conference on Axiomatic Design, Cambridge, 10–11 June 2002
22. D. Kowaltowski, S. Pina, R. Ruschel, V. Silva, L. Labaki, D. Moreira, Creating an axiomatic evaluation method for low income housing in the State of São Paulo, Brazil, in Proceedings of the International Conference "Methodologies in Housing Research", Royal Institute of Technology, Stockholm, 22–24 Sept 2003
23. A. Kankey, M. Ogot, Improving the acoustics in a historic building using axiomatic design and TRIZ. TRIZ J., 16 June 2005
24. M. Cavique, A. Gonçalves-Coelho, An energy efficiency framework for design of HVAC systems, in Proceedings of the Fifth International Conference on Axiomatic Design, Campus de Caparica, 25–27 March 2009
25. E. Ibragimova, M. Pena, M.K. Thompson, The evolution of Sihwa Dam: a formal design theory perspective, in Proceedings of the Eighth International Conference on Civil and Environmental Engineering, Busan, South Korea, 28–29 Oct 2009
26. J.S. Peck, T. Lee, E.M.W. Kolb, S.-G. Kim, Lessons learned by applying axiomatic design to an emergency department, in Proceeding of the Fifth International Conference on Axiomatic Design, Lisbon, 25–27 March 2009
27. M.K. Thompson, O.H. Kwon, M.J. Park, The application of axiomatic design theory and conflict techniques for the design of intersections: Part 1, in Proceeding of the Fifth International Conference on Axiomatic Design, Lisbon, 25–27 March 2009
28. M.K. Thompson, M. J. Park, O.H. Kwon, E. Ibragimova, H. Lee, S. Myung, The application of axiomatic design theory and conflict techniques for the design of intersections: Part 2, in Proceeding of the Fifth International Conference on Axiomatic Design, Lisbon, 25–27 March 2009
29. J.B.R. Pastor, E.M. Benavides, Axiomatic design of an airport passenger terminal, in Proceedings of the Sixth International Conference on Axiomatic Design, Daejeon, 30–31 March 2011

30. Y. Yi, M.K. Thompson, Quantifying the impact of coupling in axiomatic design: calculating the coupling impact index for traffic intersections, in Proceeding of the Sixth International Conference on Axiomatic Design, Daejeon, 30–31 March 2011
31. R.G. Lindsey III, M. Omar, A.M. Farid, An axiomatic design based approach for the conceptual design of temporary modular housing, in Proceeding of The Seventh International Conference on Axiomatic Design, Worcester, 27–28 June 2013
32. M. Marchesi, S.-G. Kim, D. Matt, Application of the axiomatic design approach to the design of architectural systems, in Proceedings of the Seventh International Conference on Axiomatic Design, Worcester MA, 27–28 June 2013
33. A.M. Farid, Multi-agent system design principles for resilient operation of future power systems, in Proceedings of the International Workshop on Intelligent Energy Systems, San Diego, CA, 5–8 Oct 2014
34. R.G. Lindsey III, M. Omar, A.M. Farid, An integrated QFD & axiomatic design methodology for the satisfaction of temporary housing stakeholders, in Proceedings of the Eighth International Conference on Axiomatic Design, Lisbon, 27–29 Sept 2014
35. W.N. Lubega, A.M. Farid, A reference system architecture for the energy-water Nexus. IEEE Syst. J. **01**, 1–11 (2014)
36. M. Marchesi, J. Fernandez, D. Matt, S.-G. Kim, Axiomatic design approach for the conceptual design of sustainable buildings, in Proceedings of the Eighth International Conference on Axiomatic Design, Universidade Nova de Lisboa
37. A.M. Viswanath, E.E.S. Baca, A.M. Farid, An axiomatic design approach to passenger enumeration in reconfigurable transportation systems. IEEE Trans. Intell. Transp. Syst. **15**(3), 915–924 (2014)
38. A.M. Viswanath, A. Farid, A hybrid dynamic system model for the assessment of transportation electrification, in Proceedings of the American Control Conference 2014, Portland, Oregon, 4–6 June 2014

Chapter 7
Applying Axiomatic Design to Prefabricated Building Design in the Housing Industry: A Case Study Analysis

Marianna Marchesi, John E. Fernández and Dominik T. Matt

Abstract Since housing market demands customized performance-effective buildings at affordable costs, prefabrication combined with mass customization is worthy for satisfying these requirements, but demands robustness and flexibility of design solutions with respect to the architect's viewpoint. Crucial decisions that affect these aspects are made during the conceptual design phase, but in this stage suitable tools are not widely available. Since Axiomatic Design approach (AD) has been able to support designers' decision-making process for the development of product concepts that would have the best chance to provide the specified requirements as well as for the analysis of ideas with respect to their capability to satisfy these requirements, AD is applied to the examination of contemporary well-appreciated prefabricated houses in order to identify crucial design decisions that have affected robustness and flexibility in their conceptual design. Subsequently, strategies adopted by their architects during the design generation are reviewed, and by comparison, similarities between AD and the architects' approaches are identified. These results prove that AD can be effectively applied to prefabricated building design in the housing industry with the goal of proving effective designs in terms of robustness and flexibility from the architect's viewpoint and therefore satisfying the current housing market demand.

M. Marchesi (✉) · D.T. Matt
Faculty of Science and Technology, Free University of Bozen-Bolzano,
Piazza Università 5, 39100 Bolzano, Italy
e-mail: marianna.marchesi@outlook.com

D.T. Matt
e-mail: dominik.matt@unibz.it

J.E. Fernández
Department of Architecture, Massachusetts Institute of Technology,
77 Mass. Ave, Cambridge, MA 02139, USA
e-mail: fernande@mit.edu

© Springer International Publishing Switzerland 2016
A.M. Farid and N.P. Suh (eds.), *Axiomatic Design in Large Systems*,
DOI 10.1007/978-3-319-32388-6_7

7.1 Introduction

In Italy, as in Europe as a whole, the construction sector has a significant impact on economy and environment. It plays an important role emerging from the current crisis and moving toward a more sustainable future growth path. In this way, the development of this sector is addressed toward performance-effective buildings. Moreover, design customization is crucial to satisfy a segmented and high varied demand. As the demand for customized performance-effective buildings continues to increase, prefabrication represents a suitable form of providing these requirements. Customers demand personalized products, and prefabrication is able to construct infinite possibilities of variations in design and production thanks to customization, but this high customization is currently achieved by tailor-made building systems. Each piece is designed and built ad hoc; therefore, costs remain high. On the other hand, building industry is able to offer prefabricated houses at affordable costs. In this case, standardization guarantees lower costs and the consequent maximization of the customer value, but high levels of standardization cause limited product variations and a consequent lack of customer appreciation. In order to satisfy the current demand for affordable customized solutions, the building industry should focus on achieving maximum variety within partially standardized solutions.

Nowadays, building design has to be optimized with consideration for a large number of different (often conflicting) requirements and constraints, and it has to be selected from different available alternatives. The early design phase provides the greatest opportunity for design team to influence the project outcomes. The concept design is the most important and challenging phase of the building design process. Usually in this phase, architects propose a rough design concept on the basis of a simplified problem statement defined using previous experiences and knowledge. Through reiterative conjecture–analysis cycles, problem and solution are iteratively refined in parallel, until they are explicitly defined. On the other hand, interdisciplinary design team lacks a common understanding of the manner in which the design process is being performed and of the direction of the project progressing. Despite the decisive role of the conceptual design phase, only few tools are available to support this stage. In order to support architect and design team's decision making for addressing early the solution to the accomplishment of the expected outcomes and for optimizing decisions with respect to a varied and complex set of requirements and conflicting parameters, conceptual design needs suitable tools. In engineering design, appropriate methodologies are available: the systematic approaches. They provide procedures, rules, and principles for the accomplishment of attended outcomes without confining designer's creativity. AD, one of them, is a formal design theory differentiated from other systematic approaches by an axiomatic scientific basis that underlies the decision-making process. It guides the decision-making process through principles of functional independence and complexity minimization. Since AD principles are axioms that can be generalized, this method can be effective and powerful in different design areas. It can be applied to all situations of design problem solving especially in the conceptual phase, from synthesis to analysis of the synthesized idea, and to the selection of good ideas from plausible solutions. In AD, like in architecture, the creation process is a sequenced process of going to-and-from problem and solution, and the formulation of problem and solution is developed in parallel with constant shuttling to-and-from problem (what) and solution (how). On the other hand, with AD, in contrast to architectural design, the process is systematic and hierarchical and two design axioms are provided to avoid unwanted couplings and complex designs. Therefore, AD may be helpful in architectural building design to minimize the trial-and-error activity found in its design process to achieve the expected outcomes and, in the case of interdisciplinary design team, to jointly manage the design process. Since it is proved that AD is able to support the development of product concepts with the best chance to provide robustness and flexibility from the designer's viewpoint as well as the analysis of designs with respect to the specified requirements, in this study AD is applied to examine the conceptual design of successful prefabricated houses in order to identify crucial design decisions that have affected the accomplishment of these two aspects. Then, the design approaches adopted by the architects during their development are reviewed through the architects' writings. Comparing the architects' approaches and AD, similarities are identified, and potential benefits of applying AD to prefabricated building design in the housing industry are deduced and discussed. These results show that AD can be effectively applied to prefabricated building design to support architect's decision-making and address the design development toward solutions better satisfying the current demand for customized performance-effective housing at affordable costs.

7.2 Literature Review

Building design is a process of creating built solutions that are able to provide technical, social, and aesthetic functions within existing constraints and available resources. The main characteristics valued in a built work consist of the space suitability for human usage, its adaptability to specific activities, the capability of the construction to protect and supply resources, and the aesthetic aspect [1]. Design is constrained by site features and finite budget, time, resource, and whole-life value. In addition, due to the increasing demand for sustainable buildings, various aspects regarding sustainability, especially energy efficiency, have to be introduced and evaluated during the design process. In conclusion, a design has to satisfy various demands in a utilitarian and aesthetic way within given socio-economic constraints and also with respect to available resources and existing context characteristics. Therefore, numerous, different, and sometimes conflicting requirements and constraints have to be early considered during the design process in order to develop suitable solutions.

7.2.1 Conceptual Building Design

In the building design process (shown in Table 7.1), conceptual design is the most challenging stage. In this phase, architects make design decisions that mostly affect the project outcomes with respect to functionality, performances, appearance, costs, time, and also sustainability, and the opportunity to influence them is highest. In contrast, decisions made in the later stages have a minimal influence on the project outcomes, and the opportunity to influence those decreases rapidly over time during the process. Moreover, poor decisions made initially cannot always be corrected in the later stages [2]. Therefore, the conceptual design phase has a decisive role on the understanding of the problem, as well as on the production of adequate design concepts.

Despite the inherent complexity of building design, in the initial phases a solution conjecture is usually generated very early by architects based on a selection of a small set of design objectives and a simplified problem statement depending on architects' subjective judgment, acquired knowledge and heuristics in order to confine the potential solutions to a manageable set. Then, it is delivered as a first

Table 7.1 Building design process [3]

Strategic definition	Identification of client's business case, strategic brief and other project requirements	
Preparation and brief	Preparation of an initial project brief on the basis of project objectives, project outcomes, sustainability aspirations, project budget, site information, and other constraints	
Conceptual design	Preparation of a concept design including proposals for structural design, building services systems, specifications, and preliminary cost plan along with project strategies in accordance with the design program	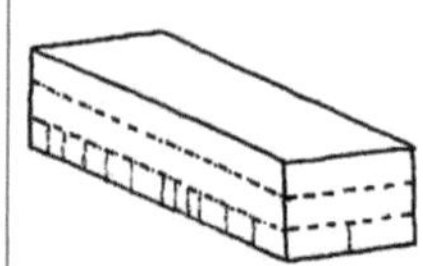
Developed design	Preparation of a developed design including coordinated and updated proposals for structural design, services systems, outline specifications, cost plan, and project strategies in accordance with the design program	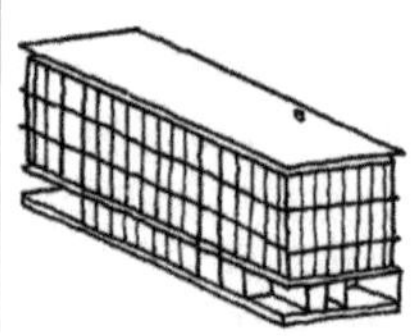
Technical design	Preparation of technical design information including all architectural, structural, and building services information and specifications in accordance with the design program	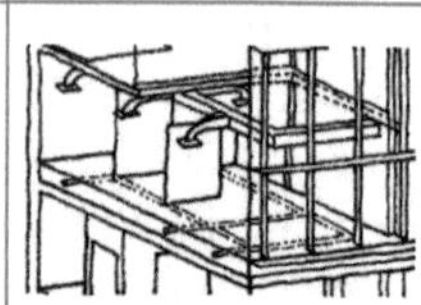

proposal to the client to increase the depth of information concerning the design [4]. As a consequence, an iterative process emerges and both problem and solution are reformulated multiple times and coevolved until all relevant issues of the design task are progressively recorded. Participants go back and forward between problem and solution through spiral and cyclic stages [5] while the opportunity to influence the design outcomes decreases rapidly over time. Meanwhile in interdisciplinary design teams, confusion often arises regarding the direction of progression due to the lack of common design approaches between architects and engineers. Team members expect that all requirements can be equally satisfied without considering the possible conflict among them [6]. Despite the decisive role of the initial design phases, there is little or no guidance in these phases on what should be done and how it should be achieved [6]. These phases are not well understood and only few tools are available to sustain them while most of the available tools for building design are concerned with detailed phases [7]. Traditional design tools, such as design-by-drawing, cannot always manage adequately the current complexity frequently imposed on design team [8]. In addition, in sustainable building design, available tools rely on detailed design information provided at the later design stages to assess the expected level of sustainability [9]. Architects and interdisciplinary design teams need adequate supports to jointly address early decision-making to the accomplishment of the expected design outcomes.

7.2.2 Building Prefabrication

Currently, housing market demands customized performance-effective buildings at affordable costs, and prefabrication represents a suitable way to achieve all these requirements. Building prefabrication consists of linear, planar, or spatial building elements that are premade in factory, and then assembled and installed permanently on the building site [10]. Before the development of information and digital technologies, building manufacturing processes were limited to mass production. Mass production consists in the creation of large amounts of identical parts in order to reduce costs significantly, but limiting individual choice. Thanks to the introduction of information and digital technologies and their advances, building industry is currently able to rapidly respond to individual customer's needs [11]. This strategy, called customization, allows providing a unique product built according to specific customer's requirements, but costs are typically high. In order to provide personalized artifacts at affordable costs, it is available another strategy based on the mix between mass production and customization in which prefabricated mass-produced artifact parts are combined to customized parts [12]. Mass customization is distinguished in four types: collaborative, adaptive, transparent, and cosmetic customization. In the collaborative customization, firms define with individual customers the precise product that satisfies their needs. In the adaptive customization, product is standardized, but it can be personalized according to the customer's preferences. In the transparent customization, individual customers are provided with unique products without the customer's awareness about it. Finally, cosmetic customization means marketing the same product to different target audiences in different ways [12].

This study asserts that mass customization, especially adaptive customization, is a suitable approach for the improvement of the building industry in order to satisfy the current housing demand. By this approach, building industry is able to personalize building parts that are decisive for clients, and limit costs by the mass production of the others. In this approach, buildings are realized in such a way that their components can be easily varied and disassembled and any modifications, such as additions, are easy to realize [13]. In order to develop customized prefabricated buildings at affordable costs, artifact robustness and flexibility with respect to the architect's viewpoint appear crucial design requirements. In general, artifact robustness from the designer's viewpoint is the ability of an artifact to produce the expected functionalities and performances despite being subjected to uncertainties and disturbances (e.g., changing customers or functions or physical components) [14]. Artifact flexibility from the designer's viewpoint expresses the ability of an artifact to be adapted in terms of functionality and performance features in order to yield similar design families with little effort, time, or penalty in response to market demand. Flexibility is meaningful if the functionality of the artifact varies in someway in terms of set of functional requirements implemented by the artifact or in terms of specific artifact performance features [15]. According to Barrow et al. [16], these abilities depend mainly on decisions made by architects at the conceptual design phase. Since literature shows engineering design methods and approaches effectively employed in the conceptual phase in product and manufacturing design as well as in the design of the built environment, and since product/manufacturing industry and building industry share several similarities, these methodologies may be successfully transferred and applied to prefabricated building design in the housing industry [17].

7.3 Axiomatic Design Theory

Crucial decisions that affect the project outcomes are made during the conceptual design phase. In this phase, architect and design team's decision-making process requires support. In engineering design, systematic approaches are suitable procedures to address effective designs while reduce workload, save time, and prevent human error, without restricting the designer's creativity [18]. AD, one of these approaches, is differentiated by including basic principles of design analysis, synthesis, and decision-making. AD is a design theory developed in engineering by Nam P. Suh and successfully applied to many different problems in various design fields including manufacturing design [19] and built environment design. AD proposes a rational framework, a systematic procedure, and principles to support early designer's decision-making from synthesis to analysis of ideas and to selection of the best idea among valid alternatives [20, 21]. In AD, designers are guided to develop the design in a specific structure through the mapping and decomposition process across four domains: the customer, functional, physical, and process domains. The process of creating this structure consists of defining what is required and how is reached through the design domains and then to develop the design in levels of detail from general to specific through a hierarchical decomposition [20, 21]. This process continues down between the domains decomposing the design in finer levels of detail until it is developed adequately [20, 21]. Initially, designers must decide what they want to achieve in terms of functions before considering how to achieve it in terms of physical components. Functions of an artifact, also called functional requirements (FRs), are what the artifact should perform to satisfy client's needs. They concern the exchange of signals, information, materials, forces, and energy. Designers define the expected functions in terms of FRs [20, 21]. According to Thompson [22], in addition, there are desirable qualities or attributes that the artifact should have (e.g., robustness, flexibility, energy efficiency, affordability, pleasantness). They imply the definition of constraints (Cs) on the product or on how the product must be designed, and affect the mapping process from FRs to physical components (so-called design parameters—DPs), but they rarely are directly translated to physical features. Therefore, they are not subject to the mapping between FRs and DPs [22]. Cs establish limits on qualities (e.g., cost, size) that design has to be observed or final artifact has to be included. FRs are then mapped into DPs that implement physically the defined functions. In AD, the definition of FRs and the subsequent assignment of DPs are both dependent on the independence axiom. The independence axiom or axiom one states that the independence of the FRs as well as the one-to-one mapping between FRs and DPs must be maintained to minimize coupling between FR/FR and FR/DP pairs and avoid conflicts [20, 21]. Such decoupling warrants that a variation of one DP or one FR will not destabilize the whole solution. In this way, it is fostered the artifact robustness from the designer's viewpoint [14]. Couplings are identified by the check of the design matrix (DM), so they can be reduced or eliminated. The second axiom fosters the artifact robustness from the user's (consumer or manufacturer) point of view [14]. It states that a decoupled design should also follow the principle of minimum information for the user or manufacturer [20, 21]. This means that the user should not have to adjust any DP in order to benefit from the functions of the system. Axiom two will not be applied in this study. When the mapping and decomposition process is completed, all the DP components need to be physically integrated into one entity, and interacting components are connected by interfaces. In AD, every DP should be combined without introducing unwanted couplings between FRs and DPs and between DPs. DP–DP couplings can be checked by a physical integration matrix in which the DPs are related to the DPs. In this matrix, the diagonal can be ignored since each physical component is always coupled with itself. Some off-diagonal couplings are required and others are undesirable [23]. On the other hand, couplings between FRs and DPs should have been addressed previously in the mapping and decomposition process. Therefore, in this phase the degree of undesired existing FR–DP couplings should not be increased as well as new FR–DP couplings should not be introduced. When each DP implements one

FR and the interfaces between connecting DPs are decoupled, the artifact architecture is defined modular [15]. In this manner, the artifact results in being robust and flexible from the designer's viewpoint.

7.4 Applying Axiomatic Design to Conceptual Building Design

According to Roozenburg and Cross [5], a design approach is appropriate for practitioners and interdisciplinary design teams when it provides a sequence of stages and varying cognitive procedures to perform the design process. It should avoid emphasis on problem analysis and specification and at the same time emphasis on early solution conjectures [5]. Moreover, it should be able to reproduce the iterative activity between problem and solution and sub-problems and sub-solutions typically found in practice [5]. AD shows being able to support the co-evolution of problem and solution providing a sequenced creation process of going to-and-from problem and solution, and sub-problems and sub-solutions and also decision-making principles. In AD, like in architect's design approach, formulation of problem and generation of solution are developed together with constant shuttling to-and-from problem and solution. On the other hand, in AD, unlike in the architect's design approach, problem and solution are systematically and consistently specified in parallel, moving down along the hierarchy, and the design decisions are made in an explicit form maintaining data of the decision-making process. In addition, general principles of decision-making support the creation and analytic processes in order to define effective designs with respect to specified requirements, to evaluate the synthesized ideas and to select the most feasible solution among valuable alternatives. Benefits of applying AD to building design are illustrated below by the comparison between the design approach adopted by architects for the concept generation of the Lord's Cricket School roof (Fig. 7.1) [24] and the AD approach applied to the design development of the roof concept (Fig. 7.2).

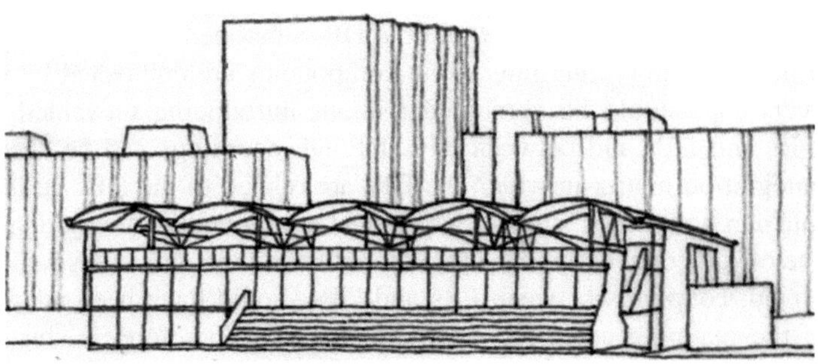

Fig. 7.1 Lord's cricket school roof [24]

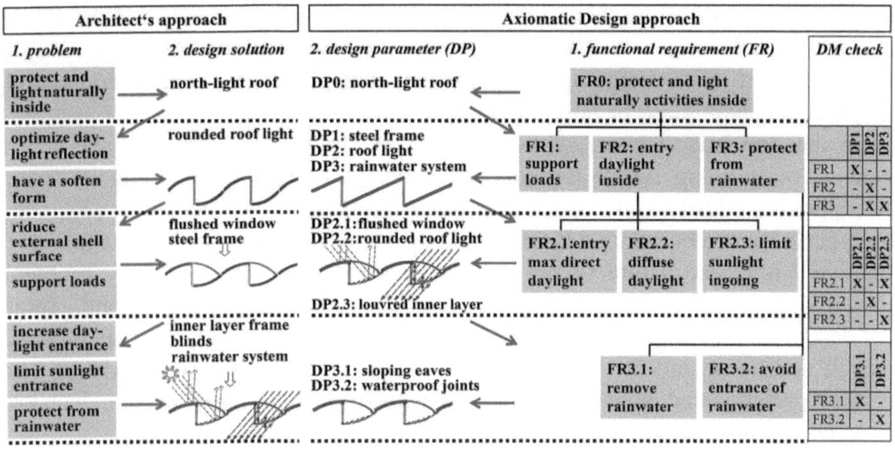

Fig. 7.2 Comparison between architects' approach and AD on the roof case study [25]

The crucial goal of this project was a high and steady degree of natural lighting through the roof inside the building. In order to guarantee this requirement, a saw-tooth north-light roof was proposed. Then, designers proposed to soften the roof-light shape for improving the internal reflections of daylight. Afterward, the design solution was redefined moving the light glass line in order to reduce the external shell surface. For favouring the improvement of daylight entrance, the inner layer was redefined as a frame. Since in this last solution the sunlight enters directly, an internal solar blinds layer was suggested that obstructs the sunlight, but allows the daylight entrance. Using AD, a minimum set of independent FRs, that fully represents the problem of designing a building roof, is defined: supporting loads, entering daylight inside, and protecting from rainwater. Then, a solution is conceptualized by mapping between domains, from the functional domain to the physical domain, and it is expressed in terms of DPs that satisfy the before-mentioned FRs. Then, by checking the DM, couplings between FRs and DPs are searched to identify undesirable iterations in the proposed design. Later, returning to the functional domain, a lower level is generated, and the process is pursued until the design is completed. Applying AD to the concept development of the building roof, it has been observed its ability to complete multiple design objectives homogeneously during the design process. By this approach, some requirements are engaged very early such as the protection from rainwater, and tardive decisions and consequent trial-and-error activity are avoided. Moreover, by the DM check, unwanted interferences have been shown: the adopted solution for daylight entry through the roof affects the function of protecting from rainwater, and the internal blinds influence the daylight entry. In particular, the last conflict observed by the DM stresses a contradiction. Initially, architects have proposed a north-light roof in order to provide steady degree of natural lighting inside the building, but then they have adjusted the design for increasing the daylight entry by flushed windows. This change causes the entrance of sunlight and therefore, contradicts the initial problem statement and the related concept initially proposed. On the basis of these observations, AD may result in being an appropriate design approach for performing decision making consistently in the conceptual phase of building design.

In the previous chapter, applications of AD to building design have shown benefits of applying AD to the conceptual design of airport regarding its spatial configuration [26] and also temporary prefabricated housing [27] and high-performance buildings [25]. On the other hand, AD has been also applied for the analysis of the nature of pre-existing systems in order to identify problems linked to their conceptual design. Since if a design theory can identify and explain the design problems with an existing design, it can also be used to identify effective design decisions, and this study intends to perform an analysis of existing successful prefabricated houses in order to identify crucial early design decisions associated with their conceptual design.

7.5 Axiomatic Analysis of Prefabricated Houses

Since AD has proven being able to guide the analysis of designs with respect to their capability to provide robustness and flexibility from the designer's viewpoint and due to the building industry's need to develop robust and flexible designs in order to satisfy the current demand for customized performance-effective houses at affordable costs, this study analyses by AD the design concept of award-winning prefabricated houses. Case studies are selected on the basis of their connection to prefabricated construction processes, the availability of the adopted architects' strategies by publications, and their recognized technical innovation and capabilities in terms of robustness and flexibility in order to guarantee attended performances and to satisfy different clients' preferences or the architects' creative freedom desire. The identified case studies are made by different construction technology materials. The aim of this analysis is to identify by AD crucial design decisions that have influenced the accomplishment of robustness and flexibility in their conceptual design. The identified strategies are then validated through the comparison with the architects' choices documented by their writings.

7.5.1 Case Study #1: Dwell Home

The first case study is an L-type two-story house, the Dwell Home (Fig. 7.3).

The Dwell Home was designed by Resolution: 4 Architecture (Joseph Tanney and Robert Luntz), in 2003, and it was awarded by the Dwell Home Design Invitational for modern prefabricated housing [28]. Moreover, architects won several awards for their design approach called "Modern Modular". The Dwell Home was the first prefabricated home prototype based on this approach. This approach focuses

Fig. 7.3 Dwell Home by Resolution: 4 Architecture [28]

on combinable user activities space modules allowing the generation of a wide range of dwelling space solutions capable of being adapted to meet variable clients' needs. This approach defines a series of space modules distinguished in communal, private, and accessory modules by the identification of sets of compatible living user's activities (Table 7.2) [28].

Table 7.2 Spatial modules [28]

Communal modules (first row), private modules (second row) and accessory modules (third row)

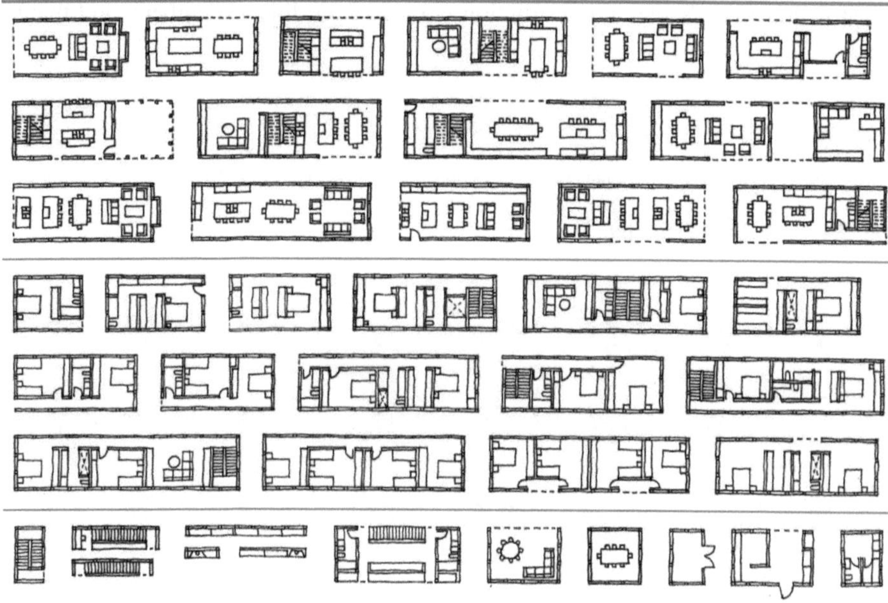

These spatial units are freely combined providing customized house configurations in order to satisfy different needs and to adapt design to diverse locations and climates, but nonetheless at moderate costs by the mass production of some modules. Homes are easily expandable and transformable allowing them to grow and be adapted to different residents' needs during the building lifetime [28]. Seven main spatial typologies are proposed by architects (Fig. 7.4). Starting from these, numerous variants are available (Fig. 7.5) adapting types to different clients' needs, site features, and available budget.

The selected case study is analyzed using AD. Design intents are expressed in terms of functional requirement (FR) and corresponding design parameter (DP):

FR0 provide a comfortable, energy efficient, robust, flexible, affordable, and pleasant isolated house

DP0 well-oriented, energy performance, renewable energy gathering, modular architecture, mass-customized and off-site prefabricated, pure geometry villa

Critics assert that generally architects identify architectural form as the constantly evolving interplay of three converging vectors, "topos," "typos," and "tectonic" [29]. "Topos" is related to context, site, and orientation; "typos" is related to activities, spaces, and their relationship, and "tectonic" is related to construction for creating spaces. Site provides design inputs and constraints (Cs) to the evolution of the architectural form. Therefore, the definition of design solution is constrained by context and site features and also by urban and building regulations. Construction is generally differentiated as skeleton construction, massive construction, and hybrid construction [30]. Skeleton construction is made from linear members, and thanks

Fig. 7.4 Spatial typologies [28]

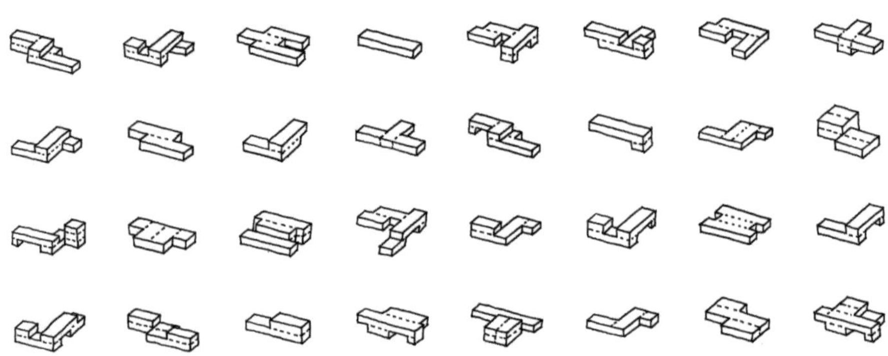

Fig. 7.5 Spatial configuration variety [28]

to this nature it is able to provide support without conditioning the creation of interior spaces and without separating interior from exterior. On the contrary, massive construction is made from walls. Since walls perform both loadbearing and enclosing functions, they create the interior space directly, and interior and exterior are distinctly separated. Hybrid construction is a combination between skeleton construction and massive construction [30]. According to the defined basic factors influencing architectural form, a minimum set of independent FRs that the artifact should perform is defined as follows:

FR1 accommodate client's living activities

FR2 support client's living

The proposed solution should satisfy the defined FRs and observe existing Cs. Design inputs and Cs are provided from site, context, and urban-building regulations. Urban and building regulations establish limits regarding border distances, maximum built volume, maximum cover surface, and others that have to be observed. Also the disposable client's budget usually fixes a hard bound on the acceptable solution. Cs bound the architectural form evolution. Moreover, desirable qualities or attributes, which describe how the artifact should be, are specified: it should be comfortable, energy efficient, robust, flexible, affordable, and pleasant. They affect the definition of Cs and the mapping of the FRs to DPs at this level and at the lower levels. The proposed solution consists in a spatial volume and a construction type plus related material (Fig. 7.6).

The adopted construction system is the platform frame made from timber frames of squared section linear members with an inner sheathing that carries loads and provides rigidity and an outer sheathing that closes the frame in which the thermal insulation is embedded. It is a hybrid construction system. In fact, similarly to massive constructions, this system separates interior from exterior because loading and separating functions are united in the same plane. This means that this construction system interferes with the defined space. At the same time, similarly to skeleton constructions, each individual layer performs essentially just one function thanks to its

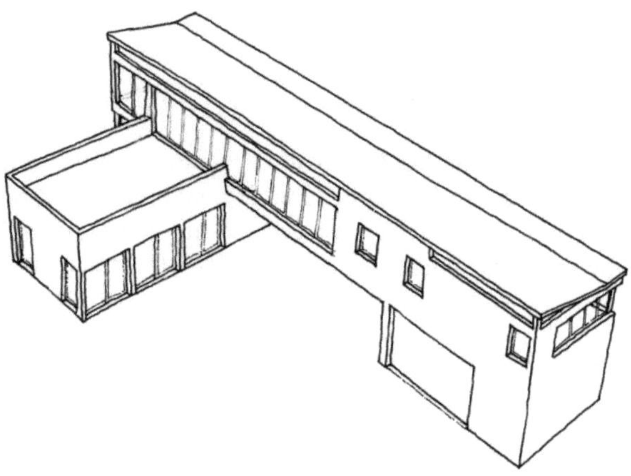

Fig. 7.6 Design concept [28]

linear members providing design freedom concerning interior plan layout and openings positioning. In order to provide a comfortable and energy-efficient building, the construction is well-oriented, well-insulated and energy gathering. The proposed volume is optimized with respect to the sun orientation in order to maximize the passive and active uses of the solar energy. The volume is defined observing local urban and building plan regulations. Robustness and flexibility are achieved by a modular architecture while affordability is obtained by the mass customization and off-site prefabrication of modules. Regarding esthetic aspects, architects propose pure geometries, simple forms, and natural-exposed materials.

This design concept is expressed in terms of DPs satisfying the FRs defined above:

DP1 pure geometry, modular architecture, two story, L-type volume composed of two interlocking rectangular volumes

DP2 well-oriented, well-insulated, energy gathering, modular architecture, mass-customized and off-site prefabricated construction made by the timber platform frame

Then, the check of the DM is performed (Table 7.3) in order to identify unwanted couplings. Strong coupling is indicated by a large X and weak coupling by a small x.

There is an unwanted weak coupling between the defined space and the construction system due to the hybrid construction. It distinctly separates interior from exterior space. However acceptable design freedom concerning plan layout and openings positioning is guaranteed thanks to its linear members. The resulting DM is triangular. The FRs are decoupled which implies that they can be satisfied only when FR1 is determined before FR2. Therefore, if the planned space is adjusted first, the function of providing protection and comfort could be met without affecting the planned space.

Then, this study analyses the proposed spatial solution using AD. The FR1 is decomposed into a lower level in order to define more specific spatial requirements:

FR1.1 accommodate communal activities

FR1.2 accommodate private activities

FR1.3 accommodate service activities

FR1.4 connect activities

FR1.5 place car

The solution has to provide adequate spaces; each space has to be sized in order to allow the fulfillment of the defined activities. Moreover, since the prefabricated modules are entirely produced off-site and delivered to the building site by truck, the modules should respect size limits for transport. According to the user requests

Table 7.3 First level DM

	DP1	DP2
FR1	X	x
FR2	–	X

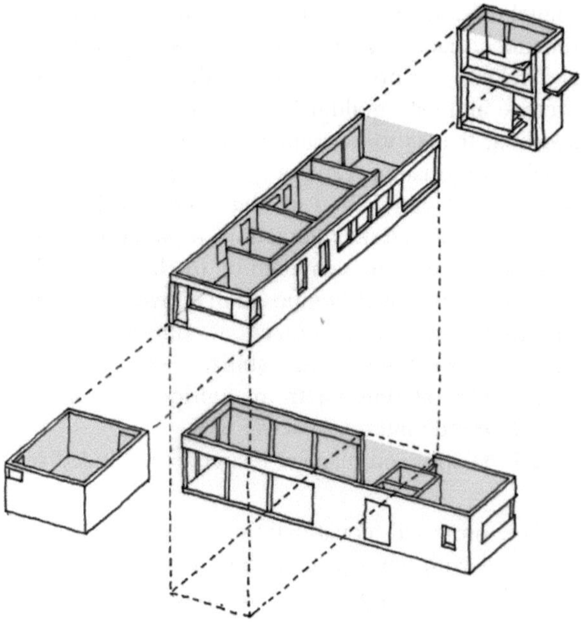

Fig. 7.7 Dwell Home's spatial modules configuration [28]

and defined Cs, the solution consists in four spatial modules arranged in two interlocking rectangular volumes and an empty space (Fig. 7.7). The ground floor is a rectangular module in which car parking and communal activities (living, dining, and cooking areas) are placed. The communal module can be entirely open to outside or enclosed by a curtain. Private activities (sleeping, cleaning, and studying areas) are located in the perpendicular rectangular module at the second floor [28].

The DPs that satisfy the FRs listed above are defined by mapping from FRs to DPs:

DP1.1 communal spatial module (living-dining room, kitchen, and services) at the ground floor
DP1.2 private spatial module (bedrooms, bathrooms and study room) at the first floor
DP1.3 accessory spatial module (service room) at the ground floor
DP1.4 accessory spatial module (staircase)
DP1.5 carport area at the ground floor

The DM is shown in Table 7.4.

The resulting DM is diagonal. The axiom one is observed, and unwanted couplings are avoided. The one-to-one mapping between FRs and DPs allows that each

Table 7.4 Second level DM—FR1

	DP1.1	DP1.2	DP1.3	DP1.4	DP1.5
FR1.1	X	–	–	–	–
FR1.2	–	X	–	–	–
FR1.3	–	–	X	–	–
FR1.4	–	–	–	X	–
FR1.5	–	–	–	–	X

spatial module provides the expected functionality without interferences with other functions. Each spatial module can be adjusted independently in terms of function and related performance during the design phase and over during the building lifetime allowing building variations in order to satisfy changed circumstances. In addition, thanks to decoupled module interfaces, spatial modules are freely combinable originating various spatial configurations according to different clients' needs, site features, and available budget. In this solution, since each DP implements one FR and the interactions between DPs are decoupled, spatial modules can be produced off-site, delivered by truck, and finally assembled on the building site.

Also the FR2 is decomposed in a lower level into a minimum set of independent FRs to provide safety, protection, comfort, privacy, and resources supply.

FR2.1 support loads and stabilize
FR2.2 separate interior from exterior
FR2.3 divide interior spaces
FR2.4 supply and manage resources (included renewable energy resources)

The proposed construction solution consists of an energy-efficient system equipped with renewable energy gatherings and a high energy-performance structural shell that supports and separates inside from outside without interferes with interior partitions (Fig. 7.8) [28].

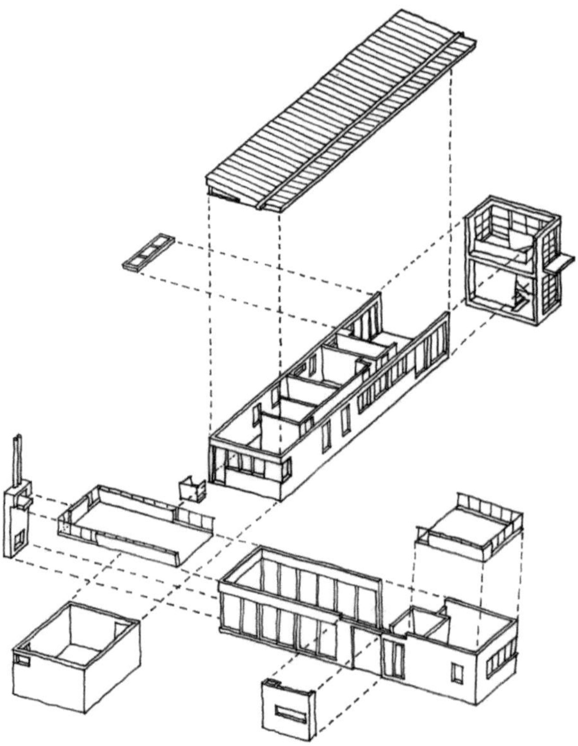

Fig. 7.8 Dwell Home's construction system [28]

The proposed construction solution is expressed in terms of DPs:

DP2.1 off-site prefabricated squared section frames with inner sheathing

DP2.2 off-site prefabricated outside sheathing that closes the frame in which a high level of thermal insulation is embedded and energy-efficient windows with low levels of air infiltration

DP2.3 interior layout (partitions and intermediate floorings)

DP2.4 active solar energy system, geothermal heating and cooling system and services system

The design is evaluated by the DM check (Table 7.5).

It is observed that the solution is decoupled. Unwanted couplings are noticed. Due to the platform frame system, the loadbearing and separating functions are united in the same plane within the wall determining an unwanted coupling. The thickness of insulation in the shell is influenced by the size of the squared sections frame. Moreover, the function of providing and managing resources is affected by the shell configuration due to photovoltaic panels integrated into the roof and by the interior partition layout. These couplings affect performances and flexibility of the construction system.

In AD, the defined spatial and construction systems can be further analyzed in detail at the lower levels by the decomposition of the specified FRs, the identification of corresponding detailed solutions, and the check of the functional independence. Finally, all the defined DPs should be analyzed with respect to their reciprocal combination into one entity. In AD, the physical integration process should be performed avoiding the introduction of unwanted DP–DP couplings or the exacerbation of existing FR–DP couplings that compromise artifact functionality and controllability. Moreover, the physical integration needs to consider how the building artifact will be assembled in order to avoid additional operations that reduce the likelihood of success. Dwell home is composed of combinable and independent spatial modules and planar building elements easy to be transported and built on-site. They are designed making the assembly of each individual component independent from other components; they are fabricated off-site and delivered on the site. Each spatial module contains external walls, flooring, and partitions. This artifact architecture allows that approximately 80 % of spatial modules, completed with external walls, flooring, and partitions, is built in factory and then provided on the building site. Only terraces and roof are managed separately [28]. By this architecture, it is also possible to minimize building costs through the mass production of those modules that clients do not need to personalize such as accessory

Table 7.5 Second level DM—FR2

	DP2.1	DP2.2	DP2.3	DP2.4
FR2.1	X	–	–	–
FR2.2	x	X	–	–
FR2.3	–	–	X	–
FR2.4	–	x	x	X

modules. High benefits in terms of cost and construction time as well as building quality are guaranteed. In addition, the environmental impact on the local ecosystem, the waste typically deposited on or near site, and the transport of crews and materials of environmental factors result in being minimized [28].

On the basis of the outcomes obtained by the case study analysis, it is observed that through the comparison between the architects' approach and AD, the architects' approach, similarly to AD, identifies at the higher level a minimum set of independent compatible user activities, and it fosters the functional independence between spatial modules. This approach shows to support the fulfillment of expected spatial functionalities and performance level in spite of uncertainties (changing customers or functions or physical components) and without compromises. Moreover, thanks to functional-independent spatial modules and independent interfaces between connecting modules, design results in being easily adaptable to different client's needs and site conditions during the design process and the building lifetime. In addition, this artifact architecture fosters the off-site production and on-site assembly of the spatial modules. Further benefits in terms of design robustness and flexibility from the architect's viewpoint may be achieved by the resolution of the identified unwanted couplings in the construction system in order to fulfill a wholly functional independence between the identified construction lower systems.

7.5.2 Case Study #2: Cellophane House

The second case study is a five-story single-family house, the Cellophane House (Fig. 7.9).

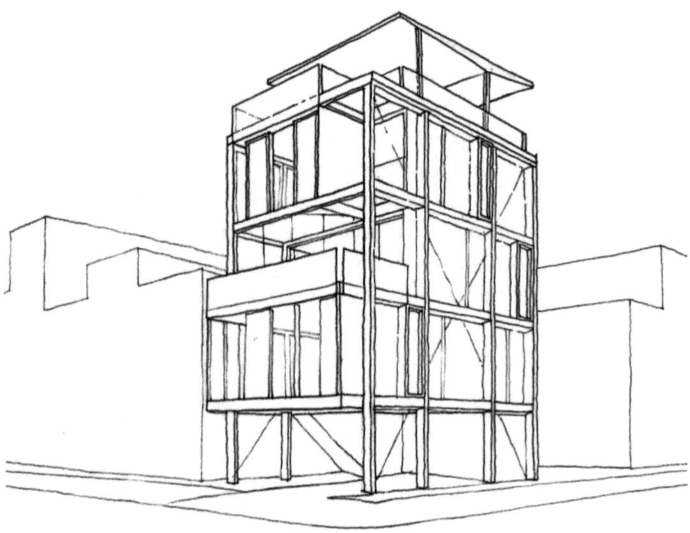

Fig. 7.9 Cellophane House by Kieran and Timberlake [31]

The Cellophane House was designed by Stephen Kieran and James Timberlake in 2008. According to the architects' approach, reported in their writings, this prototype is conceived as an energy performance and energy-gathering building focused on speed of on-site assembly, design for disassembly, and a holistic approach to the life cycles of building materials [31]. The solution consists in an "integrated components assembly". The structural frame is a standard aluminum frame to which other off-the-shelf building components (exterior shell, interior layout, and service systems) are connected. It enables the mass customization of the construction to conform it to its context with massing, material selection, and placement of the house. Through simple modifications, the house can be adapted to different climatic factors, solar orientations, slopes, and adjacencies. The customizable nature of the assembly shows that the adopted solution can be adapted to different clients' budgets and needs. Since all structural loads are carried by the frame, it is also simple to rearrange interior floor plans. Material options, easily substituted due to the ease of connection to the aluminum frame, allow the house to accommodate different clients' needs, tastes, and budgets [31]. In addition, building components are designed in order to be fabricated off-site, rapidly assembled (in six weeks), and then easily disassembled for material reuse and recycling. This approach allows that envelope materials may be disassembled instead of demolished, and they can be recycled instead of wasted in order to preserve the embodied energy in the recyclable house materials. Thanks to the effectiveness of this approach, in 2007 the Museum of Modern Art (MoMA) selected this project for an exhibition on contemporary prefabricated houses and successful design approaches [31].

The second case study is analyzed by AD in terms of FRs and DPs starting from the upper levels of the process (Fig. 7.10).

Desirable qualities that the artifact should perform are specified. It should be comfortable and energy efficient, robust, flexible, affordable, and recyclable. These attributes affect the definition of Cs and the mapping from FRs to DPs, but they are not subject to the one-to-one mapping between FRs and DPs. In order to provide a comfortable and energy-efficient building, the construction solution is highly

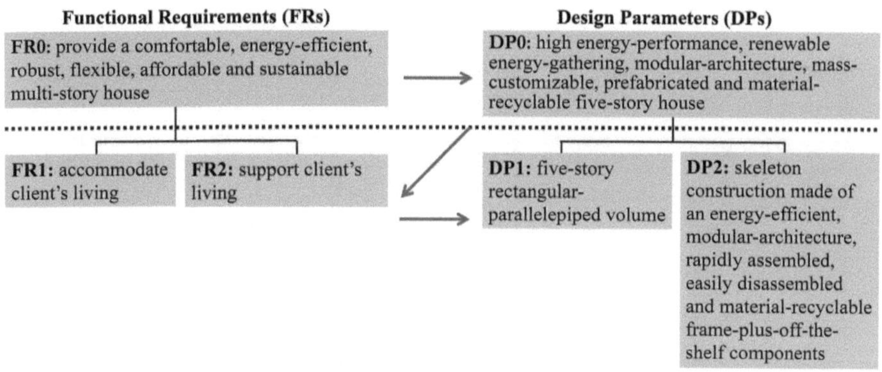

Fig. 7.10 First levels FRs and DPs

energy performing and renewable energy gathering. Robustness and flexibility are achieved by a modular architecture and affordability by mass customization, off-site prefabrication, and speed of on-site assembly. Regarding sustainable aspects, architects propose an easily dissembled construction, possibly made of recyclable materials.

The check of unwanted couplings is shown by the DM in Table 7.6.

The resulting DM is diagonal, and the axiom one is satisfied. Thanks to the adopted skeleton construction, this design is uncoupled. The separation between exterior and interior and between interior spaces is not directly created by the construction system because in this construction type the structural frame performs the loadbearing function independently with respect to the enclosing elements (facade and interior layout). This design shows that, at the high level of the decomposition process, the reduction of couplings between space and construction allows high chance to satisfy the expected functionalities in terms of space and construction and guarantees high flexibility on the design of both. The design is easily adjustable, and in the case of change, it is easily controllable.

The FR1 is decomposed into a lower level in order to define more specific spatial requirements. The spatial solution is a five-story single-family house with entrance and carport at the ground floor, daytime living activity areas at the first floor, night-time living activity areas at the two last floors, and a roof garden on the roof (Fig. 7.11). It is expressed in terms of DPs by the mapping shown in Fig. 7.12.

The design is evaluated by the DM (Table 7.4).

The resulting DM is diagonal. Thanks to the one-to-one mapping between FRs and DPs, each spatial module provides the expected functions without interferences with other functions. Each of them is easily adjustable in terms of function during the design phase and during the building lifetime in order to satisfy changed

Table 7.6 First level DM

	DP1	DP2
FR1	X	–
FR2	–	X

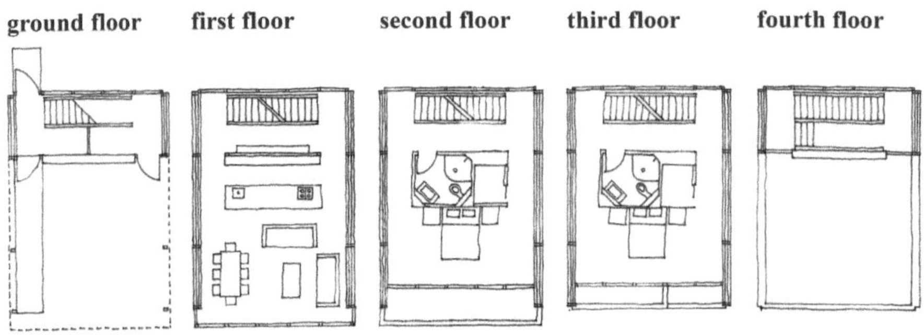

Fig. 7.11 Spatial solution [31]

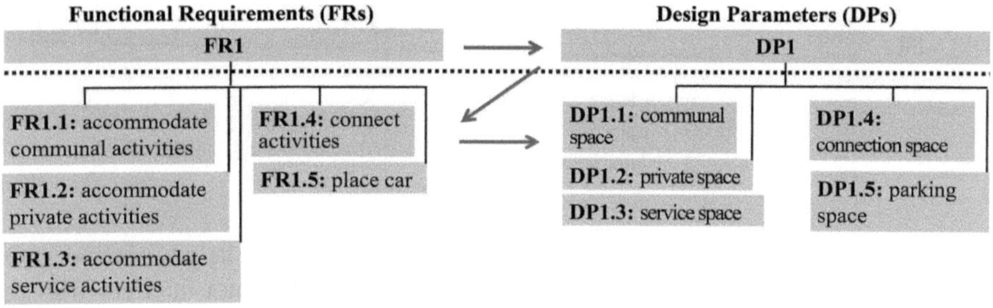

Fig. 7.12 Second levels FRs and DPs—FR1

circumstances. Moreover, identifying independent spatial modules supports the adopted assembly process based on the combination of spatial modules and planar elements.

Then, the FR2 is decomposed into a minimum set of independent FRs and the solution (Fig. 7.13) is then expressed in terms of DPs by mapping (Fig. 7.14).

It is then evaluated by the DM check in Table 7.7.

The DM shows that FRs are satisfied independently with the exception of the shell. In fact, the function of separating interior from exterior is affected by the integrated services system applied on the double glass-wall system through a thin-film PET

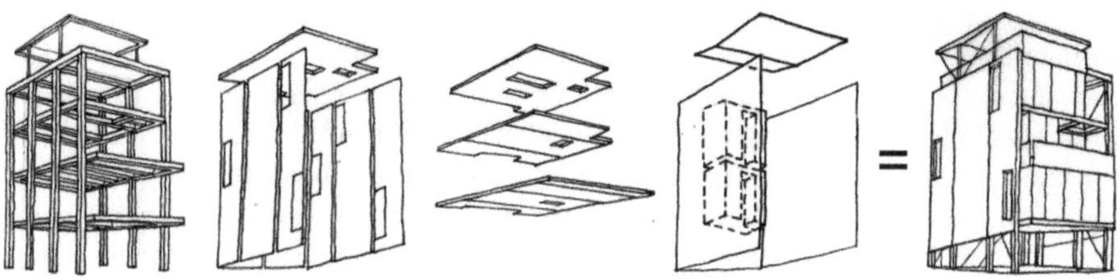

Fig. 7.13 Construction lower systems and their combination in a main construction system [31]

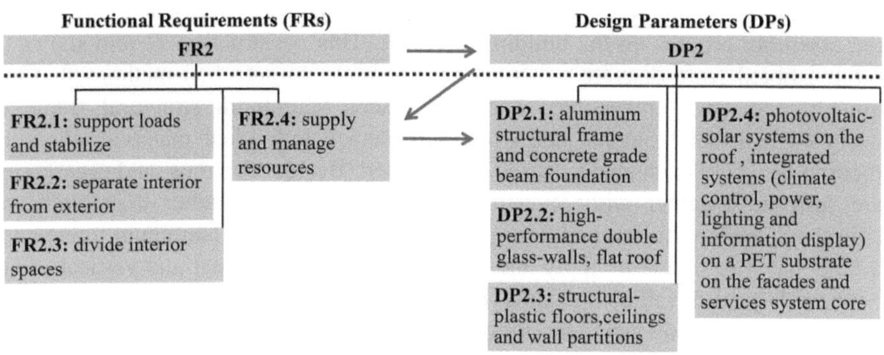

Fig. 7.14 Second levels FRs and DPs—FR2

Table 7.7 Second level DM—FR2

	DP2.1	DP2.2	DP2.3	DP2.4
FR2.1	X	–	–	–
FR2.2	–	X	–	X
FR2.3	–	–	X	–
FR2.4	–	–	–	X

membrane. This coupling influences the shell configuration and its performances. On the other hand, it is observed that the loadbearing frame does not interfere with the other sub-systems. This condition allows adapting and sizing it (foundation and elevation structure) according to the ground features, climate (wind, snow), and seismic risk typical of a site without limitations due to unwanted interferences. The functional independence of the shell with respect to the loadbearing frame determines benefits on the shell performances. In fact, in the case of high energy-efficient shell, thermal bridges can be easily avoided. Moreover, the functional independence between facades and interior layout allows them to be freely configured without reciprocal influences according to clients' needs and preferences. In addition, the identification of functional-independent construction lower systems and independent interfaces between connecting components supports the adopted assembly system.

By AD, the construction system as well as the spatial system may be specified in detail in the lower levels by the decomposition of the identified FRs, the definition of corresponding detailed solutions, and the check of the functional independence. Then, all DPs are physically combined with the others and integrated into one entity. This process should be performed without compromising function, controllability, or introducing unwanted consequences. This means that unwanted couplings between FRs and DPs and between DPs should be avoided. In prefabricated building design, the physical integration needs to consider how the building artifact will be assembled. Cellophane house has a modular architecture based on integrated assemblies (spatial and planar elements) easy to be transported and built on-site. These integrated assemblies, called chunks, are designed making the assembly of each individual component independent from other components and are fabricated off-site and delivered on the site. Each chunk contains the structural frame, floor, and external walls. Moreover, chunks are broken down further into sub-assemblies. The chunking strategy increases speed and precision of the assembly process on the building site [31]. This construction system shows to be assembled from discrete materials held in place by attached methods that are quickly and easily reversible. The off-the-shelf aluminum structural frame is designed in order to attach and eventually detach individual materials allowing them to be easily exchanged, reused, or recycled. Beams, columns, and accessories are fastened together with reversible connections (bolts) rather than welds or adhesives. Wall partitions and floor panels are attached to the aluminum frame with a tape that it is simple to apply and easy to remove. Horizontal and vertical joints between floor and wall components (Fig. 7.15) are bracketed by easily de-bracketed steel connectors allowing every component to remain reusable or easily dismantled at the end of the building or component lifetime [31].

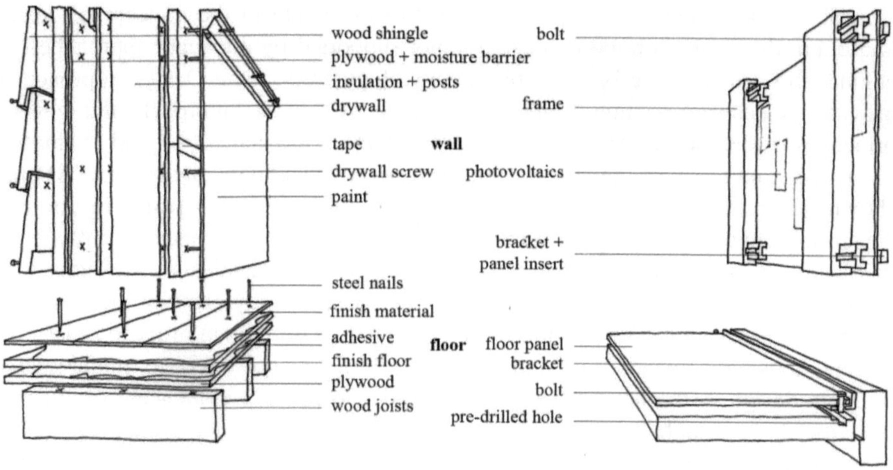

Fig. 7.15 Permanently fixed construction (*left*) versus temporarily held assembly (*right*) [31]

By the AD analysis, it is observed that the accomplishment of the functional independence allows defining better designs, able to satisfy intended requirements without limitations. Since each DP implements one FR, each FR can be changed according to specific clients' needs or changed circumstances. The design is adaptable to different demands and transformable over time. Moreover, decoupled physical interfaces between connecting DPs (structure, shell, interior layout, and services system) allow managing their different longevity during the building lifetime and applying sustainable strategies. During the building lifetime, construction lower systems can be easily disassembled, replaced on the basis of their longevity, and reassembled subsequently without disrupting the whole. Disassembled elements can be dismantled or reused on the basis of material according to sustainable strategies. In addition, by this approach, modules that do not need to be customized can be mass-produced in order to minimize building costs. In conclusion, by comparing the architects' approach and AD, it is observed that similarly to AD, the architects' approach is based on the identification of independent set of FRs, and it fosters functional independent physical components and decoupled interfaces between connecting components in the design solution in order to achieve robustness and flexibility.

7.6 Conclusions

Nowadays, prefabrication and mass customization represent suitable ways for construction trends to satisfy the current housing demands for customized performance-effective buildings at affordable costs. These strategies require robustness and flexibility of design solutions with respect to the architect's viewpoint.

Crucial decisions that affect these requirements are made in the conceptual design phase, but this stage in building design is not supported by adequate approaches. In engineering design, there is a suitable methodology: Axiomatic Design approach. It provides a decision-making framework with a systematic approach and general principles of decision-making to address designers on the generation and evaluation of the idea and to select the best design from several candidate alternatives. AD has shown being adequate for architects and interdisciplinary design teams in building design to support them in performing decision-making in the conceptual phase. In this study, it has emphasized the AD ability to guide the analysis of designs with respect to robustness and flexibility from the designer's viewpoint. Due to the building industry's need to develop robust and flexible designs in order to satisfy the current housing demand, this study has proposed to apply AD to prefabricated building design in the housing industry. In order to provide evidences of potential benefits, this research has performed an analysis of two contemporary successful prefabricated houses. By AD crucial early design decisions that have affected the specified requirements have been identified. In parallel, the design strategies adopted by their architects during the design phase have been reviewed through the architects' writings. Subsequently, each Axiomatic analysis has been compared with the related architects' design strategy. This comparison has shown that the design approaches developed by these architects' teams find a correspondence to AD. Functional independent physical components as well as decoupled interfaces between connecting components have been purposely pursued in the approaches adopted by Tanney J. and Luntz R. and by Kieran S. and Timberlake J. Therefore, by using this comparison, potential benefits of applying AD to the prefabricated building design in the housing industry have been proven. In particular, it has been asserted that the functional independence between space and construction observed at the upper level of the design process in the Cellophane House allows adapting independently space plan and construction configuration to context and client's preferences. At the lower level in the spatial system, it has been noticed that the functional independence between sets of compatible users' living activities emphasized in the Dwell Home guarantees the existence of spatial solutions with diverse floor plans and sizes. Any variations on a spatial module do not influence the other spatial modules. Modules can be easily varied satisfying different clients' needs, site features, and the available budget, and they can be changed during the building lifetime satisfying altered circumstances. In the construction system, the functional independence between construction lower systems emphasized in the Cellophane House allows each of them performing its functionality autonomously without interferences with the others. Each construction lower system can be adjusted according to context features, users' needs, and local regulations. In addition, when each DP implements one FR and the interfaces between connecting DPs are decoupled, the assembly system is easily defined in the form of linear, planar, and spatial building elements and their combination. Each layer can be easily replaced in the course of time according to varied customer's needs or exhausted lifetime. This artifact scheme also supports the application of sustainable strategies at the end of building or component lifetime by dismantling or recycling the building components sorted according to material.

Some couplings between services system and shell or between services system and interior layout have been noticed in both the case studies. Most of the time, these links are hardly avoidable, but architects and design teams can reduce the intensity of couplings with consequent benefits on the design effectiveness in terms of robustness and flexibility. By AD, designers are in a position to be aware of couplings and to do the best to avoid or reduce unwanted interferences.

In addition, this application has shown the attempt to include also attributes regarding aesthetic, social, and economic aspects in the design process. The first case study has shown to address artifact attributes such as energy efficiency, robustness, flexibility, affordability, and aesthetic features. The second case study has also included social aspects such as material reuse and dismantling. This has implied the distinction between FRs and other types of requirement as proposed by Thompson [22]. Further benefits may be achieved by the specification of the relationship between these different types of design information in order to support their inclusion and management during the design process by AD. Future studies for this purpose are recommended.

In conclusion, AD can be effectively applied to prefabricated building design in the housing industry in order to address decision making toward the development of robust and flexible design concepts. Thanks to design robustness and flexibility from the architect's viewpoint, building industry is able to develop solutions based on the combination of customized building components and mass-produced components. In this manner, clients have the chance to personalize parts that are crucial for them, and building industry can mass-produce the other parts in order to limit building costs. This strategy allows building industry being better able to satisfy the current housing demand.

In the future, this study intends to directly test AD on housing industry applications for addressing the design innovation and for developing prefabricated building systems adequate for the present building demand.

References

1. J.S. Ackerman, Architecture, in *Encyclopædia Britannica Online Academic Edition* (Encyclopædia Britannica Inc., 2013)
2. American Institute of Architects, *Integrated Project Delivery: A Guide* (AIA, 2007), p. 21
3. Royal Institute of British Architects RIBA, Plan Work (2013), www.ribabookshops.com
4. J. Darke, The primary generator and the design process. Des. Stud. **1**(1), 36–44 (1979)
5. N.F.M. Roozenburg, N.G. Cross, Models of the design process: integrating across the disciplines. Des. Stud. **12**(4), 215–220 (1991)
6. S. Macmillan, J. Steele, S. Austin, P. Kirby, R. Spence, Development and verification of a generic framework for conceptual design. Des. Stud. **22**(2), 169–191 (2001)
7. L. Wang, W. Shen, H. Xie, J. Neelamkavil, A. Pardasani, Collaborative conceptual design: state of the art and future trends. Comput. Aided Des. **34**(13), 981–996 (2002)
8. N. Cross, *Engineering Design Methods: Strategies for Product Design* (Wiley, Chichester, 2000)

9. G.K.C. Ding, Sustainable construction: the role of environmental assessment tools. J. Environ. Manage. **86**(1), 451–464 (2008)
10. M. Rosenthal, A. Dörrhöfer, G. Staib, *Components and Systems: Modular Construction: Design Structure and New Technologies* (Birkhäuser, Basel, 2008)
11. D. Schodek, M. Bechthold, K. Griggs, K.M. Kao, M. Steinberg, *Digital Design and Manufacturing: CAD/CAM Applications in Architecture and Design* (Wiley, New Jersey, 2004)
12. B.J. Pine, *Mass Customization: The New Frontier in Business Competition* (Harvard Business School Press, Boston, 1993)
13. R.A. Niemeijer, B. De Vries, J. Beetz, Designing with constraints. Towards mass customization in the housing industry. Proceedings of the 10th international conference on design & decision support systems in architecture and urban planning, Eindhoven, 2010
14. A. Hatchuel, P. Le Masson, Y. Reich, B. Weil, A systematic approach of design theories using generativeness and robustness. Proceedings of the eighteenth international conference on engineering design, Copenhagen, 15–19 August 2011
15. K.T. Ulrich, *Design: Creation of Artefacts in Society* (University of Pennsylvania, Philadelphia, 2005)
16. L. Barrow, S. Al Arayedh, S. Kumar, Performance house 1—a CADCAM modular house system. Proceeding of the 2006 annual conference of the association for computer-aided design in architecture, Louisville, 2006
17. S. Kumar, L. Barrow, Architecture and industrial design: a convergent process for design exchange. Proceedings of the 2006 ACADIA annual international conference, Kentucky (Ky), USA, 2006
18. G. Pahl, W. Beitz, *Engineering Design: A Systematic Approach* (L. Blessing, F. Bauert, trans.) (Springer, London, 1996)
19. O. Kulak, S. Cebi, C. Kahraman, Applications of axiomatic design principles: a literature review. Expert Syst. Appl. **37**(9), 6705–6717 (2010)
20. N.P. Suh, *The Principles of Design* (Oxford University Press, New York, 1990)
21. N.P. Suh, *Axiomatic Design: Advances and Applications* (Oxford University Press, New York, NY, 2001)
22. M.K. Thompson (2013) A classification of procedural errors in the definition of functional requirements in axiomatic design theory. Proceedings of the 7th international conference on axiomatic design, Worcester MA, USA, 26–28 Jun 2013, pp. 107–112
23. C.A. Brown, Elements of axiomatic design: a simple and practical approach to engineering design. (2006). Retrieved from http://www.axiomaticdesign.org
24. R. Spence, S. Macmillan, P. Kirby, *Interdisciplinary Design in Practice* (T. Telford, London, 2001), pp. 85–89
25. M. Marchesi, J. Fernandez, D. Matt, S.-G. Kim, Axiomatic design approach for the conceptual design of sustainable buildings. Proceedings of the 8th international conference on axiomatic design, Lisbon, 27–29 Sept 2014
26. J.B.R. Pastor, E.M. Benavides, Axiomatic design of an airport passenger terminal. Proceedings of the 6th international conference on axiomatic design, Daejeon, 30–31 March 2011
27. R.G. Lindsey III, M. Omar, A.M. Farid, An axiomatic design based approach for the conceptual design of temporary modular housing. Proceeding of the 7th international conference on axiomatic design, Worcester, 27–28 June 2013
28. Resolution: 4 Architecture (2015). Retrieved from http://re4a.com/
29. K. Frampton, *Studies in Tectonic Culture* (MIT Press, Cambridge, 1995), p. 3
30. A. Deplazes, *Constructing Architecture: Materials Processes Structures* (Birkhäuser, Basel, 2013), pp. 14, 96–98
31. S. Kieran, J. Timberlake, *Cellophane House* (KieranTimberlake, Philadelphia, 2011)

Chapter 8
An Application of Quality Function Deployment and Axiomatic Design to the Conceptual Design of Temporary Housing

Lindsey R. Gilbert III, Mohammed A. Omar and Amro M. Farid

Abstract The interdisciplinary complexity of modern construction projects has made meeting customer needs and requirements a difficult task. Under the current model, decisions affecting the early stages of design when designers have the largest impact on the final cost and functionality of a given product are approached in an informal and non-homogeneous manner. This study proposes an alternative approach, combining quality function deployment (QFD) and axiomatic design (AD) methodology as a systematic way to approach the conceptual design phase of construction projects, specific to temporary housing. This methodology would help to ensure the designer meets the customers needs and requirements, as well as satisfies the design objectives in a homogeneous manner. More precisely, the QFD–AD method proposed herein is considered novel because it combines two prevalent design methodologies in a way that allows a smooth transition from the translation of customer needs into a formal and methodical design approach. The method also allows for an effective framework to help evaluate and compare conceptual design decisions, including the complex process of material selection. The design of a refugee housing unit is presented as an illustrative case study of temporary housing.

Keywords Quality function deployment (QFD) · Axiomatic design (AD) · QFD–AD · Refugee housing · Civil engineering design

L.R. Gilbert III · M.A. Omar
Engineering Systems and Management, Masdar Institute of Science and Technology, Abu Dhabi, UAE
e-mail: lrgilbertiii@gmail.com

M.A. Omar
e-mail: momar@masdar.ac.ae

A.M. Farid (✉)
Thayer School of Engineering at Dartmouth, Hanover, NH, USA
e-mail: amfarid@dartmouth.edu

© Springer International Publishing Switzerland 2016

A.M. Farid and N.P. Suh (eds.), *Axiomatic Design in Large Systems*,
DOI 10.1007/978-3-319-32388-6_8

List of Acronyms

TH	Temporary house or temporary housing
QFD	Quality function deployment
AD	Axiomatic design
VoC	Voice of customer
CN	Customer needs
CA	Customer attributes
TR	Technical requirements
FR	Functional requirements
nFR	Non-functional requirement
C	Constraint
OC	Optimization criteria
SC	Selection criteria
DP	Design parameter
PV	Process variable
UNHCR	United Nations High Commissioner for Refugees

8.1 Introduction

Construction projects are rapidly growing in complexity. Modern designers in the construction industry need to consider more encompassing view including the life-cycle issues (such as constructability, durability, life-cycle maintenance, energy efficiency, the cost of maintenance, environmental impact, and social-economic impact) and the more traditional concerns (such as aesthetics, structural integrity, and initial cost) [1]. According to Marchesi et al. [2], the intricacy of modern architectural design demands a more rational approach to the design phase when decisions with fundamental and extensive effects on appearance, performance, and costs are made. This is perhaps even truer in the case of temporary housing (TH in the reminder of this text), which faces the broader challenges of a typical construction project as well as the need to satisfy a diverse range of stakeholders. To handle the growing complexity and find a more rational approach to design, many designers are looking outside of the traditional domain for solutions, particularly during the early design phase [1, 3–8].

Throughout a construction project timeline, the decisions impact decreases as the project progresses such that earlier decisions have greater importance. However, "rigorous analytical methods and optimization systems are used for decisions that impact project costs by plus or minus 7 % (detailed design phase), while decisions that impact project costs plus or minus 30 % (conceptual design phase) are internalized" [1]. Civil engineering and architectural work typically begin with a broad conceptual design performed by experienced experts who have received input from key stakeholders. However, the mounting intricacy of the conceptual design phase makes it difficult for even the most experienced engineer to effectively capture and understand the diverse range of customer demands, much less ensure all of their needs are met during preliminary design phase. Temporary housing, a field awash with different stakeholders is even more liable to have trouble capturing the customer demands. Design of temporary housing is equally, if not more, complex as a traditional construction project, particularly given the diverse contexts, environments, and stakeholders they are subject to. Therefore, it is critical to have robust, rigorous, and methodical approaches to early conceptual design for said TH projects.

Traditional design methods typically include building to code, formal/informal discussions with the clients and/or iterative design stages; however, some find these methods lacking in their ability to capture client needs and requirements and so other methods not typically applied in construction design may be useful. "In the construction industry, usually the client needs and requirements are not treated systematically. Even if they are collected before the design phase, they tend to be disregarded and finally vanish as the construction phase goes on" [3]. This has forced the construction industry to turn to other fields for direction. Newer fields, such as manufacturing engineering, have developed a number of methods to improve product design and development projects based on customer requirements. Literature has demonstrated that manufacturing new product development (NPD) and construction share a number of similarities [9]. Due to this similarity, methods used in NPD are easily adaptable to the construction industry. Two popular NPD mythologies are quality function deployment (QFD) and axiomatic design (AD), both of which are used in this study.

AD was developed by Nam P. Suh in the 1980s and, like QFD, has quickly grown in popularity because of its ability to improve the conceptual design stage of a variety of different products. It has been used to develop products as complex as an autobus or refrigerator, to simple products like an efficiently designed soda can or bottle opener [10, 11]. AD works by creating a systematic approach to decomposing the design in a series of steps that takes it from a high-level view to a low-level view, while simultaneously encouraging adaptability. While AD is a strong design methodology, currently it assumes that the designer has identified the users' functional requirements "well" before beginning. In earlier works, AD was used specifically for temporary housing conceptual design [12, 13].

The QFD methodology was developed in Japan in the 1960s by Mitsubishi Heavy Industries to improve the design of ships in the Kobe shipyards. It was adopted by Toyota in the 1970s and since has been used by car manufacturers worldwide to increase customer satisfaction [6]. Over the past forty years, QFD has continued to grow in popularity and use in other industries as a means to systemically assure that customer needs and wants are clearly specified and drive the product design and production process [14]. QFD translates the difficult to understand customer requirements into measurable technical characteristics through a cascading series of relationship matrixes. The relationship matrix ensures that every customer need is addressed by at least one element in the design and further helps designers better understand the most important design elements.

In light of the QFD's ability to capture the voice of customer (VoC) and map it into requirements, and the AD's ability to guide the design process from high-level requirements into a conceptual design, combining the two processes seems a beneficial match. While this idea has been explored in the past by Taglia and Campatelli [15] and El-Haik and Said [16], neither work strongly demonstrates how to use the two simultaneously. Also, unlike previous work, this paper proposes a slight change in the QFD method to allow it to seamlessly join with the AD process, thus taking the strengths of both methods and creating a new streamlined process.

This paper seeks to address the conceptual design of TH using a QFD–AD methodology where the two have been seamlessly connected. Refugee housing has many stakeholders that need a formal approach to address their needs. Since no formal methodologies exist in the construction industry to both assess customer requirements and systematically approach the conceptual design of a construction project, this methodology is well-suited to fill this gap and to improve the design of complex projects. The methodology is applied to a TH-illustrated example, but it has potential to find other construction project applications, or may possibly be utilized in entirely different fields.

The remainder of this paper will proceed as follows. Section 8.2 introduces the QFD and the AD inner workings and explains how the use of QFD at the start of AD is beneficial to the conceptual design process. Section 8.3 presents a case study to demonstrate the application of this theory to the conceptual design of a temporary housing unit. Ultimately, Sect. 8.4 provides a discussion of the results and a conclusion.

8.2 A QFD–AD

Methodology QFD is a well-known methodology for mapping customer needs into technical requirements and determining the most important features to ensure customer satisfaction with a product. Section 8.2.1 provides a brief introduction into the theory and literature on QFD. Axiomatic design (AD) is proposed as a methodology to develop a conceptual design for a civil engineering project. Section 8.2.2 briefly introduces the fundamental axioms that govern AD.

8.2.1 Quality Function Deployment (QFD) to Ensure Customer Needs Drive Design

QFD is composed of a series of "quality tables" that move a design from the voice of customer (VoC) down to the detailed operations level. The House of Quality (HoQ) is the first phase and arguably the most important phase of the QFD process. In fact, most QFD studies focus almost exclusively on the HoQ phase of design

[17]. The HoQ displays the VoC and translates them into technical requirements (TRs), using the importance of different customer needs values to help determine the most important TRs to ensure customer satisfaction with the product. Typically, QFD is used in product development, quality management, or customer needs analysis; however, in recent years it has been expanded into other fields of study like engineering, management, teamwork, planning, design, costing, timing, and decision making [18].

The advantages of using the QFD process in the construction industry have been strongly presented in literature. Some researchers have discovered additional benefits beyond "creating a more enhanced customer orientation," "more effective product development," and "improved communications and teamwork" that are typically discussed in QFD literature [17, 19]. Kamera et al. [16] and Griffin and Hauser [20] both found QFD to be extremely beneficial in improving communication in project teams and subsequently the success or failure of a project. One company found the use of QFD has resulted in 30–50 % reduction in engineering changes, 30–50 % shorter design cycles, 20–60 % lower start-up costs, and 20–50 % fewer warranty claims [21]. Although the benefits of QFD are highly proven in the construction industry, with dozens of papers written on the matter, the methodology has still not gained hold in the field [3–8]. However, the trend is slowly changing.

In order to seamlessly integrate the QFD and AD design process, a few adjustments need to be made to the QFD matrix. The new process works by first filling in an adjusted house of quality like the one in Fig. 8.1. The key difference between this QFD and a traditional QFD is the TRs and the roof of the house (boxes 5–13 in Fig. 8.1). The technical requirements are split into constraints (Cs), non-functional requirements (nFRs), and functional requirements (FRs), three of five essential elements of AD decomposition as highlighted by Thompson [22]. In this paper, optimization criteria (OCs) and selection criteria (SCs) are considered to be parts of constraints for simplicity. Projects that are more complicated may find it worthwhile to include these two items in addition to Cs, nFRs, and FRs. The FRs should come from the second-level decomposition. The roof of the house is done identically to a typical QFD by specifying the direction and strength of the relationship between the different TRs. However, the information provided in the roof will be used to guide the AD process. The roof provides the designer a compact and rapid view of the different Cs and nFRs that will affect the decomposition of the FRs in the AD zigzag design process. The QFD will provide the designers with important information, such as the most important FRs to ensure clients' satisfaction, and which Cs are most likely to hinder the realization of the project. From this information, designers can determine the most important areas to invest resources. When the QFD is completed, the designer moves to AD to complete the design.

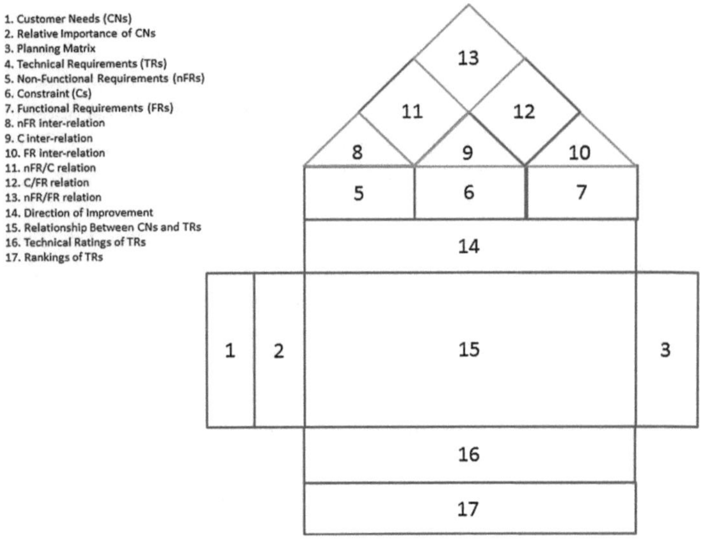

Fig. 8.1 Modified QFD

8.2.2 Fundamental Concept of Axiomatic Design

Similar to QFD, AD is used to guide the design process from the VoC down to a final design and manufacturing technique. However, literature on AD usually focuses on the process of mapping and guiding the designs FRs to design parameters (DPs) through the zigzag process. In addition, while AD does consider the customer needs, it does not have a methodical process of translating the customer needs into FDs.

The heart of AD is the two fundamental axioms upon which it is built, the independence axiom and the information axiom, where an axiom is a "truth that cannot be derived for which there are no counter examples or exceptions" [11]. These are formally stated by Suh [11]:

Axiom 1 *The independence axiom.* Maintain the independence of the FDs.

Axiom 2 *The information axiom.* Minimize the information content of the design.

For additional information regarding either axiom, the reader should refer to [1, 10, 11].

AD is a rigorous design tool and has been applied in many areas. In its relatively short history, AD has been used in fields ranging from industrial design to aerospace engineering. It has even been used in the construction industry. It helps designers start with the statement of "what we want to achieve" and ends with a clear idea of "how we want to achieve it" AD was established to create a systematic, scientifically based process that would make "human designers more creative, reduce random search process, minimize iterative trial-and-error processes, and select the best designs among those proposed" [11].

Axiomatic design has been applied in architecture by Marchesi et al. [2], structural engineering by Albano and Suh [1], and transportation engineering by Baca and Farid [23]. In earlier works, it was also applied to the design of a modular temporary housing unit, where it was found to be beneficial in making the design process more systematic and flexible to changes in requirements or resources [12, 13].

In this integrated methodology, very few changes were made to the AD process. The key difference was that QFD was used to capture the customer needs and transform them into FDs, non-functional requirements, and constraints which can then be used in the AD zigzag process, where the Cs and FRs from the QFD guide the decomposition of the FRs and DPs.

8.3 Case Study: Design of a Refugee House

In the following section, a case study is used to demonstrate the application of the combined QFD–AD methodology to the design of a refugee temporary housing unit. A brief introduction into refugee housing is provided in Sect. 8.3.1. While Sect. 8.3.2 demonstrates how the QFD can be used to capture customer needs and convert them into ranked FRs. Section 8.3.3 takes the TRs from Sect. 8.3.2 and converts them into DPs using AD. Section 8.3.4 concludes the case study by demonstrating how the use of the AD information axiom helps assess and select the most appropriate solutions (DPs) to the given FRs. The case study focuses on one branch of the AD decomposition, a broader scope has been done in an earlier paper [12].

8.3.1 Case Study Brief

Every year, large numbers of people are forced to relocate from their homes due to wide spread violence, ethnic persecution, natural disaster, war, and other forms of political instability and natural disasters. According to National Geographic, there are over 44 million forcibly displaced people in the world today [24]. Of these, "*15.4 million are refugees, 27.5 million are internally displaced persons (IDPs), and 800,000 are asylum-seekers hoping to achieve refugee status*" [24]. Many of these refugee's and IDP's end up living in camps, where they spend an average of 12 years waiting for conditions to improve in the area they are fleeing from, or a better option to open up. Many children spend their entire childhoods in these camps.

The Syrian conflict alone highlights the number of people who demand better access to temporary shelters. As of December 2013, the Syrian conflict has resulted in over 2 million refugees (1.8 million registered), and 4.25 million IDPs within Syria itself. Over a million of the refugees are children under the age of 18, and nearly three-quarters of a million are under the age of 11. Zaatari is the largest refugee camp in Jordan, the second largest in the world, and the fourth largest city in Jordan. Zaatari houses nearly 150,000 refugees and grows by up to 2000 people daily. There are nearly 30,000 shelters in the camp, 1700 administration buildings, and nearly 4000 shops and restaurants. To stem the continued growth of Zaatari, another camp named Azraq is being designed as an overflow camp. The majority of the buildings in the camp are tents provided by groups like United Nations High Commissioner for Refugees (UNHCR) [25]. There are no plans to make any of the buildings permanent, as the Jordanian government hopes all residents will return to their homes when the Syrian conflict is resolved. However, before it is resolved they will continue to experience large influxes of refugees into their country.

It is easy to contend that a safely built infrastructure and adequate housing conditions are among the most elemental human needs. Yet still a large proportion of refugees live in terrible and inhumane conditions [26]. The camps are often overcrowded, and housing within the camps filled beyond capacity [27]. Not only are the housing units overcrowded, they are poorly designed with little thought in mind for meeting the occupants needs. In a study of Sri Lankan Refugee camps, typical housing was found to be poorly ventilated, overcrowded, with no chimney to vent smoke from cooking with wood [4]. In another study of housing in the Palestinian refugee camp, Jalazonee, dampness was present in 72.5 % of the houses, while 50.5 % had mold, 37 % had leaks, and only 41.5 % were exposed to the Sun [27]. In addition to the above problems, residents of many of these shelters have to deal with the constant threat of contagious diseases, especially Malaria.

Many organizations provide temporary housing for these refugees; however, the limited funds shift the focus to speed and quantity over quality and functionality. This typically involves the distribution of tents. In fact, currently more than 3.5 million people worldwide live in tents provided by agencies like UNHCR. The tents are compact, easy, and cheap to manufacture, store, and ship. However, the technology behind the tents has not changed in years, and they provide little security and perform poorly in hot and cold conditions. Their inadequacy demonstrates a strong need for better designed housing options for refugees. Realizing this, the Ikea foundation and UNHCR recently joined in a collaborative project to design a new type of refuge shelter. The new design is built to have a lifetime of several years (compared to the current tent lifetime of 6 months), better thermal resistance, more privacy, and access to solar power for lighting. It is also designed to be compact for easy storage and transportation, and inexpensive to manufacture (expected cost of $1000 per unit). They are not alone, and a range of other groups have been founded to address this growing problem.

While the work done by Ikea foundation is a step toward improving the housing situation faced by refugees, there are still millions of refugees in need of better housing. Currently, temporary housing camps are unsafe; thus, it is essential to provide safe homes that are free of physical hazards. A number of studies on the effect of poor housing on health conditions have found that crowded-cramped conditions in conjuncture with inadequate housing can lead to anxiety stress, high-blood pressure, acute respiratory infections, and poor mental health among children [26, 27]. If dampness and mold are present, these problems may expand to include aches and pains, digestive disorders, and respiratory tract infections [28]. The crowded conditions of the camps also encourage the spread of communicable and contagious diseases such as tuberculosis. New housing needs to address the health and safety issues of the refugees while simultaneously meeting the shipping, storage, manufacturing, and cost requirements of agencies providing the structures. What is more, since the refugee's status is fundamentally temporary, their housing needs a temporary solution. However, it is clear that do to the tremendous heterogeneity and diversity of voices of stakeholders, an integrated one size fit all approach will not work.

8.3.2 Assessing Customer Needs

The first step of creating a QFD is obtaining the voice of customer (VoC). This information can be obtained from a range of sources including, but not limited to surveys, interviews, focus groups, and observation. Often customers are ambiguous with their description of needs and may confuse a physical object for functional requirement. For example, a customer may specify they need an A/C unit (an object); however, what they mean is a way to regulate the internal temperature (a functional requirement). Customers may also provide vague (subjective) specifications or provide very general ideas. Affinity trees and diagrams can help clarify and assist in the completion of the list of needs.

In this study, the customer needs (CNs) are determined from the open literature published on the subject. Table 8.1 shows the CNs found based on the work of Gilbert et al. [13], Arnold [29] and Ballerino [20]. This was determined by first specifying the higher-level CNs, and then determining the components that compose said high-level needs. The importance of each low-level element to the user was determined and averaged to find the importance of the high-level elements to the customer. Note that unlike a typical VoC, the table gathers CNs from multiple stakeholders.

Similar to the CNs, the TRs are determined from the literature, as well as an extensive review of the attributes highlighted by temporary housing as proposed on habitat.com, morethenshelters.org, and the Ikea foundation home. Like with the customer needs, the higher-level TRs where further decomposed into the Cs, nFRs, FRs. Each of these three is then further decomposed into high-level TRs for the QFD, and low-level TRS to capture a more complete view. This is shown in Table 8.2.

The QFD in this case study is created around the VoC of the people who will purchase and provide the temporary structure (groups like UNHCR or the Red Cross and Red Crescent), not just the end users (IDPs and refugees). This is different from a typical product designed using QFD. This is not to say the end users' requirements are not taken into account, but rather, they are taken into account alongside the other tradeoffs made by the purchaser/owner. For example,

Table 8.1 High- and low-level customer needs and level of importance

VOC				
Who it matters to	High level	Low level	Importance low level (1–9)	Importance high level (1–9)
End user	Be climatically comfortable	Insulate from fluctuations in external temperature	9	7.00
		Insulate from external noise	4	
		Prevent penetration of precipitation	8	
		Resist incoming air flow	7	
		Maintain internal humidity	7	
	Support health and safety	Resist rot and corrosion of materials	7	7.25
		Resist fire	9	
		Resist intruders	6	
		Prevent entrance of insects (mosquitos, etc.)	7	
	Support user activity	Facilitate cultural specific activities	7	5.60
		Provide area for sleeping	8	
		Provide area for food prep	8	
		Provide area for work	3	
		Provide area for personal hygiene	5	
		Provide area for entertainment	2	
		Provide area for storage	4	
		Provide privacy	7	
		Provide access to electricity	7	
		Provide access to water	5	
	Be aesthetically pleasing	Have aesthetically pleasing interior	5	4.67
		Maximize natural light inside	7	
		Have aesthetically pleasing exterior	2	

(continued)

Table 8.1 (continued)

VOC				
Who it matters to	High level	Low level	Importance low level (1–9)	Importance high level (1–9)
End user/provider	Function and performance	Last for 3 + years with minimal maintenance	8	8.25
		Be expandable/customizable	7	
		Resist local hazards	9	
		Meet international standards	9	
	Be easy to assemble	Need little or no experience to assemble	6	6.33
		Need minimal tools to assemble	6	
		Require little time to assemble	7	
Provider	Be easy to manufacture	Minimize shape complexity	7	5.67
		Use readily available materials	6	
		Use scalable process	4	
	Match site	Accommodate high density	5	5.67
		Connect to available services	3	
		Function independent of infrastructure	9	
	Be sustainable	Minimize embodied energy	5	6.00
		Ensure reusability of units or materials	7	
		Use local resources	5	
		Limit use of hazardous materials	8	
		Limit site disturbance	5	
	Minimize cost	Minimize cost to store materials/units	9	8.00
		Minimize cost to manufacture	9	
		Minimize cost to ship	9	
		Support local economy	5	
	Be easy to transport and store	Use minimal weight	9	8.50

Table 8.2 High- and low-level constraints (Cs), non-functional requirements (nFRs) and functional requirements (FRs)

High level	Low level
Constraints	
Environmental impact	Minimize depletion of natural resources
	Minimize soil and land degradation
	Maximize recyclability/reusability
Volume during transportation/storage	Minimize volume during transportation
	Minimize volume during storage
Number of components	Minimize total number of components
Number of materials	Minimize total materials used
Design/volume when built	Maximize efficiency of layout
	Hit target area of built structure
	Hit target height for roof
Complexity of assembly	Minimize required experience to assemble
	Minimize equipment required to assemble
	Minimize complexity of assembly
Complexity of manufacturing	Minimize required experience to manufacture
	Minimize equipment required to manufacture
Modularity	Make easy to customize
	Make modular connections possible
Material physical properties	Weight
	Thermal insulation properties
	Stability/expect lifetime
	Fire resistance
	Odor
	Chemical activity
	Thermal exchange properties
	Moisture resistance
	Corrosion resistance
	Biological resistance
	Deformation due to heat
	Absorption
	Embodied energy
	etc.
nFRs	
Inexpensive	–
Lightweight	–
Aesthetically pleasing	–
Functional requirements	
Passively protect and maintain internal climate	All FRs are further decomposed in the AD zigzag process
Actively protect and maintain internal climate	
Maintain structural integrity	
Support user activity	

the end user does not care about the amount of energy required to ship, store, and manufacture the shelter. However, they do care about the internal temperature of the shelter during the peak of summer. As can be seen in the list of CNs, both of these factors are acknowledged. This is due to the fact that from the provider's point of view, the end users comfort and the embodied energy of the structure are both important.

The displayed QFD in Fig. 8.2 demonstrates that only the high-level CNs and TRs are used. The reasons to approach this from a high-level instead of leaf-level (lowest level) view are twofold: improve clarity and eliminate unintentional bias toward high-level elements that have more leaf elements. Projects of smaller scale do not require this top level decomposition, but projects as complex as construction projects do.

The QFD above also provides a benchmark analysis of 3 different existing and proposed temporary housing solutions. Specific information was not available for all aspects of the units, so ratings are based on literature about each unit. *UNHCR tents* are the units typically used for refugee housing today. As can be seen in the benchmark, they are inexpensive to produce, store, and ship; however, they are not very effective at addressing the comfort or activity needs of the users. UNHCR is

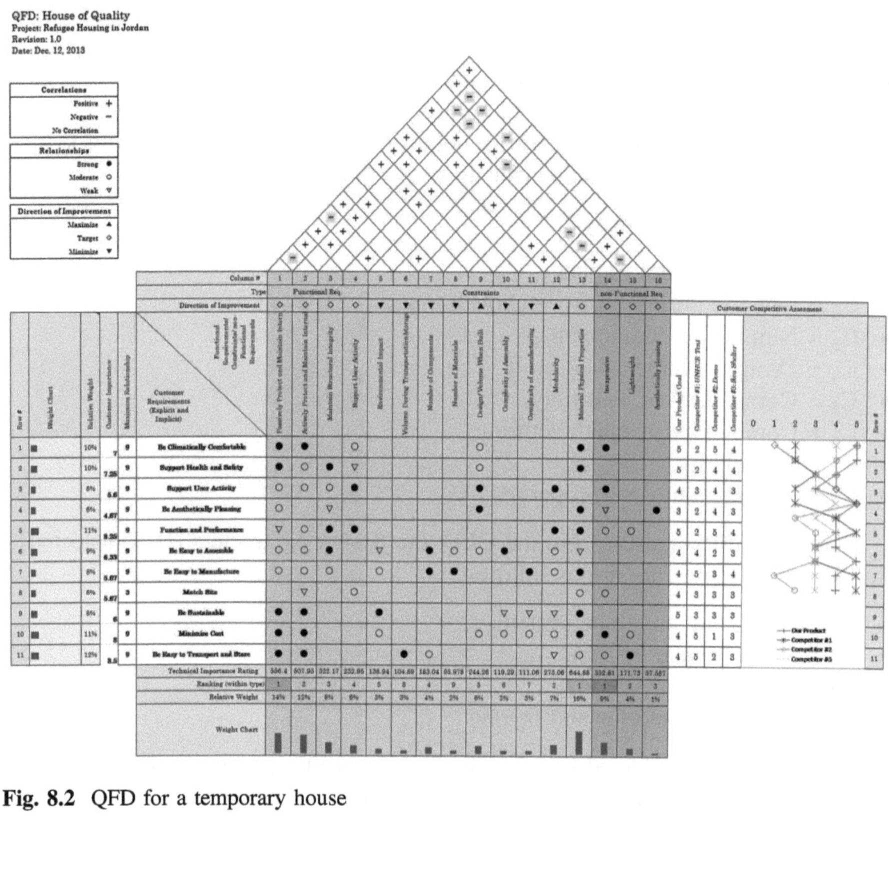

Fig. 8.2 QFD for a temporary house

looking at addressing these issues in the near future [30]. The second unit, *Domo*, was designed by a German group called More Than Shelter. It is a conceptual design that has been proposed to improve the quality of life of people living in refugee housing and slums by creating spaces to empower people. These units will be more expensive than a UNHCR tent, but are much more adept at meeting the user needs [31]. The final solution, the *IKEA shelter*, was unveiled in 2013 and is considered a promising solution to improve the quality of housing for refugees and IDPs [32]. Ikea seems to be a more middle of the road solution between the UNHCR tent and the Domo, providing less versatility then the Domo, but better at meeting user needs then the UNHCR tent. Using these different units as benchmarks helps to recognize where opportunities exist and can help designers to better understand how other designers address, or do not address, the VoC.

8.3.3 Decomposition of a Refugee Shelter

After the customer needs were used to highlight the high-level FRs, nFRs, and Cs of the system, the design of the temporary housing system was done using the AD zigzag methodology.

As can be seen in the QFD, the high-level FRs are follows:

- FR_1 = Passively protect and maintain internal climate,
- FR_2 = Actively maintain internal environment,
- FR_3 = Maintain structural integrity (against static and dynamic loading),
- FR_4 = Support user activities (for up to 5 ± 2 people).

Which are constrained by:

- C_1 = Environmental impact,
- C_2 = Volume during transportation/storage,
- C_3 = Number of components,
- C_4 = Design/volume when built,
- C_5 = Complexity of assembly,
- C_6 = Complexity of manufacturing,
- C_7 = Modularity,
- C_8 = Material physical properties.

Using the nFRs and Cs from the QFD, the design parameters (DPs) selected to fulfill each of these FRs are as follows:

- DP_1 = Building envelope system,
- DP_2 = Mechanical system,
- DP_3 = Structural system,
- DP_4 = Building interior and layout.

The DPs that are selected to fulfill the high-level FRs provide insights about the form of the shelter. The selected DPs may also change depending on designer's point of view and previous experiences. For example, designers more comfortable working with structural insulated panels (SIPs) may have chosen to combine the structural and envelope system into a single DP. In short, the decomposition of the same system by two different designers will nearly always be different. This is considered an advantage, because it highlights that the methodology does not impede creativity in the design process.

During the AD design process, the conceptual design should start to take form in the designers mind. Each continuous step of the zigzag process and expansion of the design matrix (DM) will further develop the shelter form. A design matrix, like the one displayed in Eq. 8.6 below, needs to be formulated for each level of the decomposition to avoid violating the independence axiom. In this case, the choice of a building envelope system will have an effect on the mechanical system used and the choice of a structural system. For example, if the building envelope is designed as to be load bearing, it will be part of the structural system; likewise, if the building has high-thermal resistance, the demand for a mechanical system to ensure the air is properly tempered will be lessened.

$$\begin{Bmatrix} FR_1 \\ FR_2 \\ FR_3 \\ FR_4 \end{Bmatrix} = \begin{bmatrix} X & 0 & 0 & 0 \\ X & X & 0 & 0 \\ X & 0 & X & 0 \\ 0 & 0 & 0 & X \end{bmatrix} \begin{Bmatrix} DP_1 \\ DP_2 \\ DP_3 \\ DP_4 \end{Bmatrix} \tag{8.1}$$

Equation 8.6 shows that the design is decoupled at the highest level and the independence axiom is not violated. Next, each of the FRs will be further decomposed. For brevity, only FR_1s decomposition will be shown; however, the other FRs will follow a similar decomposition format. FR_1 was chosen because it provides the primary function a refugee house needs to afford based on the Maslow hierarchy of needs [33]. The other three FRs all provide secondary, albeit important, functions for the users.

The building envelope system is perhaps the most important part in ensuring the good health and safety of its occupants. It is responsible for a number of very important functions related to the internal climate of the structure. While the mechanical system may play an important role in this function in a typical building, most refugees have limited or no access to electricity or driving power that allow most mechanical systems to function. This means that majority of the control of the internal climate will be done passively with the external envelope. The envelope of the structure must maintain a reasonable internal temperature throughout the entire day and should resist fluctuations in external temperature from the summer to winter seasons or from day to night. The envelope should also prevent excessive moisture and water ingress. Condensation due to excess moisture is one of the leading problems of health issues in the refugee camps. Safety of the occupants and their belongings is also an essential FR. Crime is often a major problem in large camps. It is essential that refugees' security is maximized, and they can help protect the few belongings they have left. Protection from mosquitos is also important since malaria is a rampant problem in refugee camps.

Because a refugee house needs to be simple by nature, the decomposition of the FRs is also relatively simple. Systems that are more complex may require more iterations of the zigzag process. FR_1s decomposition is shown below:

$FR_{1.1}$ = Allow controllable interaction with external environment,
$FR_{1.2}$ = Passively control indoor climate,
$FR_{1.3}$ = Prevent entrance of insects and pest.

Which are solved using the following DPs:

$DP_{1.1}$ = Fenestration (door/window),
$DP_{1.2}$ = Curtain wall and floor,
$DP_{1.3}$ = Insect-resistant features.

Again this can be mapped into a design matrix to ensure the second axiom is not violated. The design matrix (7) below shows that the decisions regarding the both curtain wall and floor as well as the fenestration affect the ability of the structure to "passively control indoor climate." This intuitively makes sense since the door and window will be important in passively cooling the building in hot weather and will be one of the main sources of heat leakage from the structure in cold weather. Likewise, choices of door and window will affect the buildings ability to prevent the entrance of insects (in addition to other insect-resistant features).

$$\begin{Bmatrix} FR_{1.1} \\ FR_{1.2} \\ FR_{1.3} \end{Bmatrix} = \begin{bmatrix} X & 0 & 0 \\ X & X & 0 \\ 0 & X & X \end{bmatrix} \begin{Bmatrix} DP_{1.1} \\ DP_{1.2} \\ DP_{1.3} \end{Bmatrix} \tag{8.2}$$

Since the independence axiom is not violated, the third level of decomposition can be created by following the zigzag process. First $FR_{1.1}$ is decomposed into the following:

$FR_{1.1.1}$ = Allow controllable entrance to structure,
$FR_{1.1.2}$ = Allow Entrance of natural light into structure,
$FR_{1.1.4}$ = Remove smoke from cooking/heat fires,
$DP_{1.1.1}$ = Door,
$DP_{1.1.2}$ = Window,
$DP_{1.1.4}$ = Closable cooking vent.

$$\begin{Bmatrix} FR_{1.1.1} \\ FR_{1.1.2.} \\ FR_{1.1.3} \end{Bmatrix} = \begin{bmatrix} X & 0 & 0 \\ X & X & 0 \\ X & X & X \end{bmatrix} \begin{Bmatrix} DP_{1.1.1} \\ DP_{1.1.2} \\ DP_{1.1.3} \end{Bmatrix} \tag{8.3}$$

Next, $FR_{1.2}$ is broken down into the following:

$FR_{1.2.1}$ = Regulate air flow/quality,

$FR_{1.2.2}$ = Regulate moisture in air and prevent accumulation of free standing water within unit,
$FR_{1.2.3}$ = Maintain internal temperature of 23 ± 6 °C.
$DP_{1.2.1}$ = Natural ventilation,
$DP_{1.2.2}$ = Water-resistant barrier,
$DP_{1.2.3}$ = Passive cooling and heating techniques.

$$\begin{Bmatrix} FR_{1.2.1} \\ FR_{1.2.2.} \\ FR_{1.2.3} \end{Bmatrix} = \begin{bmatrix} X & 0 & 0 \\ 0 & X & 0 \\ X & 0 & X \end{bmatrix} \begin{Bmatrix} DP_{1.2.1} \\ DP_{1.2.2} \\ DP_{1.2,3} \end{Bmatrix} \tag{8.4}$$

In the last step of the level 2 zigzag decomposition, $FR_{1.3}$ is broken into the following:

$FR_{1.3.1}$ = Prevent entrance of insects from openings,
$FR_{1.3.2}$ = Prevent Entrance of bugs and pest from under structure,
$DP_{1.3.1}$ = Screen on all openings with mesh size < 1 mm,
$DP_{1.3.2}$ = Impenetrable base.

$$\begin{Bmatrix} FR_{1.3.1} \\ FR_{1.3.2} \end{Bmatrix} = \begin{bmatrix} X & 0 \\ 0 & X \end{bmatrix} \begin{Bmatrix} DP_{1.3.1} \\ DP_{1.3.2} \end{Bmatrix} \tag{8.5}$$

To demonstrate how this can be continued to be broken down into the leaf elements, $FR_{1.1.3}$ (regulate internal temperature) is brought down to the fourth level of decomposition.

$FR_{1.2.3.1}$ = Regulate internal temperature from convection,
$FR_{1.2.3.2}$ = Regulate internal temperature from solar radiation,
$FR_{1.2.3.3}$ = Regulate internal temperature from conduction.

While convection, radiation, and conduction are all highly correlated, in this case some assumptions were made to minimize correlation. The worst-case scenario for regulating internal temperature occurs at night during the winter. During this time, no energy is gained from solar radiation, so this correlation is negated (though this FR does matter in the middle of the day in the summer and, therefore, should not be removed). It is further assumed that only a fraction of energy (~30 %) is lost to convection through gaps between parts. The remainder of energy loss in winter is expected to occur by conduction through the ceiling and walls.

In another branch of the AD decomposition that is not shown in this paper, it was determined that a propane heater (that produces around 2.8 kW of energy per day) can be installed in the housing unit. The housing unit has been designed to have total area of 20 square meters with an average height of 2.5 m. The heat produced by the heating unit and body radiation (0.1 kW × 5 people) must be sufficient to counter the rate of heat loss through conduction through the walls and floor. Any additional heat energy gained by the house (from solar radiation, additional people, etc.) can be countered by reducing heat produced by propane heater. The average winter low for this case studies' location is expected to be around 0 °C.

$DP_{1.1.3.1}$ = Adjustable vent on top of ceiling,
$DP_{1.1.3.2}$ = External coating,
$DP_{1.1.3.3}$ = Thermal insulation panel.

$$\begin{Bmatrix} FR_{1.2.3.1} \\ FR_{1.2.3.2.} \\ FR_{1.2.3.3} \end{Bmatrix} = \begin{bmatrix} X & 0 & X \\ 0 & X & 0 \\ 0 & 0 & X \end{bmatrix} \begin{Bmatrix} DP_{1.2.3.1} \\ DP_{1.2.3.2} \\ DP_{1.2.3.3} \end{Bmatrix} \quad (8.6)$$

The use of a thermal mass is a popular passive heating/cooling method typically implemented to modulate daily temperature variations; however, in the presented case this is a difficult proposition to implement given the importance of keeping the units light-weight for shipping. This means that an alternative effective passive heating and cooling techniques needs to be devised. This highlights the role of solar radiation, because home users might capture the heat of solar radiation when it is cold, but block it or reflect it when it is warm. Typically, this is done passively by orienting the house in a way that allows the solar heat to enter directly through windows during the winter (when the Sun is lower in the sky) and be blocked by the roof in the summer, or through some form of thermal mass. Developing a method of regulating internal temperature will be a difficult task and will be highly dependent on selecting an appropriate material.

8.3.4 Checking Appropriateness of the Solution

The information axiom of AD is used to select the best design parameter to meet the given functional requirement within the system constraints. In a complete design, this should be done for every leaf-level component; however, for brevity a single sample has been selected to demonstrate the concept. In this particular case, the best DP option to satisfy $FR_{1.1.3.3}$, regulate internal temperature from conduction, is being selected from 3 different alternatives. Table 8.3 provides the three options along with their material properties. The "goal" line also provides the acceptable design range for each property for the given FR, data for both selection criteria and the design alternatives were assumed to be uniformly distributed.

Table 8.3 Material properties of three alternatives and the AD information content for each alternative

	Thickness	*R*-value	Weight	Flame spread index	Vapor transmission rate	Recycled content	Sum of information content
Unit	m	M^2/W	Kg/m^2	–	SI Perm	%	–
Goal	0–0.08	1.2–5	0–0.8	0–25	57–285	10–50	–
Product 1	0.07–0.085	1.6–2	0.56–0.816	20–23	240–378	8–12	–
Information	0.585	0	0.093	0	1.616	1	3.29
Product 2	0.065–0.07	1–1.4	0.488–0.60	24–25	265–400	30–50	–
Information	0	1	0	0	2.755	0	3.755
Product 3	0.05–0.06	2–2.2	0.475–0.63	18–19	100–120	0–5	–
Information	0	0	0	0	0	∞	∞

The information axiom equation

$$I = \log_2(1/p) \tag{8.7}$$

is used to determine the I value for each property, where p is the probability that satisfying the goal. This value is then summed to determine the DP with the lowest information content. In this example, the information content of $DP^a_{1.1.3.3}$, $DP^b_{1.1.3.3}$, and $DP^c_{1.1.3.3}$, was found to equal 3.29, 3.75, and infinity, respectively. This signifies of the three options, $DP^a_{1.1.3.3}$, or Product 1, is the best possible option. This is interesting considering Product 3 is the best in every way except recycled content.

8.4 Conclusion and Future Work

The same tents have been utilized for most natural disaster and refugee camps for the past twenty years. While many design ideas for refugee shelters have been proposed, none have been able to completely replace the tent. This is because they are unable to adequately meet the stakeholder requirements, either from a design or cost point of view. Recently, many designers, including the IKEA foundation, have attempted to address this problem; however, only time will tell if their proposed designs will be successful.

This paper proposed a new method to systematically guide the process of creating a temporary house conceptual design based on stakeholder needs. The method is built on two proven design methodologies, QFD and AD. QFD has already proven its use to construction projects in literature, and AD has developed wide acceptance due to its ability to improve creativity, minimize the iterative process, and quickly optimize for the best solution [11, 21]. The new method integrated these two methods by using the strengths of both of approaches. After introducing the changes made to the QFD and AD process, a case study was provided to demonstrate the use of the process in the design of a temporary housing shelter. Although the case study does not present the complete design, it does demonstrate the methodology's ability to capture the VoC in a systematic design process.

The case study found that the combination of QFD–AD method streamlined the design process and helped to ensure that the VoC directed the entire conceptual design creation. This new approach to combining the QFD–AD methodology may be applicable in other industries as well. Future work may include developing a more efficient way of combining the process to maximize the impact of the VoC on the design process, and perhaps expanding the QFD to include other design variables beyond nFRs, FRs, and constraints. This includes selection criteria (SCs) and optimization criteria (OCs). Material selection using the AD information axiom also proved less efficient then desired.

There is complex coupling that occurs at the material level. This pairing makes material selection a complex process of managing a number of different tradeoffs. In the above example, the density, cost, thickness, and thermal conductivity all are highly coupled. An extremely dense and expensive material may have a much higher thermal conductivity and ultimately be the most inexpensive option for the same thermal performance. While the information axiom presents a method of handling these tradeoffs, it creates a bias by weighing all factors effecting the decision equally. Suh [11] recommends applying weight factors to prevent this bias. Although this eliminates bias, it is difficult to correctly determine what value of weights to assign. The method also does nothing to directly differentiate between required material characteristics and preferred material characteristics, and it is poor at dealing with subjective criteria. Lastly, if a material option does not meet one of the FRs for the material, it will immediately be eliminated because it has infinite information content (as was the case with product three in the case study). Future work will investigate if material selection may be carried out through alternate decision-making frameworks such as AHP. This may serve to handle subjective criteria and allow the comparison of fundamentallly different parameterizations.

Acknowledgements The International Renewable Energy Agency (IRENA) is acknowledged for their support.

References

1. L.D. Albano, N.P. Suh, Axiomatic approach to structural design. Res. Eng. Design **4**, 171–183 (1992)
2. M. Marchesi, S-G. Kim, D.T. Matt, in *Application of the axiomatic design approach to the design of architectural systems: a literature review,* ed. by M.K. Thompson. Proceedings of the 7th International Conference of Axiomatic Design (Worcester, MA 2013)
3. I. Dikmen, M. Talat Birgonul, S. Kiziltas, Strategic use of quality function deployment (QFD) in the construction industry. Build. Environ. **40**, 245–255 (2005)
4. R.L. Armacost, P.J. Componation, M.A. Mullens, W.W. Swart, An AHP framework for prioritizing customer requirements in QFD: an industrialized housing application. IIE Trans. **26**, 72–79 (1994)

5. Y.Q. Yang, S.Q. Wang, M. Dulaimi, S.P. Low, A fuzzy quality function deployment system for buildable design decision-makings. Autom. Constr. **12**, 381–393 (2003)
6. D.J. Delgado-Hernandez, K.E. Bampton, E. Aspinwall, Quality function deployment in construction. Constr. Manag. Econo. **25**, 597–609 (2007)
7. L.S. Pheng, L. Yeap, Quality function deployment in design/build projects. J. Architectural Eng. **7**, 30–39 (2001)
8. L.A. Gargione, *Using quality function deployment (QFD) in the design phase of an apartment construction project,* in Proceedings IGLC (1999), p. 357
9. C.T. Formoso, P. Tzortzopoulos, R. Liedtke, A model for managing the product development process in house building. Eng. Constr. Architectural Manag. **9**, 419–432 (2002)
10. N.P. Suh, Design and operation of large systems. J. Manuf. Syst. **14**, 203–213 (1995)
11. N.P. Suh, *Axiomatic Design Advances and Applications* (Oxford University Press, New York, 2001)
12. L.R. Gilbert III, A.M. Farid, M. Omar, in *An Axiomatic Design Based Approach for the Conceptual Design of Temporary Modular Housing*. ed. by Thompson M.K. Proceedings of the 7th International Conference of Axiomatic Design (Worcester, MA, 2013)
13. L.R. Gilbert III, A.M. Farid, M. Omar, in *An Axiomatic Design Based Approach to Civil Engineering,* ed. by Thompson M.K. Proceedings of the 2nd International Workshop on Design in Civil and Environmental Engineering (Worchester, MA, 2013)
14. L Cohen, *Quality function deployment: how to make QFD work for you* (Addison-Wesley Reading, MA 1995)
15. A. Del Taglia, G Campatelli, *Axiomatic design & Qfd: a case study of a reverse engineering system for cutting tools*, in Proceedings of ICAD2006, 4th International Conference on Axiomatic Design 2006
16. B. El-Haik, Axiomatic quality: integrating axiomatic design with six-sigma, reliability, and quality engineering: Wiley.com, 2005
17. L.-K. Chan, M.-L. Wu, A systematic approach to quality function deployment with a full illustrative example. Omega. **33**, 119–139 (2005)
18. L.-K. Chan, M.-L. Wu, Quality function deployment: a literature review. Eur. J. Oper. Res. **143**, 463–497 (2002)
19. L.-K. Chan, M.-L. Wu, Quality function deployment: a comprehensive review of its concepts and methods. Qual. Eng. **15**, 23–35 (2002)
20. C.C. Ballerino, Building Materials & Engineering Design: Low-Income Housing Projects, Port Elizabeth-South Africa. (Kungl Tekniska Hogskolan (Royal Institute of Technology), Stockholm, Sweden 2002)
21. A. Zakarian, A. Kusiak, Forming teams: an analytical approach. IIE Trans. **31**, 85–97 (1999)
22. M.K. Thompson, in *A Classification of Procedural Errors in the Definition of Functional Requirements in Axiomatic Design Theory*, ed. by Thompson M.K. Proceedings of the 7th International Conference of Axiomatic Design. (Worcester, MA, 2013)
23. E.E.S. Baca, A.M. Farid, in *An Axiomatic Design Approach to Reconfigurable Transportation Systems Planning and Operations,* ed. by Thompson M.K. Proceedings of the 2nd International Workshop on Design in Civil and Environmental Engineering (Worcester, MA, USA, 2013)
24. Schleeter R, in *Mapping Displaced People Around the World*, ed. by Caryl-Sue, O'Connor S. P. National Geographic: Education (2013)
25. A. Qattawi, A. Mayyas, M. Abdelhamid, M.A. Omar, Incorporating quality function deployment and analytical hierarchy process in a knowledge-based system for automotive production line design. Int. J. Comput. Integr. Manuf. **26**, 839–856 (2013)
26. N. Brifcani, R. Day, D. Walker, S. Hughes, K. Ball, D. Price, A Review of Cutting-edge Techniques for Material Selection (2012)
27. F. De Felice, A. Petrillo, A multiple choice decision analysis: an integrated QFD–AHP model for the assessment of customer needs. Int. J. Eng. Sci. Technol. 2 (2011)
28. E.S.S.A. Ho, Y.-J. Lai, S.I. Chang, An integrated group decision-making approach to quality function deployment. IIE Trans. **31**, 553–567 (1999)

29. Arnold C, Building Envelope Design Guide—Introduction. Des Guidance (2009)
30. M. Lolachi, *High Tech Solar Tents, and How to Design a Refugee Camp: An Interview With Manoucher Lolachi* (Senior Physical Planner, UNHCR, 2010)
31. morethanshelters. More than shelters: Domo http://www.morethanshelters.org/eng/domo/
32. Zimmer L, IKEA Unveils Solar-Powered Flat Pack Shelters for Easily Deployable Emergency Housing
33. J.A. Simons, D.B. Irwin, B.A. Drinnien, Maslow's hierarchy of needs. Retrieved October **9**, 2009 (1987)

Part IV
Manufacturing Systems

Chapter 9
Design and Implementation Approach for Distributed Manufacturing Networks Using Axiomatic Design

Dominik T. Matt and Erwin Rauch

Abstract Rising logistics costs, mass customization, and market-specific product variants are the reason for the current trend toward decentralized and geographically distributed manufacturing systems. Often, these mini-factories are organized and managed in production networks. While there exist already many scientifically discussed methods for the design of single manufacturing systems, there are only few works discussing the design of production networks with distributed manufacturing units. This chapter provides an Axiomatic Design-based approach for the design, planning, and implementation of distributed manufacturing systems using the example of a franchise network with geographically distributed, changeable, scalable as well as replicable manufacturing units. The aim of the proposed approach was to derive a set of appropriate design parameters (DP), which supports practitioners in their work to design the manufacturing system. The presented approach has been successfully applied in a real case study of an Italian franchise company.

Keywords Axiomatic Design · Manufacturing systems · Franchising · Distributed manufacturing · Production networks

9.1 Introduction

The success of an enterprise no longer only depends on their own performance, but on the performance of the entire production network they are situated in. With a rising number of such networks, the inter-organizational coordination becomes to a success factor for these companies [63]. Currently, the trend shows an increase of

D.T. Matt (✉) · E. Rauch
Faculty of Science and Technology, Free University of Bozen-Bolzano,
Piazza Università, 5, 39100 Bolzano, Italy
e-mail: dominik.matt@unibz.it

E. Rauch
e-mail: erwin.rauch@unibz.it

© Springer International Publishing Switzerland 2016

A.M. Farid and N.P. Suh (eds.), *Axiomatic Design in Large Systems*,
DOI 10.1007/978-3-319-32388-6_9

distributed productions with the aim of a global market development as well as for meeting local needs. A so-called global production combines the goals of global market development and the fulfillment of local customer requirements [75]. With the rising demand for individual products and product variants, the shift from mass production toward a more personalized “mass customization” becomes more and more realistic [38]. In order to stay competitive in today’s global market, manufacturing companies need to be flexible. To ensure flexible production, shorten processing times, and reduce time to market, companies are utilizing the distributed manufacturing system paradigm, wherein geographically distributed local resources are used for product development and production [11]. In a global market, production systems must also consider business processes of geographically distributed manufacturing units to take advantages of local facilities [18].

In the future, long-established paradigms of production will still have to continue to change in order to meet the demand for even more individuality, customer-specific product variants, and shortest delivery times within the meaning of the term “production on demand.” Therefore, in the future, the concept of the distributed manufacturing by geographically distributed manufacturing systems plays an increasingly important role. New and innovative ways of organizing production operations will be necessary. It needs modern organizational models for small, flexible, and scalable production units in decentralized production networks, which take into account the local and individual customer needs and produce as possible locally [60]. The main advantages of decentralized production structures are a higher flexibility to reflect local customer, lower logistics costs, and shorter delivery times. Besides these advantages, there are also negative aspects, such as the high investment costs for a decentralized structure and the lower efficiency of decentralized production in comparison with usually highly automated central production factories [61]. The following are the key trends and reasons for the development toward distributed manufacturing systems [60]:

1. Megatrend Sustainability (reduction of transport and therefore CO_2 emissions).
2. Rising logistics costs (reduction of physical transports).
3. Individuality and mass customization (individual and local products).
4. Democratization of design and open innovation (involvement of the customer in product development).
5. Proximity to the market and point of consumption (just-in-time delivery and shorter delivery times).
6. Production at the place of critical resources (e.g., raw materials or highly qualified human resources).
7. Regionalism and authenticity (in special cases such as “authentic food”).

The concept of distributed manufacturing includes many possible forms for their design. Figure 9.1 summarizes different forms of decentralized manufacturing [60]. The classification summarizes eight identified forms for distributed manufacturing:

Evolution stages of distributed model factories			Special forms of distributed production		
Type	**Classification**	**Description and characteristics**	**Type**	**Classification**	**Description and characteristics**
1	**Standardized and replicable model factory**	Replicable and standardized model factories for geographically distributed production of defined products with a defined number of units.	5	**Service model of industrial contract manufacturing**	Production service providers and intermediaries ("Production Provider") for distributed industrial contract manufacturing of different products with similar manufacturing steps and varying quantities
2	**Modular and scalable model factory**	Modular model factories for geographically distributed production of defined products with flexibility in relation to item quantity and thus scalability of the manufacturing system.	6	**Mobile and non-location-bound model factories**	Mobile non-location-bound and highly flexible as well as scalable model factories for temporary production requirements reducing procurement and/or distribution transports.
3	**Flexible and reconfigurable model factory**	Flexible and rapidly reconfigurable model factories for geographically distributed manufacturing of products in different variants (product flexibility) and various quantities (quantity flexibility).	7	**Production-Franchise**	Model factories, operated independently by franchisees, with more or less flexible and adaptable production units for geographically distributed production of products in a franchise network.
4	**Changeable and „smart" model factory**	Intelligent and self-optimizing model factories with a high degree of adaptability to geographically distributed production of different products with similar manufacturing steps and varying quantities.	8	**Additive manufac-turing (Cloud Production)**	Highly flexible and geographically capillary distributed labs for the production of various products with generative manufacturing processes (3D printing) by digital transmitted CAD data (Cloud)

Fig. 9.1 Forms of distributed manufacturing systems [60]

A distinction is made between decentralized model factories with their individual stages of evolution and special forms of distributed manufacturing. All of the shown concepts are based on collaborative production networks for managing the decentralized manufacturing units.

Innovative and modern production networks following a globally standardized production combined with local and flexible manufacturing units are used very often in franchising [70]. Franchising has lately become more and more important. Under franchising, we mean in its broadest sense to build a "best practice" business model and the subsequent transfer of licenses for the replication or duplication of the concept in different target markets [13, 66]. By franchising, manufacturers can establish facilities in new markets with a minimum of delay and capital outlay [33].

Besides the traditional pure franchise sales or service license (e.g., Burger King or Subway), franchising is also possible in the form of a production franchise or license to assign the production of goods to a franchisee [87]. Often, these companies produce not in a central location, but in a decentralized structure, because of the individual customer requests in the various destination countries or especially in the case of food products with a short shelf life. The individuality of products is sometimes given by ethnic-, religious-, or cultural-based differences in the markets [57]. For the above-described reasons, franchising models in the form of geographically distributed production franchises or mixed forms (production franchise with simultaneous sales or service franchise) are increasingly used to expand into new markets. This work puts this special type of production company in focus, which will become increasingly important due to actual and future growth of franchise business models [70].

A manufacturing system should not only produce high-quality products at the lowest possible price; it should also quickly adapt to market changes and react to consumer behavior and trends. Geographically distributed production facilities composed of reconfigurable manufacturing systems allow these quick adjustments of production capacity and functionality with respect to local customer needs [10]. Given the promising development in the past and the anticipation of further growth in franchising brands and their significant share of total economic output [4], it becomes important to develop specially adapted changeable and agile manufacturing systems also for this sector.

In many research works about distributed manufacturing, the main focus has been given often to isolated aspects of layouts, organizational structure, process design, or collaboration mechanisms [58]. A comprehensive approach for the design of scalable modular manufacturing systems that promote distributed production in collaborative networks in a highly dynamic environment is still missing.

Therefore, the main objectives of this research and the development of the illustrated approach in this chapter can be summarized as follows:

- Changeability through a modular and scalable expansion of the manufacturing systems capacity.
- Replicability of the manufacturing system in the rollout phase and expansion of the franchise system.
- Identification of needs for manufacturing systems for franchise models using a systematic methodology.
- Derivation of an appropriate guideline with a set of design parameters for manufacturing system designers.
- Development of a holistic approach to design and implement a franchise system with decentralized production units, which includes not only technical but also organizational and strategic content.
- Ensuring the practical applicability and validation using a case study.

Axiomatic Design provides a systematic approach to derive in a first step, the functional requirements (FR) and in a second step a set of design parameters (DP) for a changeable and modular manufacturing system for franchising models. By applying the Axiomatic Design methodology [79] and the MSDD approach [16] in this work, the requirements and specific DP could be achieved in a systematic and structured way.

The research in this chapter is based on a real case study with a north Italian start-up franchise brand. The aim of the collaboration in this case study was to design and implement a modular and scalable manufacturing system for a network of distributed franchise production facilities. The application of the AD-based approach in the case study was very useful and effective for the systematic investigation of the requirements as well as for the elaboration of a concept for scalable and modular manufacturing systems for distributed manufacturing in franchising networks.

9.2 Literature Review

In the current state of the art, great attention is paid to changeability in distributed manufacturing systems. There exist countless articles and research papers to this argument [2, 22, 32, 35, 54, 64, 65, 68, 70, 72, 78, 89–92]. Changeable systems are able to make anticipatory adjustments in addition to reactive interventions [89]. The design principles of reconfigurable module-based manufacturing systems are as follows: convertibility, flexibility, scalability, modularity, integrability, and diagnosability [40, 41]. Dove [21, 22] describes, in his research concrete practical examples, how plant and machinery can be designed and constructed in a flexible and changeable manner.

As more research emerged, additional approaches and philosophies around the topic of distributed manufacturing such as Holonic Manufacturing Systems [46, 86], Bionic Manufacturing Systems (BMS) [67, 85], Fractal Factory and multi-agent systems [74, 88], and the networked organization [84] were developed. However, the impact of this expansion has been little discussed because of the traditional focus on information technology [19]. All of the concepts have basic properties in common, including autonomy, distribution, decentralization, flexibility, adaptability, and agility [73]. Research into industrial networks has mostly neglected the dynamic forms of communication and coordination [19]. Collaboration is a hot topic in industrial networks and needs expansion beyond the current concepts to arrive at a more grounded theory [6, 25, 45]. Manufacturing systems should present self-organization features and reorganization techniques in order to adapt to the external changes. The way to represent the reorganization techniques and the responsibilities associated with the trigger of the reorganization is a complex task, which requires additional research in order to develop a standard model to represent those techniques [47].

9.2.1 *Changeable, Scalable, and Distributed Production in Franchise Models*

The above-mentioned approaches usually have a universal and general character and hardly respond to special operational or organizational forms such as distributed manufacturing and franchising. In recent decades, the topic of franchising was addressed almost exclusively from the business and legal side [1, 8, 17, 20, 23, 42, 44, 49, 50, 62, 76, 77, 82]. Manufacturing aspects were highlighted only very superficially. While there are a number of practical guidelines on the introduction of franchising and the creation of franchise manuals (e.g., [1, 39]), it is missing entirely a guideline for the planning, design, and implementation of geographically distributed manufacturing systems within franchise networks.

Only a few authors have done research on production franchising and/or geographically distributed production. The following literature review summarizes the most important works on this argument.

Hayfron et al. [33] developed firstly rough approaches for the design and implementation of production franchising networks. The authors show, however, only partially the requirements of the technical and organizational design of appropriate manufacturing systems.

Unlike licensing systems, a franchise system consists of the transfer of an entire business model and production concept from the franchisor to the franchisee [34]. Carrie et al. [12] present in their research a few basic requirements for the successful implementation of production franchise models:

- The applied technologies and work processes must be established and tested (preferably by means of a pilot production facility).
- The model must be easily replicable.
- The franchisor has the ability and expertise to transfer its know-how and knowledge to its franchisees.
- The staff of the franchisee must be able to be trained in an efficient, fast, and economical manner.

Hildebrand et al. [36] developed a so-called PLUG+PRODUCE concept, which could be applicable also for franchise models. The research aims were to develop a modular factory concept, which should enable particularly for small and medium enterprises, to expand production without much effort and to move the production facility also to a new location. The research focuses on the design of a standardized "type factory" with the aim of duplicating it without great effort. However, the approach is based on a specific example of the industrial partner in the research project and can therefore be used only as a very limited guide for the design of manufacturing systems for franchising models.

Zäh and Wagner [93] developed in their research project named "Market-oriented production of customized products" a concept of mini-factory structures. The objective of the project was similar to the project PLUG +PRODUCE, to develop a modular concept of a mini-factory for the purposes of mass customization [71]. The design of the mini-factory is based on a modular kit which differentiates in necessary basic modules and optional modules. The requirements for the mini-factories are similar to those from the task of this work, but it is strongly focused on the topic of mass customization. The concept therefore has significant weaknesses to apply for franchising models as there are no recommendations regarding the integration and refinement in a franchise network.

9.2.2 Axiomatic Design Approach for the Design of Manufacturing Systems

Axiomatic Design is a systematic approach for design by the top-down decomposition of "what we want to achieve," to "how we can satisfy the requirements." The theory of Axiomatic Design is applicable to many different kinds of complex systems [80]. A manufacturing system can be defined as a dynamic and complex system, because it is subject to temporal variation and must be reconfigurable and adaptable. In such cases, Axiomatic Design shows a suitable and helpful method to reduce complexity in the manufacturing systems design [56]. Axiomatic Design is based on four domains to transform so-called customer needs or customer attributes (CA) into functional requirements (FR), design parameters (DP), and process variables (PV). Through its top-down approach, Axiomatic Design is a very systematic and structured methodology. Starting from a main goal, a hierarchically structured catalog of requirements (FR) with proposed solutions (DP) is developed. By breaking down (decomposition of) the top goals and design proposals can be identified specific DP at operational level.

Cochran developed an approach for the design of manufacturing systems, which is based on the principles of Axiomatic Design [15, 16]. Cochran's methodology "Manufacturing System Design Decomposition" (MSDD) is the graph of the derivative FR-DP tree and very clear and easy to understand. In the background are analyzed the interactions between the individual requirements and DP in a mathematical way. This results, ultimately, in an ideal sequence or path to implement the DP at the lowest level. AlGeddawy and ElMaraghy [2] describe Axiomatic Design as a very suitable and frequently used method to derive the target system as well as the requirements and evaluate the interactions of the identified requirements in a systematic way. Bergmann applies the MSDD methodology and thus the Axiomatic Design approach for the derivation of requirements for a sustainability-oriented holistic manufacturing system [7]. The work of Bergman proves once again that the application of the Axiomatic Design methodology is suitable for a systematic and structured derivation of requirements and DP. Also, other authors [3, 5, 9, 27, 29–31, 37, 43, 51–54, 56–58, 69, 70, 81, 83] and Matt and Rauch [60] suggest the application of Axiomatic Design for the design of manufacturing processes and systems in different research papers and case studies.

The application of Axiomatic Design for designing distributed as well as reconfigurable and changeable manufacturing systems is particularly treated by [5, 14, 26–30, 48, 55], also in alternative ways.

9.2.3 Research Gap and Need for Action

None of the shown approaches in literature, to achieve changeability and reconfigurability in manufacturing, provides information on the specific application in distributed manufacturing and franchising networks. All the discussed approaches show important and relevant findings for this work, but they are often only partially suitable and/or only generally formulated. Thus, it was important to develop a comprehensive approach to the design of changeable and modular manufacturing systems with geographically distributed production based on the example of franchise models.

9.3 Axiomatic Design-Based Derivation of Parameters for the Design of Distributed Franchise Manufacturing Systems

The AD-based approach for the determination and derivation of the DP can be basically divided into the following five usual steps in Axiomatic Design [79, 80]:

1. Identification of customer attributes (CAs).
2. Transfer of customer needs into functional requirements (FRs) at the highest level.
3. Assignment ("mapping") of solutions or design parameters (DP) to the respective functional requirements (FRs). In the assignment, the two axioms of Axiomatic Design have to be considered:
   - the Independence Axiom in order to reduce the coupling of the system (avoid dependencies between the DPs and other FRs).
   - The Information Axiom for the selection of solution alternatives (choose always the "simplest" solution with the least information content).
4. Decomposition ("zigzagging") into several hierarchical levels (top-down) to move from abstract requirements to concrete design parameters (FR-DP tree).
5. Development and revision of the design matrix.

The following sections describe the approach for the derivation of a comprehensive catalog of appropriate DP by the use of the above-mentioned steps in the Axiomatic Design methodology. It should be noted that the presented work has been built primarily on the approach of AD for fixed systems. Suh distinguishes between AD for fixed systems and AD for flexible systems [80]. While it is assumed that requirements in fixed systems do not change for a certain time period, the FR in flexible systems are changing over time. This means for flexible systems that different FR sets are used at different points in time. It is quite difficult and comparable to a look in the crystal ball to derive at what time which combinations of FR are substituting previous defined FR combinations. Thus, in this work is used the approach for fixed systems, and the functional requirements are defined based on a five-year outlook with the ability to adapt the system. For regular adjustments and adaptations of the system, a corresponding feedback loop with a redesign process has been foreseen at the end of the presented approach.

9.3.1 Customer Needs and Functional Requirements on the Highest Level

The customer needs in this case study were identified through interviews with management and executives of the Italian franchise company. Based on these interviews, the functional requirement at the highest hierarchical level (level 0), which is the main objective of the manufacturing system, was determined:

FR_0 Build a network of changeable, scalable, and economic franchise production facilities.

To satisfy this functional requirement (FR_0), the following solution (DP_0) was defined on the physical design domain:

DP_0 Changeable and efficient manufacturing system for franchising models.

The proposed solution DP_0 is formulated very abstractly, and as expected, it could not be a sufficient design parameter for the manufacturing system. Therefore, it is necessary to split the top functional requirement FR0 into more detailed functional requirements at the next level.

9.3.2 Mapping and Decomposition Process

In this section, the first level FR-DP pair is broken down by the application of the decomposition process in further and more detailed levels. By "zigzagging" from one level to the next and simultaneous assignment of DPs to the respective associated FRs, the FR-DP tree is created. The review of the interactions, and therefore of the Independence Axiom, occurs simultaneously in the corresponding design matrix. Result of the decomposition process is a set of DP of a changeable and modular production system for franchising models. The mapping and decomposition process, starting from FR_0, shows at the first hierarchical level five basic requirements, henceforth called the design fields (DF) of the manufacturing system:

FR_1 Provide franchise-suitable and high qualitative products.
FR_2 Find a franchise-suitable network structure of distributed production facilities.
FR_3 Maximize scalability, adaptability, and cost efficiency of production.
FR_4 Enable an affordable supply and logistics.
FR_5 Use optimized and standardized processes.

The corresponding solutions to meet these functional requirements are as follows:

DP_1 Definition of products and services (**DF1—Assortment**);
DP_2 Franchise model and network structure (**DF2—Franchise model**);
DP_3 Changeable, scalable, replicable, and profitable production units (**DF3—Production unit**);

DP_4 Efficient supply structure (**DF4—Supply**);
DP_5 Franchise process organization (**DF5—Process**).

The design matrix on level 1 shows the influence of the solutions (DPs) on the functional requirements (FRs):

$$\begin{Bmatrix} FR_1 \\ FR_2 \\ FR_3 \\ FR_4 \\ FR_5 \end{Bmatrix} = \begin{bmatrix} X & 0 & 0 & 0 & 0 \\ X & X & 0 & 0 & 0 \\ X & X & X & 0 & 0 \\ X & X & 0 & X & 0 \\ X & X & X & X & X \end{bmatrix} \cdot \begin{Bmatrix} DP_1 \\ DP_2 \\ DP_3 \\ DP_4 \\ DP_5 \end{Bmatrix} \quad \text{(decoupled)} \tag{9.1}$$

The design matrix shows a decoupled design. The functional requirements are not clearly distinguishable from each other, but can be uncoupled ordering them in a proper sequence. Therefore, they show a useful or "good" system design. Figure 9.2 illustrates the graphical form of the FR-DP tree structure on hierarchy level 1. In their MSDD approach, Cochran et al. [16] visualize the dependencies between FRs and DPs in the form of arrow connections and align the structure of the FR-DP tree based on the principle that the picture is read from top to bottom (top-down) and from left to right (recommended sequence or path for iterating the DPs). Because those FR-DP pairs with most interactions with other elements are always located to the left, in the presence of a decoupled matrix, the correct path is necessarily the reading see "from left to right."

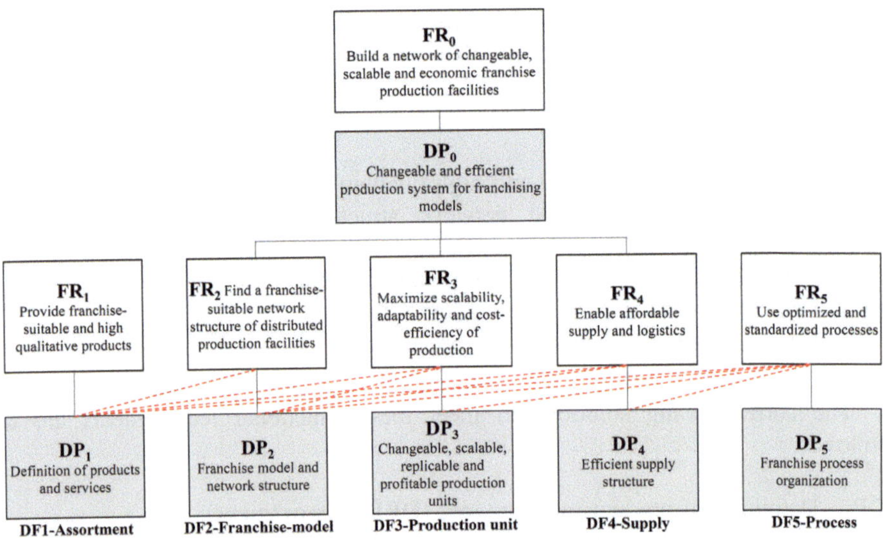

Fig. 9.2 FR-DP tree—hierarchy level 1

Starting from the decomposition of the first hierarchy level, the decomposition process continues to the next levels. The complete decomposition process comprises a total of 50 identified DP at the operational level. The above-shown five design fields were broken down to the levels 3, 4, and 5 obtaining concrete DP for the design of the franchise manufacturing system. For a better understanding of the approach, the decomposition process is shown exemplary on a part of the identified design fields (DF3-Production unit) analyzing the coupling of FRs and DPs from level 2 to level 5.

The functional requirement FR_3 (maximize scalability, adaptability, and cost efficiency of production) can be subdivided into three further functional requirements (see Table 9.1).

The design matrix shows a decoupled matrix.

$$\begin{Bmatrix} FR_{3.1} \\ FR_{3.2} \\ FR_{3.3} \end{Bmatrix} = \begin{bmatrix} X & 0 & 0 \\ 0 & X & 0 \\ X & X & X \end{bmatrix} \cdot \begin{Bmatrix} DP_{3.1} \\ DP_{3.2} \\ DP_{3.3} \end{Bmatrix} \quad \text{(decoupled)} \tag{9.2}$$

$DP_{3.1}$ is concerned with the adaptability and replicability of the production units, but needs a further decomposition to be broken down into more concrete proposals for solutions (see Table 9.2).

The design matrix for $FR_{3.1}$–$DP_{3.1}$ is thus a triangular matrix and must be decoupled by the correct sequence.

$$\begin{Bmatrix} FR_{3.1.1} \\ FR_{3.1.2} \\ FR_{3.1.3} \end{Bmatrix} = \begin{bmatrix} X & 0 & 0 \\ X & X & 0 \\ X & X & X \end{bmatrix} \cdot \begin{Bmatrix} DP_{3.1.1} \\ DP_{3.1.2} \\ DP_{3.1.3} \end{Bmatrix} \quad \text{(decoupled)} \tag{9.3}$$

Table 9.1 Decomposition FR_3—level 2

FR	Functional requirement	DP	Design parameter
$FR_{3.1}$	Maximize changeability of the production units	$DP_{3.1}$	Changeable and replicable production units
$FR_{3.2}$	Minimize production costs	$DP_{3.2}$	Elimination of non-value added activities
$FR_{3.3}$	Minimize overhead costs	$DP_{3.3}$	Reduction of assets, fixed capital, and overheads

Table 9.2 Decomposition $FR_{3.1}$—level 3

FR	Functional requirement	DP	Design parameter
$FR_{3.1.1}$	Maximize changeability and flexibility of machines	$DP_{3.1.1}$	Design guidelines of changeable machines
$FR_{3.1.2}$	Enable a gradual expansion of the production capacity	$DP_{3.1.2}$	Modular expansion levels (capacity, resources, layout)
$FR_{3.1.3}$	Minimize the effort for the realization of a new production	$DP_{3.1.3}$	Replicability of the production unit without effort

Table 9.3 Decomposition $FR_{3.1.1}$—level 4

FR	Functional requirement	DP	Design parameter
$FR_{3.1.1.1}$	Enable an easily shifting and movement of machines	$DP_{3.1.1.1}$	Mobility by locally unrestricted machines (wheels, …)
$FR_{3.1.1.2}$	Enable universal use of the machines	$DP_{3.1.1.2}$	Universal and flexible machines and work processes
$FR_{3.1.1.3}$	Enable simply linking the machines	$DP_{3.1.1.3}$	Compatibility with standard interfaces
$FR_{3.1.1.4}$	Allow reusability and extensibility	$DP_{3.1.1.4}$	Modular and scalable structure

The design guidelines for changeable manufacturing systems and equipment ($DP_{3.1.1}$) are based fundamentally on the changeability enablers: universality, mobility, scalability, modularity, and compatibility [24]. Table 9.3 shows the decomposition of $FR_{3.1.1}$.

The design matrix is again a triangular matrix (decoupled) and must follow a correct sequence.

$$\begin{Bmatrix} FR_{3.1.1.1} \\ FR_{3.1.1.2} \\ FR_{3.1.1.3} \\ FR_{3.1.1.4} \end{Bmatrix} = \begin{bmatrix} X & 0 & 0 & 0 \\ X & X & 0 & 0 \\ X & 0 & X & 0 \\ X & X & X & X \end{bmatrix} \cdot \begin{Bmatrix} DP_{3.1.1.1} \\ DP_{3.1.1.2} \\ DP_{3.1.1.3} \\ DP_{3.1.1.4} \end{Bmatrix} \quad \text{(decoupled)} \qquad (9.4)$$

The same procedure for top-down decomposition was applied for all other design fields and levels.

9.3.3 FR-DP Tree and Design Parameters

The result of the iterated decomposition process is the FR-DP tree with concrete DP at the lowest level (see Fig. 9.3). In this work, the software Acclaro DFSS was used to create the design matrix and the FR-DP tree as well as to do a digitally assisted review and check for the Independence Axiom. The entire FR-DP tree consists of five hierarchy levels. FR-DP pairs are marked with blue, and the blue lines between

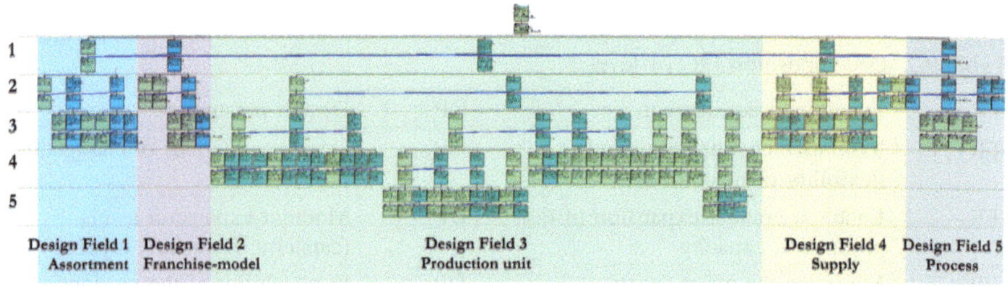

Fig. 9.3 Visualization of the full FR-DP tree with five hierarchical levels in Acclaro DFSS

DPs and FRs represent a path-dependent approach (decoupled), while green boxes and lines stand for an uncoupled design. A coupled design would be signalized with red boxes and lines. The FR-DP tree has to be read from left to right. Therefore, this AD-based sequence in the FR-DP tree is also a recommendation for the sequencing of the various DPs.

9.4 Design Fields and Design Elements

As just mentioned, to guarantee a systematic modeling of the manufacturing system, there were defined so-called design fields (DFs) (first level of the AD-based decomposition). At this design level, independent from location-specific factors (such as labor cost) in the franchise system, the system designer could create a uniform and standardized template of the manufacturing system. The identified five design fields, with their set of DP, form the normative framework for the further expansion and development of the franchise system with geographically distributed production sites.

As a result of this study, the recommended sequence of these design fields could also be determined, in order to avoid iterative loops in the design process to the extent possible and to reduce the complexity to a minimum. Figure 9.4 shows the identified design fields (DF1–DF5) and graphically describes the order in which the various fields should be treated. After determining the product or service assortment (DF1), the right franchise model (DF2) has to be defined. Once the franchise structure is clearly defined, the design of decentralized, changeable, and profitable production units (DF3) needs to be elaborated. In a next step, the supply of the production facilities and outlets has to be modeled (DF4). Ultimately, it is necessary to standardize and summarize all results acquired in the design fields in form of processes and procedures (DF5).

Within the design fields, the so-called design elements (DE) are defined. A manufacturing system is designed and assembled element by element; therefore, the design elements correspond to the derived DP in the decomposition process (concrete DP and solutions at the lowest level of the FR-DP tree). A total of 50 design elements (see Fig. 9.5) could be derived through the AD-based approach for

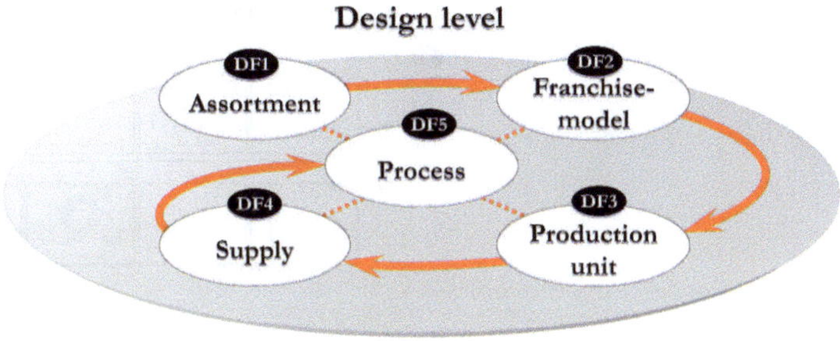

Fig. 9.4 Five resulting design fields of the franchise manufacturing systems

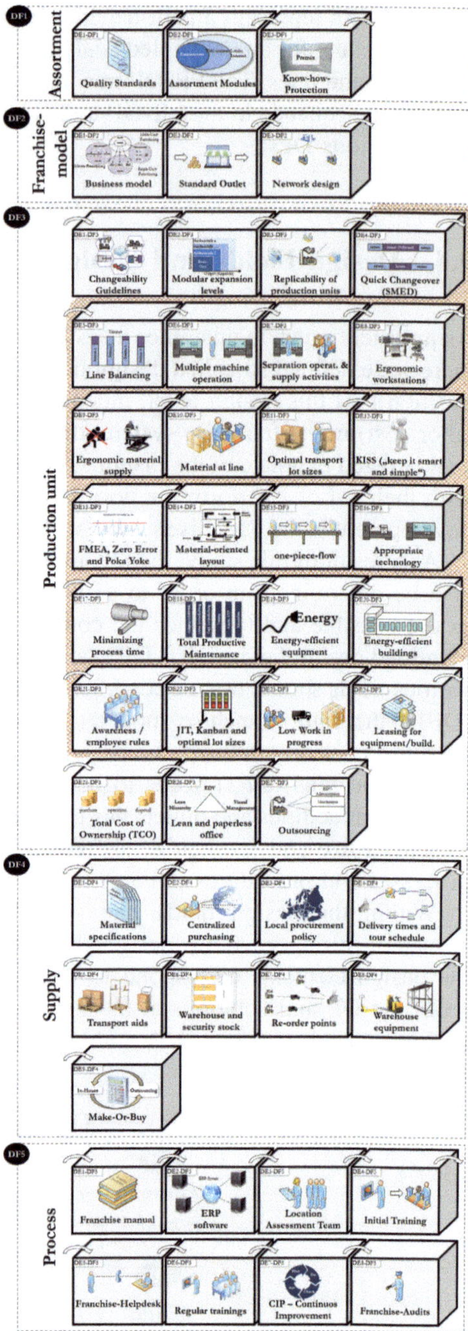

Fig. 9.5 Set of 50 design elements (DE) for the design of the manufacturing system

the design of the franchise manufacturing system, which in their totality constitutes a very useful tool for the system designer.

In design field DF1 (assortment), quality standards and a quality policy have to be defined. In addition, assortment modules should be formed defining a standardized core assortment and optional or locally individual assortments. It is also necessary to define strategies for protecting important product know-how.

The design of the franchise model (design field DF2) serves to select the right franchise model and to define the structure of the franchise network. Therefore, in a first step, an appropriate franchise model (e.g., master franchise, single-unit franchises, multi-unit franchise) has to be selected. Based on market studies, the capacity of standard outlets and the franchise network is performed.

The design of the production unit (design field DF3) offers the system designer a set of guidelines and DP for the design and modeling of the production units. Primary functional requirements of the production system are the ability to change and expand gradually the capability of the production units in order to replicate this as easy and fast as possible can. Another requirement of the production units is the resource-efficient and economical production of the products. A variety of methods and practices to eliminate waste, to optimize material flow and layout, to minimize overhead costs, etc., are listed in the design elements. Some design elements (see dashed area in Fig. 9.5) can also be combined into a macro-block "lean and green production." They include a number of known methods of lean manufacturing and energy efficiency improvement.

Design field DF4 (supply) includes various modules for structuring procurement, distribution, and warehouse logistics with the objective of efficient supply of production units and outlets. These are material specifications, a central procurement, local sourcing policy, standardized delivery, definition of transport modes and dimensioning of inventory, safety stock, and reorder points.

Design field DF5 processes are fixed in the form of a production franchise manual and suggestions for the ERP system. Central functions of business development and training are also defined in this design field as well as the establishment of a production franchise Helpdesk. Ultimately, continuous improvement (CIP teams) and responsibilities for auditing are determined.

9.5 Approach for Implementation—A Three-Level Model

The previously presented design fields with their design elements form the normative framework and the basis for the expansion and multiplication (rollout) of the franchise manufacturing system. However, for the testing of manufacturing system as well as for a systematic and prudent rollout, important elements are missing on a strategic–tactical level and the operational level. To give system designers a tool for the design and implementation of franchise manufacturing systems, a three-level

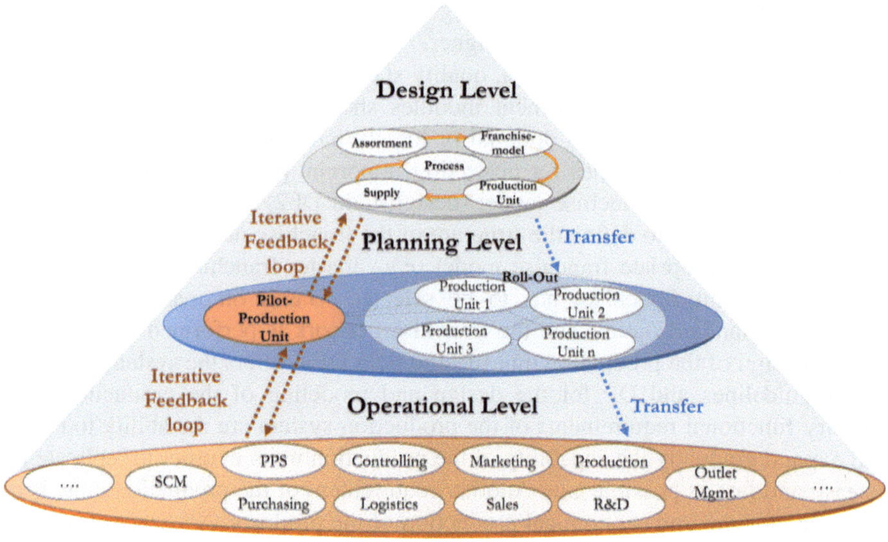

Fig. 9.6 Three-level model for the design, planning, and operation of a franchise manufacturing system

model is proposed (see Fig. 9.6). The model illustrated includes not only the design level, planning level, and the operational level, but also a feedback loop for the continuous adaptation of the manufacturing system.

9.5.1 Design Level (Normative Framework)

At the normative level, the system designer defines the design of the franchise manufacturing system. At this level, the design fields with their design elements are elaborated and defined. Thus, the modeling framework with its design templates is created. The horizon of the design level is long term and is thus over a period of five years. Periodically, the design fields and elements, however, should be checked for any necessary adjustments (trigger point for the redesign of the manufacturing system—see also [56]). At this design level, aside from the specific and location depending factors in the manufacturing system, the system designer can create a uniform and standardized template of the manufacturing system. The identified five design fields with their set of DP form the normative framework for the further expansion, adaption, and development of the manufacturing system with geographically distributed production sites.

9.5.2 Planning Level (Strategic–Tactical Framework)

Once the DP or elements for modeling the manufacturing system are developed on the design level, they have to be tested through the realization of a pilot production unit. The first step in the strategic and tactical planning level is planning and implementation of a pilot plant. The pilot production unit, which is operated by the franchisor itself, has to test and develop new products and production technologies. Once the pilot production is consolidated by iterative feedback to the design and operational level and the profitability of the business model has been proven, finally the multiplication of the production units and thus the rollout of the franchise model can be started. Before the start of the rollout, a multi-year scenario plan or business plan is being developed. This business plan includes not only the potential regions and countries, but also the number of planned outlets and production units as well as the time line for its implementation. The time horizon for this level includes the strategic planning in a time frame of three to five years and an annual, detailed tactical planning and budgeting.

9.5.3 Operational Level (Operational Framework)

The operational level comprises the implementation of the production units and the operational tasks of the franchisor with all his responsibilities. Of particular importance is that before the start of the rollout, all processes and operational issues (e.g., ordering procedure in the outlets and production units, integrated data management, process for product development, etc.) are tested and examined in the pilot production. As shown in Fig. 9.6, iterative feedback loops ensure that only a functional and viable production and franchise system is transferred to the franchisee. If not, there is a risk of failure of the franchisee and of the entire business model. The time horizon for the operational level is dominated through "daily business" and therefore shorter than one year.

9.5.4 Feedback Loop (Redesign and Replanning)

As described in Fig. 9.6, between the different levels there is an iterative feedback loop, similar to a control loop, to transfer the experiences from the pilot production unit to the other levels while "adjusting" and consolidating the manufacturing system. Between the different levels, we can distinguish two types of feedback loops or trigger points:

- Feedback loop on the design level ("*redesign*").
- Feedback loop on the planning level ("*replanning*").

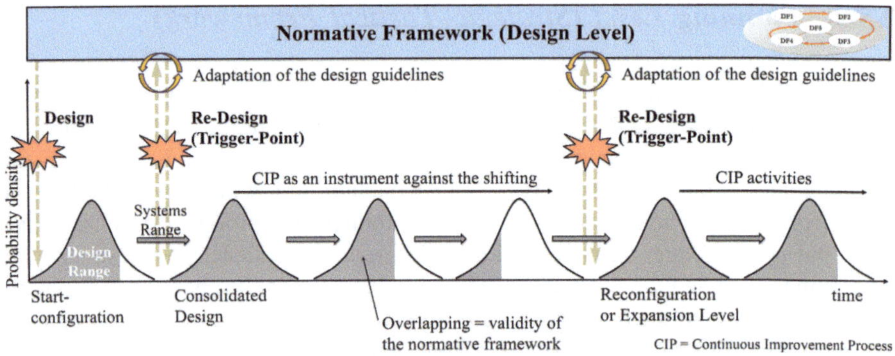

Fig. 9.7 Feedback and control loops for continuous adaption (trigger points for redesign)

The experience gained from the pilot production unit, as well as its reconfigurations, is transferred through the iterative feedback loops to new production units (rollout). The need for these control loops or feedback can be explained using the system theory. In the normal case can be assumed that the manufacturing system is based on requirements identified at a given time. At the time of initial operation of the manufacturing system, the design corresponds (more or less) to the previously defined requirements. The goal of a system design is to make the system range lies inside the design range (maximum probability density) [79]. Complexity arises if the system range (environment and external factors) lies outside the design range (designed manufacturing system). This is right for static systems or time-independent systems.

In Fig. 9.7, we can see the behavior of a manufacturing system over the time. Companies are, more than ever, subject of a turbulent environment with the result that the requirements of the production system are constantly changing. Future events are typically unpredictable and might shift the system range away from the defined design range—and thus creates complexity [54]. This ultimately carries the risk that it eventually comes to a collapse of the system, when the overlap between the production system and the requirements does not exist anymore. Once the overlap between the two areas is no longer sufficient, and thus the normative framework of the current design is no longer valid, a redesign (trigger point) is necessary [56].

By the above-described regular and systematic feedback loop and the continuous adaptation of the design level, the ability to change and adapt the entire manufacturing system can be guaranteed in the proposed approach.

9.6 Application in a Case Study

The shown approach was developed and applied in a real case study and subjected to validation. The company in the case study is a new Italian franchise brand, which began its activities several years ago with the opening of its first own outlets. The business idea is based on the concept of coffee shops with an integrated shop. The specialty of the company in the case study is the combination of coffee shop and self-made products in the shop. For the production of its own products, the company has established in advance an own pilot production unit, which first developed and produced in a traditional manner the products for the pilot market. With an increasing pilot market also the pilot production developed the industrial production methods. After the initial experience with the pilot production and outlets in the pilot market, the company pursued the vision of an international chain of franchise outlets and started a project for the development of a concept for global expansion and the related supply of the outlets. Due to the required freshness of the products and the limited shelf life and because of possible local needs of customers in the target countries, the company decided to produce with geographically distributed franchise production units. The case study showed very clearly that the implementation of such a franchise system without a suitable methodology would take very long and can be disturbed by frequent iterative loops in the planning and design phase. The approach described in this chapter was applied in the case study and was very helpful for the company. Through the approach, not only the DP for the manufacturing system could be defined, but also a simple and systematic approach for its implementation was developed. Subsequently, the application of the methodology in the case study is explained, in particular the design of the production units.

The *Design Field 1 (assortment)* initially required the definition of quality standards for products in order to ensure a consistent and standardized product quality in both the pilot production, as well as later in the franchise production units. In a second step was developed a clear structure of the product range differentiating between a standardized core assortment, an optional assortment depending on the market as well as rules for individual local products.

In the *Design Field 2 (franchise model),* the various alternatives in franchise models were investigated and evaluated for their suitability. Because of the geographically very distant target markets in the case study (Europe, Middle East, North America, and Asia) was selected master franchising as franchise form.

In the *Design Field 3 (production unit)*, the production system was designed highly flexible and adaptable. The requirements of the production system were not only a high degree of scalability in capacity expansion but also a high product flexibility to create new products or new product variants without major reconfigurability measures. With the increasing automation in the expansion stages was given particular emphasis on a high universality, mobility, and interoperability of the systems, according to the derived design guidelines for changeable production systems. For each of the defined product families have been defined production modules which, starting from a traditional operation (basic module), can be extended to industrialized production lines with correspondingly high output by stepwise and modular expansion. Figure 9.8 shows the derived production modules and the developed concept for expansion stages of a standard production unit in the franchise system. Depending on the assortment in the different outlets, the production unit is able to activate or deactivate flexible and scalable production

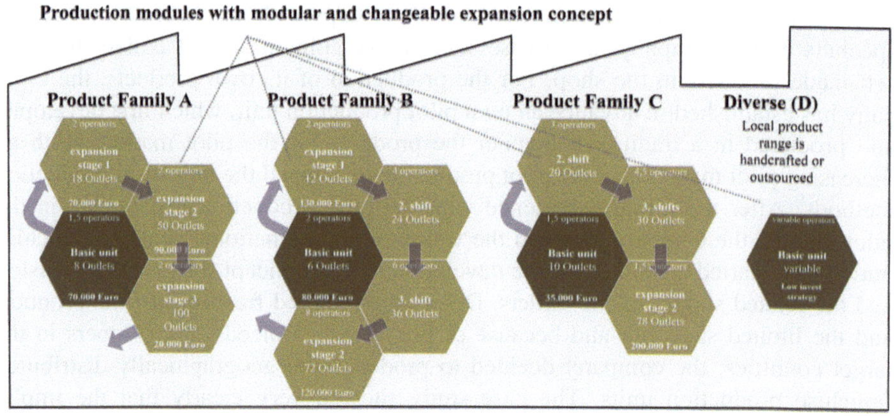

Fig. 9.8 Design field production unit: modular concept of expansion

modules for every product family. In the example, the capacity adaptation of the production unit can be achieved through either investment in a new expansion stage or through the switch from one shift on two or three shifts.

Bringing together the single modular expansion stages of every product family, a holistic scheme and development plan for the entire production unit was elaborated. Figure 9.9 shows the result with necessary information about capacity, human resources, needed space as well as investment for a manual basic unit, and a semi-industrialized and an industrialized production unit.

The plan of international expansion in the years after the consolidation of the pilot production unit shows an exponential increase (Fig. 9.10). The rollout plan gives information about the possibility to reuse machinery and equipment of the individual expansion stages at other locations. In addition, both the demand and the utilization of resources for central project planning, as well as for the training of the employees, can be aligned based on the defined rollout plan. The rollout plan gives also an idea of the necessary time for the realization of new production units and for activating the next expansion stages. It should be noted that in the industrial case

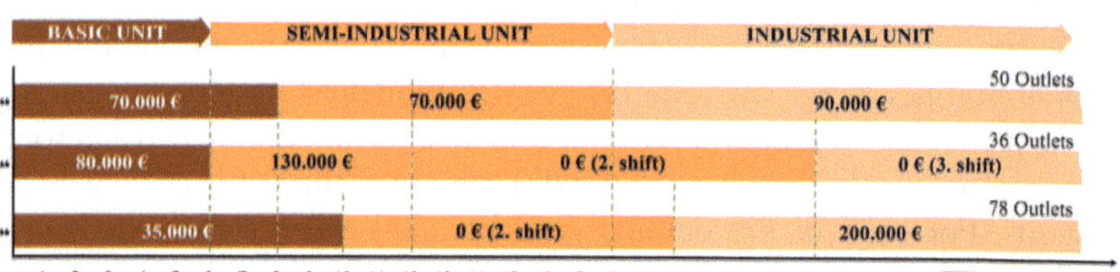

	Basic unit	Semi-industrial unit	Industrial unit
STD.-OUTLETS	6	9	36
INVESTMENT	435.000 €*	200.000 €	290.000
EMPLOYEES	5	5 - 9	9 – 9,5
SPACE	350 m²	420 m²	550 m²

* including 250.000 € of costs for initial infrastructure

Fig. 9.9 Expansion stage plan of a single production unit in the case study

market	production unit	year 1	year 2	year 3	year 4	year 5	year 6	year 7	year 8	year 9	year 10
local market	PU-pilot	5	6	8	10	12	14	16	18	20	22
Asia	PU-A1			3	5	7	9	12	15	18	22
Middle East	PU-ME1					3	6	9	13	18	23
USA	PU-US1							3	6	9	12
Total number of outlets		5	6	11	15	22	29	40	52	65	79

basic unit | semi-industrial unit | industrial unit

Fig. 9.10 Rollout planning of production units

was defined a yearly review of the scenario plan and therefore a revision of the rollout plan (replanning) in order to already react to any changes.

In *Design Field 4 (supply)* were defined the logistics supply structure of the production unit and the distribution of the products. In a first step was elaborated a concept to define, where the geographically distributed production units should order their procurement items for the production stage. Once set the procurement rules, were defined the logistical processes to supply the outlets in the production network.

In the *Design Field 5 (processes)* were defined the procedures in terms of process descriptions, work instructions, checklists, and forms. At the same time, the quality management system has been established in accordance with DIN ISO 9001. In this phase, also the roles and the responsibilities within the franchise system were defined clearly. Then, the remaining design elements (IT, training, continuous improvement, franchise manual, and audits) were defined.

9.7 Conclusion

Through the approach, not only the DP for a changeable and reconfigurable manufacturing system could be defined, but also a simple and systematic approach for its implementation was developed. By the "top-down" AD-based approach and the decomposition process was created a holistic overview of the requirements and design options. By the presented three-level model, system designers can find a complete and technically, economically as well as organizational aligned model for the design and implementation of distributed manufacturing systems. With this approach, a scientific contribution was presented to close the demonstrated research gap for the design of distributed manufacturing systems and production networks. The application in the case study showed that the one-time expense and effort in the AD decomposition, to develop the design fields and to create the normative framework on the design level, is not negligible, but then offers great benefits through a quick and high-quality design, planning, and implementation process of the manufacturing system. Through the built-in feedback loops, the internally developed knowledge can also be incorporated and transferred to other manufacturing units in the network and a continuous adaptation (changeability) of the manufacturing systems network is guaranteed. The approach was shown with the example of a franchising network, but is applicable also for other distributed manufacturing networks.

Further research will be necessary to investigate the right "trigger points" for a periodical redesign of the production system and to analyse how productivity reduces and changes if system range (environment) and design range (production system) shifts asunder.

References

1. D. Ahlert, *Handbuch Franchising und Cooperation. Das Management kooperativer Unternehmensnetzwerke* (in German) (Deutscher Fachverlag GmbH, Frankfurt, 2001)
2. T. AlGeddawy, H.A. ElMaraghy, Changeability effect on manufacturing system design, in *Changeable and Reconfigurable Manufacturing Systems*, ed. by H. ElMaraghy (Springer, London, 2009), pp. 267–280. http://link.springer.com/chapter/10.1007/978-1-84882-067-8_15
3. G. Arcidiacono, C.A. Brown, Axiomatic design of production systems for operational excellence, in *Proceedings of the 7th International Conference on Axiomatic Design (ICAD)*, Worcester (MA) 2013, ed. by M.K. Thompson
4. AZFranchises (2012) Quick Franchise Facts, Franchising Industry Statistics. http://www.azfranchises.com/franchisefacts.htm. Accessed 07 Oct 2012
5. B. Babic, Axiomatic design of flexible manufacturing systems. Int. J. Prod. Res. **37**(5), 1159–1173 (1999)
6. D. Bennett, R. Dekkers, Industrial networks of the future—a critical commentary on research and practice, in *Proceedings of the 12th International EurOMA Conference on Operational and Global Competitiveness*, Budapest (Hungary), 2005, ed. by K. Demeter
7. L. Bergmann, *Nachhaltigkeit in Ganzheitlichen Produktionssystemen* (in German) (Vulkan Verlag, Essen, 2010)
8. G. Bonani, *Guida al Franchising – come diventare Affiliato* (in Italian) (FrancoAngeli, Milan, 2004)
9. C. Brown, Axiomatic design of manufacturing processes considering coupling, in *Proceedings of the 8th International Conference on Axiomatic Design (ICAD)*, Universidade Nova de Lisboa, Lisbon, 24–26 Sept 2014
10. M. Bruccoleri, G.L. Nigro, G. Perrone, P. Renna, S.N. La Diega, Production planning in reconfigurable enterprises and reconfigurable manufacturing systems. CIRP Ann. Manuf. Technol. **54**(1), 433–436 (2005)
11. P. Butala, R. Vrabič, G. Oosthuizen, Distributed manufacturing systems and the internet of things: a case study, in *Proceedings of SAIIE 25*, Stellenbosch, South Africa, 9–11 July 2013, pp. 565-1–565-12
12. A.S. Carrie, L.E. Hayfron, U.S. Bititci, Is franchising the future of global manufacture? In *Proceedings of the IEEE Engineering Management Society*, Albuquerque, 13–15 Aug 2000, pp. 260–265
13. L.M. Castro, J. Mota, S. Marnoto, Towards a networks perspective of franchising chains. Serv. Bus. **3**, 15–30 (2009)
14. Y. Chen, S. Fan, Y. Ding, F. Guo, Design of reconfigurable manufacturing systems based on axiomatic design. Ind. Eng. Manag. **14**(5), 79–83 (2009)
15. S. Cochran, Y.-S. Kim, Teaching manufacturing system design with the axiomatic design methodology, in *Proceedings of the 1st International Conference on Axiomatic Design (ICAD)*, Cambridge, 21–23 Jun 2000, pp. 42–47

16. S. Cochran, J.F. Arinez, J.W. Duda, J. Linck, A decomposition approach for manufacturing system design. Int. J. Manuf. Syst. **20**(6), 371–389 (2001)
17. R.P. Dant, P.J. Kaufmann, Structural and strategic dynamics in franchising. J. Retail. **79**(2), 63–75 (2003)
18. R.M. Da Silva, M.F. Blos, F. Junqueira, D.J. Dos Santos Filho, P.E. Miyagi, Control architecture for reconfigurable and distributed manufacturing system, in *Proceedings of ABCM Symposium Series in Mechatronics*, vol. 6, pp. 219–230
19. R. Dekkers, A co-evolutionary perspective on distributed manufacturing, in *Distributed Manufacturing. Paradigm, Concepts, Solutions and Examples*, ed. by H. Kühnle (Springer, London, 2010), pp. 29–49
20. P. Dieses P, *Zukunft des Franchising in Deutschland* (in German) (Peter Lang Verlag, Frankfurt, 2004)
21. R. Dove, Design principles for highly adaptable business systems, with tangible manufacturing examples, in *Maynard's Industrial Handbook*, ed. by K. Zandin, H. Maynard (McGraw Hill, New York, 2001), pp. 9.3–9.26
22. R. Dove, Engineering Agile systems: creative-guidance frameworks for requirements and design, in *Proceedings of the 4th Annual Conference on Systems Engineering Research (CSER)*, Los Angeles, 7–8 Apr 2006
23. B. Elango, V.H. Fried, Franchising research: a literature review and synthesis. J. Small Bus. Manag. **35**(3), 68–81 (1997)
24. H. ElMaraghy, H.-P. Wiendahl, Changeability—an introduction, in *Changeable and Reconfigurable Manufacturing Systems*, ed. by H. ElMaraghy (Springer, London, 2009), pp. 3–24
25. B. Fagerström, M. Jackson, Efficient collaboration between main and subsuppliers. Comput. Ind. **49**(1), 25–35 (2002)
26. A.M. Farid, Facilitating ease of system reconfiguration through measures of manufacturing modularity. Proc. Inst. Mech. Eng. Part B J. Eng. Manuf. **222**(B10), 1275–1288 (2008)
27. A.M. Farid, Axiomatic design & design structure matrix measures for reconfigurability & its key characteristics in automated manufacturing systems, in *8th International Conference on Axiomatic Design (ICAD)*, Universidade Nova de Lisboa, Lisbon, 24–26 Sept 2014
28. A.M. Farid, D.C. McFarlane, Production degrees of freedom as manufacturing system reconfiguration potential measures. Proc. Inst. Mech. Eng. Part B J. Eng. Manuf. **222**(B10), 1301–1314 (2008)
29. A.M. Farid, Measures of reconfigurability & its key characteristics in intelligent manufacturing systems. J. Intell. Manuf. **1**(1), 1–17 (2014)
30. A.M. Farid, L. Ribeiro, An axiomatic design of a multi-agent reconfigurable manufacturing system architecture, in *8th International Conference on Axiomatic Design (ICAD)*, Universidade Nova de Lisboa, Lisbon, 24–26 Sept 2014
31. P. Gu, H.A. Rao, M.M. Tseng, Systematic design of manufacturing systems based on axiomatic design approach. CIRP Ann. Manuf. Technol. **50**(1), 299–304 (2001)
32. R. Hadar, A. Bilberg, *Glocalized Production—A Holistic Approach for Future Production at The LEGO Group*. Dissertation, University of Southern Denmark, 2014
33. L.E. Hayfron, A.S. Carrie, U.S. Bititci, K.V. Pandya, Manufacturing franchising: transferring the service experience, in *IEE Workshop Responsiveness in Manufacturing (Digest No. 1998/213)*, London, Feb 1998
34. L.E. Hayfron, A.S. Carrie, U.S. Bititci, K.V. Pandya, Manufacturing franchising and enterprise networks, in *Strategic Management of the Manufacturing Value Chain. International Conference of the Manufacturing Value-Chain*, Troon, Aug 1998, ed. by U.S. Bititci, A.S. Carrie (Kluwer Academic Publishers, Norwell), pp. 315–322
35. R. Hernández Morales, *Systematik der Wandlungsfähigkeit in der Fabrikplanung* (in German) (VDI-Verlag, Dusseldorf, 2002)
36. T. Hildebrand, K. Mäding, U. Günther, *PLUG+PRODUCE Gestaltungsstrategien für die wandlungsfähige Fabrik* (in German) (IBF, Chemnitz, 2005)

37. M. Houshmand, B. Jamshidnezhad, Conceptual design of lean production systems through an axiomatic approach, in *Proceedings of the 2nd International Conference on Axiomatic Design (ICAD)*, Cambridge, 10–11 Jun 2002
38. S.J. Hu, Evolving paradigms of manufacturing: from mass production to mass customization and personalization. Procedia CIRP **7**, 3–8 (2013)
39. W. Kieser, *Franchising - Wachstumschancen für KMU: Ein Leitfaden für Franchisegeber* (in German) (Gabler, Wiesbaden, 2010)
40. Y. Koren, U. Heisel, F. Jovane, T. Moriwaki, G. Pritschow, G. Ulsoy, H. Van Brussel, Reconfigurable manufacturing systems. CIRP Ann. **48**(2), 527–540 (1999)
41. Y. Koren, M. Shpitalni, Design of reconfigurable manufacturing systems. J. Manuf. Syst. **29** (4), 130–141 (2010)
42. C. Kubitschek, *Franchising. Effizienzvergleich mit alternativen Vertriebskonzepten* (in German) (Deutscher Universitätsverlag, Wiesbaden, 2000)
43. O. Kulak, S.I. Satoglu, M.B. Durmusoglu, Multi-attribute material handling equipment selection using information axiom, in *Proceedings of the 3rd International Conference on Axiomatic Design (ICAD)*, Seoul, 21–24 Jun 2004
44. M. Kunkel, *Franchising und asymmetrische Informationen. Eine institutionenökonomische Untersuchung* (in German) (Deutscher Universitätsverlag, Wiesbaden, 1994)
45. A. Lau, M. Hirsch, H. Matheis, From collaborative development to manufacturing in production networks: the SmartNets approach, in *Robust Manufacturing Control. Proceedings of the CIRP Sponsored Conference RoMaC 2012*, Bremen, June 2012, ed. by K. Windt (Springer, Heidelberg, 2012), pp 287–299
46. H. Lee, A. Banerjee, Executable design and control framework using reconfigurable manufacturing holons and its evolution algorithm. Int. J. Prod. Res. **49**(5), 1405–1423 (2011)
47. P. Leitão, F. Restivo, A framework for distributed manufacturing applications, in *Proceedings of the Advanced Summer Institute International Conference*, Bordeaux, 18–20 Sep 2000, pp. 75–80
48. P. Leitao, Agent-based distributed manufacturing control: a state-of-the-art survey. Eng. Appl. Artif. Intell. **22**(7), 979–991 (2009)
49. M. Martinek, Erscheinungsformen von Vertriebsverträgen (in German), in *Handbuch des Vertriebsrechts*, ed. by M. Martinek, F.-J. Semler (C.H. Beck, Munich, 2003), pp. 97–111
50. W. Martius, *Fairplay Franchising: Spielregeln für partnerschaftlichen Erfolg* (in German) (Gabler, Wiesbaden, 2008)
51. D.T. Matt, Axiomatic design of production modules templates, in *Proceedings of the 4th International Conference on Axiomatic Design (ICAD)*, Firenze, 13–16 June 2006
52. D.T. Matt, Achieving operational excellence through systematic complexity reduction in manufacturing system design. Key Eng. Mater. **344**, 865–872 (2007)
53. D.T. Matt, Template based production system design. J. Manuf. Technol. Manag. **19**(7), 783–797 (2008)
54. D.T. Matt, Functional periodicity as a concept for the (re-)design to agility of manufacturing systems. Prod. Eng. Res. Dev. **4**(4), 363–369 (2010)
55. D.T. Matt, Application of axiomatic design principles to control complexity dynamics in a mixed-model assembly system: a case analysis. Int. J. Prod. Res. **50**(7), 1850–1861 (2012)
56. D.T. Matt, E. Rauch, Continuous improvement of manufacturing systems with the concept of functional periodicity. Key Eng. Mater. **473**, 783–790 (2011)
57. D.T. Matt, E. Rauch, Design of a scalable modular manufacturing system for a two-stage food service franchise system: a case analysis. Int. J. Eng. Bus. Manag. **32**(4), 1–10 (2012)
58. D.T. Matt, E. Rauch, Design of a network of scalable modular manufacturing systems to support geographically distributed production of mass customized goods, in *Proceedings of the 8th CIRP Conference on Intelligent Computation in Manufacturing Engineering (ICME)*, Naples, 18–20 July 2012
59. D.T. Matt, E. Rauch, An AD based design and implementation approach for franchise-networks with distributed manufacturing units, in *Proceedings of the 7th International Conference on Axiomatic Design (ICAD)*, ed. by M.K. Thompson, Worcester (MA), 2013

60. D.T. Matt, E. Rauch, P. Dallasega, Trends towards distributed manufacturing systems and modern forms for their design, in *Proceedings of the 9th CIRP Conference on Intelligent Computation in Manufacturing Engineering (ICME)*, Naples, 23–25 July 2014
61. D.T. Matt, E. Rauch, D. Fraccaroli, A three level model for the design, planning and operation of changeable manufacturing systems in distributed manufacturing, in *Enabling Manufacturing Competitiveness and Economic Sustainability*, Munich, October 2013, ed. by M.-F. Zäh (Springer, Heidelberg, 2014), pp. 23–28
62. K. Metzlaff, Typischer Inhalt eines Franchisevertrages (in German), in *Praxishandbuch Franchising*, ed. by K. Metzlaff (C.H. Beck, Munich), pp. 43–214
63. R. Moch, R. Riedel, E. Müller, Key success factors for production network coordination, in *Enabling Manufacturing Competitiveness and Economic Sustainability*, Munich, October 2013, ed. by M.-F. Zäh (Springer, Heidelberg, 2014), pp. 327–332
64. P. Nyhuis, M. Heins, J. Pachow-Frauenhofer, G. Reinhart, P. Krebs, E. Abele, A. Wörn, Wandlungsfähige Produktionssysteme – Fit sein für die Produktion von morgen (in German). ZWF Zeitschrift für wirtschaftlichen Fabrikbetrieb **103**(5), 333–337 (2008)
65. H. Nylund, K. Salminen, P.H. Andersson, Framework for distributed manufacturing systems, in *Enabling Manufacturing Competitiveness and Economic Sustainability. Proceedings of the 4th International Conference on Changeable, Agile, Reconfigurable and Virtual production (CARV)*, Montreal, October 2011, ed. by H. ElMaraghy (Springer, Heidelberg, 2012), pp. 172–177
66. S.O. Monye, Transaction costs and the internationalization of business franchising, in *Research Papers in International Business*, South Bank University, London, 1997, pp. 1–97
67. N. Okino, Bionic manufacturing systems, in *Proceedings of the CIRP Seminar on Flexible Manufacturing Systems Past-Present-Future*, Bled, Apr 1993, ed. by J. Peklenik, pp. 73–95
68. H.-S. Park, H.-W. Choi, Development of a modular structure-based changeable manufacturing system with high adaptability. Int. J. Precis. Eng. Manuf. **9**(3), 7–12 (2008)
69. E. Puik, J. Gerritsen, E. Smulders, B. van Huijgevoort, D. Ceglarek, A method for indexing axiomatic independence applied to reconfigurable manufacturing systems, in *Proceedings of the 7th International Conference on Axiomatic Design (ICAD)*, Worcester (MA), 2013, ed. by M.K. Thompson
70. E. Rauch, *Konzept eines wandlungsfähigen und modularen Produktionssystems für Franchising-Modelle* (Concept of a changeable and modular manufacturing system for franchising models). (Fraunhofer Verlag, Stuttgart, 2013)
71. R. Reichwald, F. Piller, Der Kunde als Wertschöpfungspartner: Formen und Prinzipien (in German), in *Wertschöpfungsmanagement als Kernkompetenz*, ed. by H. Albach, B. Kaluza, W. Kersten (Gabler, Wiesbaden, 2002), pp. 27–52
72. G. Reinhart, A. Krüger, M. Prasch, Stückzahl- und Variantenflexible Montage (in German), in *Die wandlungsfähige Fabrik*, ed. by H.-P. Wiendahl (IFA, Hannover, 2003), pp. 183–215
73. K. Ryu, *Fractal-Based Reference Model for Self Reconfigurable Manufacturing Systems.* Dissertation, Pohang University of Science & Technology, 2003
74. K. Ryu, Y. Son, M. Jung, Modeling and specifications of dynamic agents in fractal manufacturing systems. Comput. Ind. **52**(2), 161–182 (2003)
75. S. Schmid, P. Grosche, *Glokale Wertschöpfung im Volkswagen-Konzern – Auf dem Weg zu mehr Dezentralisierung bei Produktion und Entwicklung* (Glocal value in the Volkswagen Group—Towards more decentralization of production and development). ESCP-EAP Working Paper Nr. 41, Nov 2008 (European School of Management, Berlin, 2008)
76. W. Skaupy, *Franchising. Handbuch für die Betriebs- und Rechtspraxis* (in German) (Vahlen, Munich, 1995)
77. W. Skaupy, Grundlagen, in *Praxishandbuch Franchising* (in German), ed. by K. Metzlaff (C. H. Beck, Munich, 2003), pp. 1–19
78. D. Spath, Lean und intelligent produzieren – aber wie? (in German), in *Modellfabrikforum 2006 – Ganzheitliche Gestaltung wettbewerbsfähiger Produktionssysteme* (IAO, Stuttgart, 2006), pp. 1–51
79. N.P. Suh, *The Principles of Design* (Oxford University Press, New York, 1990)

80. N.P. Suh, *Axiomatic Design: Advances and Applications* (Oxford University Press, New York, 2001)
81. N.P. Suh, D.S. Cochran, P.C. Lima, Manufacturing system design. CIRP Ann. Manuf. Technol. **47**(2), 627–639 (1998)
82. J. Sydow, Franchisingnetzwerke. Ökonomische Analyse einer Organisationsform der Dienstleistungsproduktion und -distribution (in German). Zeitschrift für Betriebswirtschaft **64**(1), 95–113 (1994)
83. M.K. Thompson, J.T. Foley, Coupling and complexity in additive manufacturing processes, in *8th International Conference on Axiomatic Design (ICAD)*, Universidade Nova de Lisboa, Lisbon, 24–26 Sept 2014
84. G.Y. Tian, G. Yin, D. Taylor, Internet-based manufacturing: a review and a new infrastructure for distributed intelligent manufacturing. J. Intell. Manuf. **13**(5), 323–338 (2002)
85. K. Ueda, I. Hatono, N. Fujii, J. Vaario, Line-less production system using self-organization: a case study for BMS. CIRP Ann. **50**(1), 319–322 (2001)
86. H. Van Brussel, J. Wyns, P. Valckenaers, L. Bongaerts, P. Peeters, Reference architecture for holonic manufacturing systems: PROSA. Comput. Ind. **37**(3), 255–274 (1998)
87. B. Versaevel, Coordination costs and vertical integration in production franchise networks: a common agency model. Res. Econ. **56**(2), 157–186 (2001)
88. H.J. Warnecke, *The Fractal Company: A Revolution in Corporate Culture* (Springer, Berlin, 1993)
89. H.-P. Westkämper, E. Zahn, P. Balve, M. Tilebein, Ansätze zur Wandlungsfähigkeit von Produktionsunternehmen. Ein Bezugsrahmen für die Unternehmensentwicklung im turbulenten Umfeld (in German). wt Werkstattstechnik online **90**(1/2), 22–26 (2000)
90. H.-P. Wiendahl, C.L. Heger, Justifying changeability—a methodical approach to achieving cost effectiveness, in *Proceedings of the 2nd CIRP International Conference on Reconfigurable Manufacturing*, Ann Arbor, 20–21 Aug 2003
91. H.-P. Wiendahl, H.A. ElMaraghy, P. Nyhuis, M.F. Zäh, H.-H. Wiendahl, N. Duffie, M. Brieke, Changeable manufacturing—classification, design and operation. CIRP Ann. Manuf. Technol. **56**(2), 783–809 (2007)
92. Y.Y. Yusuf, M. Sahardi, A. Gunasekaran, Agile manufacturing: the drivers, concepts and attributes. Int. J. Prod. Econ. **62**(1/2), 33–43 (1999)
93. M.F. Zäh, W. Wagner, Planning mini-factory structures for the close-to-market manufacture of individualized products, in *Proceedings of the 2nd Interdisciplinary World Congress on Mass Customization and Personalization*, Technical University of Munich, Munich, 6–8 Oct 2003

Chapter 10
Axiomatic Design of Production Systems for Performance Improvement: A Project Identification and Prioritization Model

Gabriele Arcidiacono, Christopher A. Brown, Luca Bucciarelli and Francesco Melosi

Abstract Companies should align production systems according to their overall strategy and consider the strategic goals of the organization as a whole. To be competitive and profitable, it is not sufficient to improvise, although it is necessary to consider all the variables and scenarios and accommodate the different contexts and situations as appropriate. To improve their competitive abilities and to enhance cost-reduction opportunities and process efficiency, organizations are bringing about improvements in their operations and processes, adding global operations optimization to a global manufacturing strategy. However companies' ability to develop sustainable competitive advantage from these improvements is hampered by the lack of objective approaches for targeting their improvement efforts. There are significant limitations to the approaches used for project selection and prioritization, therefore the purpose of this work is to provide a structured approach, using Axiomatic Design (AD) principles, to identify and prioritize the best projects that are conducive to process excellence and performance improvement. Through the application of the Axiomatic Design method, we identify where complexity exists within the requirements and design activities that underpin the model. Using this analysis, this work identifies the critical points within the Project Identification

G. Arcidiacono (✉)
Department of Innovation and Information Engineering, Marconi University, Rome, Italy
e-mail: g.arcidiacono@unimarconi.it

C.A. Brown
Department of Mechanical Engineering, Worcester Polytechnic Institute, Worcester, MA, USA
e-mail: brown@wpi.edu

L. Bucciarelli
Business School, Leanprove, Florence, Italy
e-mail: luca.bucciarelli@leanprove.com

F. Melosi
Department of Industrial Engineering, University of Bologna, Bologna, Italy
e-mail: francesco.melosi@studio.unibo.it

© Springer International Publishing Switzerland 2016
A.M. Farid and N.P. Suh (eds.), *Axiomatic Design in Large Systems*,
DOI 10.1007/978-3-319-32388-6_10

and Prioritization model, and suggests necessary improvements to facilitate the implementation of Process Efficiency within a company.

Keywords Axiomatic Design · Operational Excellence · Project Identification · Project Prioritization · Performance Improvement

10.1 Introduction

Success, profitability, and overall competitiveness of a manufacturing organization are closely attributed to the effectiveness and to the efficiency of its operations [1]. In order to reach Operational Excellence, many organizations strive to pursue a strategy of Continuous Improvement to reduce costs and improve productivity, with the main goal of improving overall performance [2]. To gain and sustain a competitive advantage, it is critical to identify and carry out improvement initiatives to enhance their operations. Therefore the first step is to select the right projects.

However, many companies did not have a formal project identification and selection process for performance improvement [3] and this leads to failure. In response to this difficulty, a significant amount of work has been done in the area of project selection and prioritization.

Different authors have developed and proposed a wide range of objective methods to help practitioners deal with the complexity of the selection and prioritization of improvement projects. Subjective approaches may involve brainstorming, focus groups, interviews, and customer visits, while objective methods and tools include Pareto analysis [4], Pareto Priority Index (PPI) [5], Project ranking matrix [6], Project selection matrix [7], Quality Function Deployment (QFD) [8], Project assessment matrix [9], Cost Benefit Analysis (CBA) [10], Analytical Hierarchy Process (AHP) [11], Theory of Constraints (TOC) [5].

However, there isn't a quantifiable/scientific way to identify parts of a business that are conducive to performance improvement. This often results in the application of the wrong tools/methodology to resolve a particular problem [12] and, despite that, the literature consistently shows the importance of project selection. A great number of companies have failed at performance improvement primarily because of poor project selection models.

The subsequent sections of this work describe a model to aid in projects identification, selection, and prioritization. The model, through characterization and process clustering, establishes the evaluation criteria to enable the identification of business aspects conducive for process efficiency, a topic that is extensively addressed in the literature today. Then, through a Cost Deployment Framework and project prioritization tools, it is possible to identify a list of potential projects and prioritize them.

To give consistency to the proposed methodology, Axiomatic Design (AD) is used as the tool to design the model, while a Balance Scorecard (BSC) perspective [13] is used as the theme for the decomposition. Axiomatic Design provides a framework in which the design process can be managed [14]. In particular, it

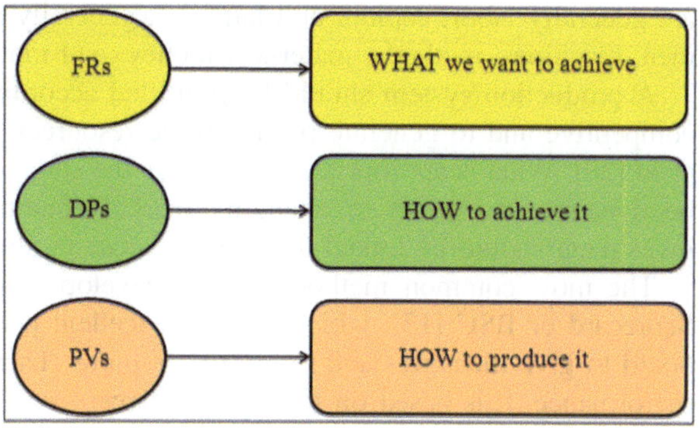

Fig. 10.1 Explanation of the different variables related to the domains

provides criteria for distinguishing bad designs from good ones [15]. The systematic bi-dimensional decomposition used in Axiomatic Design facilitates the inclusion of all the relevant variables and scenarios, as well as contexts and situations.

The first dimension of the decomposition into functional, physical, and process domains provides a clear categorization of Functional Requirements (FRs), Design Parameters (DPs), and Process Variables (PVs). These represent the domain where the concepts "WHAT we want to achieve" and "HOW we want to achieve it" lie (Fig. 10.1).

The second dimension of the decomposition is hierarchical within the domains. This analysis can be done according to equivalence relations, based on partitioning [16]. The objective is to achieve a collectively exhaustive and mutually exclusive collection of the functions [14, 17] to address the relevant business situations.

Axiomatic Design supplies companies with a disciplined design process [18]. In particular, the AD process drives the decomposition between domains and "qualitatively" defines the project structure. It provides the basis for the selection of the key physical variables (DPs) that characterize the design satisfying the FRs. The selection of the DPs is tested against the axioms. The process of matching variables in one domain (e.g., FRs) with other variables in another domain (e.g., DPs) is called mapping: to go from WHAT to HOW.

10.2 Production System Design: Theory

A system produces an output by acting on and transforming its inputs. The output is influenced by noise factors, which are generated from interactions. AD provides for control of interactions and noise factors [19].

The production system is the set of methods used in industry and the related processes that transform resources into finished goods and services. The resources are generally labor, capital, and land, but generally are called also the "six M's": men, machines, methods, materials, money, and mother nature.

A production system should be projected according to company strategy to be competitive and to generate profits, these resources should be appropriately managed [20]. What is appropriate depends on the situation. Every organization should be dynamic and adapt to changing market conditions. In addition, the capital investment should be linked to focus on areas in alignment with the strategy.

The most common method used to develop company strategy is a Balanced Scorecard or BSC [13], which uses an excellent performance measurement dashboard to give managers and executives a more "balanced" view of organizational performance. It is based on four perspectives:

1. Economic-Financial perspective
2. Customer-Market perspective
3. Processes perspective, (e.g., Operational Excellence strategy, decrease operation costs and cycle time)
4. Learning and Innovation perspective

The courses of action selected by the company should be structured so that they can be overseen from these four perspectives. This oversight would verify their efficiency in the chosen market segment. It would also establish the role by which companies are ordinarily classified. This classification is based on:

1. Product
2. Product plus (the best product compared to the competition, e.g., extra comfort in an airline)
3. Price
4. Customization

The first step is to choose the placement in the market, i.e., the first of the four categories mentioned above, and to project the subsequent business model. At the same time, it is also necessary to design an appropriate production system to optimize the processes. The objective of this design is to improve process efficiency and to introduce new products/services or new technologies.

The Production System basically consists of four general types:

1. The project (one-shot) system-for a one-off product, such as a made-to-order ship, or a prototype.
2. The batch system–variable lot sizes, depending on the kind of process/product.
3. The continuous system (assembly line)—common in mass production.
4. Any mix of the above systems.

The production system is characterized by physical flows of materials and by the flow of information in the process, depending on the previous typology of the system.

10.3 Production System Design and Axiomatic Design: Design to Target Improvement Efforts

This work focuses on production system design, using AD in order to decompose what we want to achieve (Functional Requirements) and how to achieve it (Design Parameters). Adding a global operations optimization to a global manufacturing strategy can provide cost-reduction opportunities and make the processes more efficient. In particular, focusing on building and sustaining supply chain organization and capabilities, is useful to facilitate the implementation of Process Efficiency within a company. The first step is to identify, select and prioritize the projects for Process Efficiency.

Top managers (also Chief or C-Levels) have to be focused on assessing and developing a customized global production system. CEOs of some major companies that have developed customized, global production systems have been studied in order to define the business macro aims (FRs) within the functional domain. Typical BSC perspectives are used to suggest a theme for the decomposition (see Figs. 10.3 and 10.4):

FR1 = Establish shareholders' value (Economic-Financial perspective)
FR2 = Provide competitiveness in the Market (Customer-Market perspective)
FR3 = Improve process efficiency (Processes perspective)
FR4 = Provide innovations (Learning and Innovation perspective)

To satisfy these FRs, the following DPs have been suggested by the CEOs:

DP1 = Sector selection and the placement of the company (Economic-Financial perspective)
DP2 = Business Model Design (Customer-Market perspective)
DP3 = Production System Design (Processes perspective)
DP4 = New products/services or new technologies Innovation System (Learning and Innovation perspective)

The highest level Design Matrix (DMX) is shown in Fig. 10.2. The interactions have been determined by the CEOs.

The DMX demonstrates that the project is decoupled, considering A12, A13, A23 (whose correlation value has been indicated with a dot, ".") and negligible with respect to the others values "x" as well as "X". In other words, it is possible to consider a dot as being equal to "0". Axiom 1 can also be satisfied by a decoupled design, taking into account the order in which the DPs must be adjusted (the proper sequence). It is worth noting that, for a full triangular matrix, there is only one order in which the DPs can be adjusted to satisfy the FRs without iterating. In other words, when designing from scratch, it is best to find an uncoupled design. If this is impossible, a decoupled design is acceptable. Under some circumstances, however, it might be necessary to deal with coupled designs. Even in these cases, it is important to realize that Axiom 1 can still provide guidance. Beyond the three main categories of coupling, further sub-types of coupling with variable levels of severity

FRs/DPs	**DP1** Sector/Placement	**DP2** Business Model	**DP3** Production System	**DP4** Innovation System
FR1 Add value	X	.	.	0
FR2 Competitiveness	X	X	.	0
FR3 Improve efficiency	x	X	X	0
FR4 Innovate	x	X	x	X

Fig. 10.2 Design matrix DMX

exist (e.g., full coupling is worse than sparse coupling, and stiff coupling is worse than robust coupling) [20]. In this way, the proper sequence has been identified [21], as required by the first axiom of Axiomatic Design [22]. First, select the sector, then the business model, followed by the production system, and, lastly, the innovation system.

Through the decomposition process, it is possible to study the details in the functional and physical domains (FRs in Fig. 10.3a, b, DPs in Fig. 10.4a, b) through zig-zagging. Using mapping and zig-zagging, the design can be summarized in two structures that are hierarchically arranged in levels of increasing detail and correlated by the design matrices.

The expected output of this exercise is a production system that leads the company to maximum competitiveness, considering the constraints of available resources and available capital. Competitiveness in the market requires a calculation

(a)

FR1= Establish Shareholders' Value (Economic-Financial)

- FR11= Grow revenues
- FR12= Improve EBIT margin
- FR13= Increase Return on Invested Capital
- FR14= Increase free Cash Flow
- FR15= Generate profitable business growth
- FR16= Build a solid financial structure

FR2= Provide Competitiveness (Customer-Market)

- FR21= Increase Customer Satisfaction
  - FR211= Ensure ROI and Value for the Customer
  - FR212= Deliver products on time (TTM, quantity and quality)
  - FR213= Maintain Product Quality Consistency
  - FR214= Provide effective Customer Service
- FR22= Grow in the core business
- FR23= Optimize geographic diversification

Fig. 10.3 **a, b** Functional domain

(b)

FR3= Improve Process Efficiency (Processes)
FR31= Target improvements
FR311 = Identify Processes
FR3111= Decompose the Value Chain
FR3115 = Measure Lead Time and Cycle Time
FR3112 = Identify Process Input/Output and Internal/External actors
FR3113 = Map the Process Steps
FR3114 = Identify VA and NVA activities
FR312 = Define the characteristics of the processes
FR3121= Choose the Process Parameters
FR3122= Establish an allocation Score and
Score the processes based on process parameters
FR313 = Cluster and Evaluate Processes
FR3131= Group Processes with similar characteristics
FR3122= Evaluate Overall Desirability using geometric means
FR3133=Weigh process parameters (select the shape parameters)
FR3124=Establish cut-off for Process efficiency conductive cluster
FR314 = Identify the intervention areas and the list of potential projects
FR3141 = Identify the costs of transformation and their nature and classify them in the appropriate categories
FR3142 = Allocate losses and wastes
FR3143 = Identify the possible causes of losses and wastes
FR3144 = Calculate the cost of losses and wastes
FR3145 = Identify methods for recovery losses and wastes
FR3146 = Estimate the cost of the improvement and the corresponding reduction of losses and wastes
FR3147 = Establish and Implement the improvement plan
FR315 = Prioritize Projects
FR3151 = Select the projects according to business goals and sys tem's constraints
FR3152 = Prioritize Portfolio Projects
FR32= Increase process performance
FR321= Lead and sustain processes efficient
FR322= Reduce NVA by reviewing the Value Chain
FR323= Restore basic conditions & standard Best Practice
FR322= Reduce or eliminate the Non-Value Added activities
FR33= Cut the costs ("hard" cost savings)
FR331= Reduce labor cost
FR332=Reduce material cost
FR333= Reduce products/activities portfolio
FR34= Avoid the costs ("soft" cost avoidance)
FR341= Avoid a labor's hours increase
FR342= Avoid a row material/supplier's increase
FR343= Avoid a new material purchase in the introduction of a new product

FR4= Provide Innovations (Learning & Innovation)
FR41=Build a strong corporate culture
FR42= Become innovators and customer-driven
FR43= Increase number of New Products and Service Development
FR44= Develop a new competitive business model for the Market

Fig. 10.3 (continued)

(a)

DP1= Sector Selection and the Placement of the Company (Economic-Financial)

DP11=Focusing on the market segments where product portfolio has significant competitive advantage (emerging economies; 35% of total revenues coming from after-sales activities; price increase strategy)
DP12= Positioning product and services at market value (Price) and reviewing organization & structure on a regular basis (Cost)
DP13= Thorough analysis prior to engaging in any capital investment activity and subsequently stringent project management
DP14= Three main levers: Days Sales Outstanding (DSO), Days Payable Outstanding (DPO) and Inventory
-DSO: contract negotiations, credit collections and disputes resolution
-DPO: extended payments to supplier base
-Inventory: minimize inventory levels and maximize inventory turns
DP15= Pipeline management, Quotation budgeting & follow up, Project execution management, Overall flow governance
DP16= Correct mix between debt and equity with regularity and certainty of income (minimum cost of capital, minimum risk, maximum return, commensurate to legal requirements)

DP2= Business Model Design (Customer-Market)

DP21= Customer delighters management
DP211= Kano model
DP212= Lead Time management
DP213= Surveys analysis
DP214= VOC analysis
DP22= Increasing Market share focusing the Company efforts
DP23= Creating new extendible opportunities across different countries

Fig. 10.4 **a**, **b** Physical domain

of the capacity of the system. Too much capacity could burden a company with high costs. With too little capacity, opportunities could be lost, especially if a market is developing rapidly.

Mechanisms such as hiring-&-firing workers, scheduling overtime and cutting back on work hours, changing the rate of production, adding and shutting down machines, etc., are singular important leverages to be included in a global company strategy. Some of the effectiveness of the "adjustment" in the company capability would be an important design tool.

The system capability of managing the flows in order to achieve the expected FRs depends, for example, on the quality of the goods and services, durability, functionality, and on-time delivery by the company and by the suppliers. The flexibility of the production volume, which is required to meet changes in market demand, depends on the technology to be used and on the process design. These include the choice of equipment, layout, space, and procedures. In this scenario, the process effectiveness has to be improved with the appropriate production system design.

(b)

DP3= Production System Design (Processes)

- DP31=Identification and Prioritization Model
  - DP311 = Process Definition Framework
    - DP3111 = Process Decomposition Model
    - DP3115 = Value Stream Mapping
    - DP3112 = SIPOC
    - DP3113 = Functional Flow Chart Diagram
    - DP3114 = Value-added Flow chart Diagram
  - DP312 = Process Characterization
    - DP3122= Process Parameters selection
    - DP3121= Likert Scale Analysis
  - DP313 = Process Clustering
    - DP3131= Hierarchical Clustering Algorithms
    - DP3132= Desirability function
    - DP3133= Analytic Hierarchy Process
    - DP3134= Cut-off point
  - DP314 = Cost Deployment Framework
    - DP3141= Cost Allocation Matrix
    - DP3142= Loss Allocation Matrix
    - DP3143= Loss – Causes Matrix
    - DP3144= Costs – Losses Matrix
    - DP3145= Loss – Solutions Matrix
    - DP3146= Cost-Benefit Matrix
    - DP3147= Projecting and Budgeting
  - DP315 = Project Prioritization Matrix
    - DP3151 = Portfolio Optimization Model
    - DP3152 = Project Ranking Matrix
- DP32= Operational Excellence Model
  - DP321= Lean Six Sigma (LSS)
  - DP322= Toyota Production System (TPS)
  - DP323= World Class Manufacturing (WCM)
  - DP324= Basic Lean
- DP33= Cost Reduction
  - DP331= Downsizing
  - DP332= Optimizing BOM
  - DP333= Redesigning assets and Company Strategy
- DP34= Cost Avoidance
  - DP341= Increasing job rotation and people flexibility
  - DP342= Selecting alternative raw materials/suppliers with the same quality
  - DP343= Standardizing codes and reducing the relative number

DP4= New Products/Services or New Technologies Innovation System (Learning & Innovation)

- DP41= Communicating and sharing the Vision and the responsibilities of the Company
- DP42= Deploying the VOC into VOP (through QFD)
- DP43= Using Product Design and Development
- DP44= Using Design for X (through TRIZ, Robust Design)

Fig. 10.4 (continued)

The focus should be on the strengths for value-added activities, simultaneously designing a business model that can capture the voice of the customer and increase customer satisfaction.

Cochran [23] uses AD to illustrate the differences between two different production systems (mass and lean production). More specifically, AD is an important element for defining how the production system goals are accomplished from a system design perspective, in order to increase customer satisfaction and improve performances.

In particular, the goal of this work is to design a customized production system model to improve process efficiency in order to optimize overall processes, by targeting improvement efforts. Specifically, this work is focused on the identification and the prioritization of process improvement projects, starting from an analysis that begins with business processes and up to the prioritization of the projects identified, considering both macro-economic and market prospective as well as the company prospective, which can also vary quickly. For this purpose, the model developed by Montgomery [12] has been extended, using recent methodological developments. To validate the proposed methodology, Axiomatic Design (AD) is used as the tool to design the model. From this analysis, it is possible to create process efficiency and obtain hard/soft cost savings.

Our analysis starts from the study of the design matrix DM_{3X}. Figure 10.5 shows the results of this comparison. DM_{3X} is decoupled and satisfies Axiom 1. It could be argued that the FRs 'cut cost' and 'avoid cost' are inherently coupled. If so, then this decomposition would violate the decomposition directive to be mutually exclusive. However, in this case, 'cut the cost' refers to reducing existing costs, and

FR3/DP3	**DP31** Identification and Prioritization Model	**DP32** Operational Excellence Model	**DP33** Cost Reduction	**DP34** Cost Avoidance
FR31 Target improvements	X	0	0	0
FR32 Increase performance	x	X	0	0
FR33 Cut the costs	x	X	X	0
FR34 Avoid the costs	x	X	X	X

Fig. 10.5 Design matrix DM_{3X}

'avoid costs' refers to avoiding new costs; so the two are independent and satisfy Axiom 1.

Decomposing F31 and DP31, 'Target improvements' and 'Identification and Prioritization Model', results in the following elements (see Figs. 10.3 and 10.4):

FR311 = Identify Processes
FR312 = Define the characteristics of the processes
FR313 = Cluster and Evaluate Processes
FR314 = Identify the intervention areas and the list of Potential Projects
FR315 = Prioritize Projects

And

DP311 = Process Definition Framework
DP312 = Process Characterization
DP313 = Process Clustering
DP314 = Cost Deployment Framework
DP315 = Project Prioritization

Whose design matrix (DM_{31X}) is (Fig. 10.6):

The DM_{31X} demonstrates that the project is decoupled, considering A324, A334, (whose correlation value has been indicated with a dot, ".") negligible with respect to the others values "x" as well as "X". In other words, it is possible to consider a dot as being equal to "0".

In this case, the design matrix is decoupled, indicating that Axiom 1 is fulfilled, so this is the right sequence. The proper sequence of adjustment that satisfies the FRs without iteration is indicated.

The first step (FR311) is to identify the processes to which increase effectiveness. For this purpose, it is necessary to create a process framework to enable an end-to-end value chain definition and characterization by using a Process Identification Framework (DP311). This framework is composed of five steps, each of which uses a specific tool: Fig. 10.7 shows the method and the tools used.

In this case, the design matrix is coupled, indicating that Axiom 1 is not fulfilled. Therefore it is necessary to find a proper sequence for the FRs and the DPs required for decoupling by using "Reordering" [22]. The design matrix DM_{31X} after the "Reordering" is represented in Fig. 10.8.

The first step (FR3111) is to decompose the value chain, with the goal of identifying the processes where it is possible to increase efficiency.

For the purpose of this work, a Classic Process Decomposition Model (DP3111), with its hierarchical structure will suffice [22]. This model is based on Process Classification Framework, developed by the American Productivity and Quality Center (APQC) [24]. Figure 10.9 is an example of one such model.

There are 5 levels of decomposition: Value Chain (level 1), Process Chain (level 2), Process Steps (level 3) and Process Activities and Task (levels 4, 5). Usually, level 3 process decomposition was chosen as the lowest level of decomposition for the identification model. In fact, any further process decomposition will result in projects being identified at the activity and task level instead of at the functional

FR31/DP31	DP311 Process Definition Framework	DP312 Process Characterization	DP313 Process Clustering	DP314 Cost Deployment Framework	DP315 Project Prioritization
FR311 Identify Processes	X	0	0	0	0
FR312 Define the characteristics of the processes	x	X	0	.	0
FR313 Cluster and Evaluate processes	x	X	X	.	0
FR314 Identify the intervention areas and the list of Potential Projects	x	x	x	X	0

Fig. 10.6 Design matrix DM_{31X}

process level. As a result, projects will be characterized according to a very narrow scope. Alternatively, the level 2 of process decomposition does not provide an appropriate level of process granularity.

The second step (FR3115) is necessary to measure lead time and cycle time. For this purpose, it will be used the Value Stream Mapping (VSM); VSM will help to demonstrate the flows beyond every single process by providing a common language and showing the link between material and information flows. It also detects value-added activities and not added value through the system logistics and production, which would help to assess the waste causes.

It is then necessary to identify the actors (internal/external) and the inputs/outputs impacting the process (FR3112). In this phase, the tool used is the SIPOC (DP3112), a high-level process mapping that allows the identification of the

FR311/DP311	DP3111 Process Decomposition Model	DP3112 SIPOC	DP3113 Functional Flow Chart Diagram	DP3114 Value-added Flow chart Diagram	DP3115 Value Stream Mapping
FR3111 Decompose the Value Chain	X	0	0	0	0
FR3112 Identify Process Input/Output and Internal/External actors	0	X	0	0	x
FR3113 Map the Process Steps	x	0	X	x	X
FR3114 Identify VA and NVA activities	0	0	x	X	X
FR3115 Measure Lead Time and Cycle Time	0	0	0	0	X

Fig. 10.7 Design matrix DM_{311X}

suppliers, inputs, process steps, outputs and customers. SIPOC mapping contains information useful to process improvement and problem solving by defining the perimeter of the process.

After the SIPOC a detailed process mapping is performed (FR3113), identifying all the activities and tasks to determine the business functions involved in the process. For this purpose a Functional Flow Chart Diagram will suffice (DP3113). Finally, by using a Value-added Flow Chart Diagram (FR3114), it is possible to distinguish the VA activities from the NVA activities.

The process definition framework described in the previous sections sets the landscape of processes that will be evaluated. The second Step (FR312: Define the characteristics of the processes) helps to evaluate the applicability of process

FR311/DP311	DP3111 Process Decomposi-tion Model	DP3115 Value Stream Mapping	DP3112 SIPOC	DP3113 Functional Flow Chart Diagram	DP3114 Value-added Flow chart Diagram
FR3111 Decompose the Value Chain	X	0	0	0	0
FR3115 Measure Lead Time and Cycle Time	0	X	0	0	0
FR3112 Identify Process Input/Output and Internal/External actors	0	x	X	0	0
FR3113 Map the Process Steps	x	X	0	X	x
FR3114 Identify VA and NVA activities	0	X	0	x	X

Fig. 10.8 Design matrix DM_{311X} after reordering

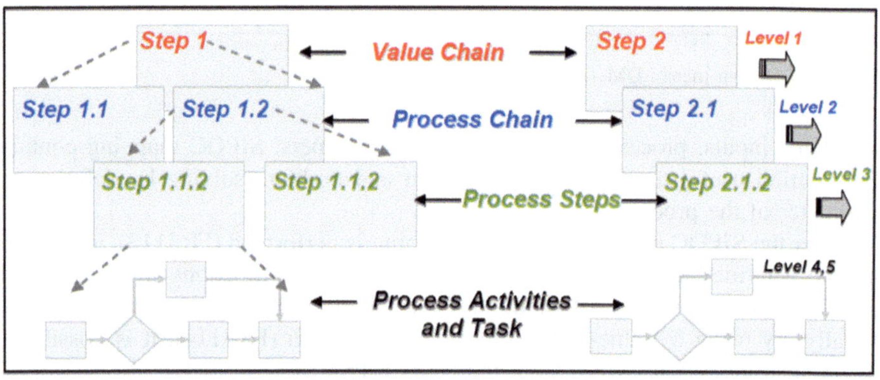

Fig. 10.9 Process definition framework [22]

FR312/DP312	DP3121 Likert Scale Analysis	DP3122 Process Parameters
FR3121 Establish an allocation Score	X	X
FR3122 Score the processes based on process parameters	X	X

Fig. 10.10 Design Matrix DM_{312X}

efficiency tolls to a business process and provides guidelines regarding which of these methodologies are best suited for each individual process. Therefore, several factors are considered while evaluating a process, and a Likert scale of 1–5 [25] is used to score each process based on these factors (Fig. 10.10).

In this case, the design matrix is coupled, indicating that Axiom 1 is not fulfilled. Therefore, it is necessary to change the FRs and find a proper sequence for the DPs required for decoupling, using "Reordering" [22] between FR/DP_{3121} and FR/DP_{3122}. The design matrix DM_{31X} after the change of functional requirements and "Reordering" is represented in Fig. 10.11.

So the first step is to choose the process parameters. For example, Lean Six Sigma is a process improvement methodology that has a strong statistical undertone. The methods and tools used are data driven and work best on structured repeatable processes that are not performing relative to customer expectations, so the frequency

FR312/DP312	DP3122 Process Parameters selection	DP3121 Likert Scale Analysis
FR3121 Chose the Process Parameters	X	0
FR3122 Establish an Allocation Score and Score the processes based on process parameters	x	X

Fig. 10.11 Design matrix DM_{312X} after reordering

of execution of a process, the process structure, and the availability of an established process metric/measurement are of crucial importance. Other parameters that can be evaluated are process cost, performance factor, strategic impact (risk)/VOC, geographical dispersion, level of process automation. Processes that are localized are good candidates for Lean Kaizen events, and processes that are extremely manual are potential opportunities for productivity type improvements. For processes with a lack of strategic impact or a lack in process structure, it is advisable to change (or create) a business process, policy, or even the IT based infrastructure.

The second step is to establish an allocation score and score the processes. For example, the frequency of a process varies from 1—High frequency of execution-daily- to 5—the Process is executed once a year. It is necessary that the scoring process is performed by a multifunctional team of experts. The allocation score tool used for this phase is the Likert Scale.

The data collected in the previous step was sanitized and validated to ensure that the scoring process was consistently applied to all processes. After defining the characteristics of the processes, it is necessary to group all processes that have similar characteristics and evaluate them (FR313 Cluster and Evaluate Processes). Figure 10.12 shows the method and the tools used. DM_{313x} is decoupled and satisfies Axiom 1.

FR313/DP313	**DP3131** Hierarchical Clustering Algorithms	**DP3132** Desirability function	**DP3133** Analytic Hierarchy Process	**DP3134** Cut-off point
FR3131 Group Processes with similar characteristics	X	0	0	0
FR3122 Evaluate Overall Desirability using geometric mean	x	X	0	0
FR3133 Weigh process parameters (select the shape parameters)	0	x	X	0
FR3124 Establish cut-off for Process improvement conductive cluster	x	0	0	X

Fig. 10.12 Design matrix DM_{313X}

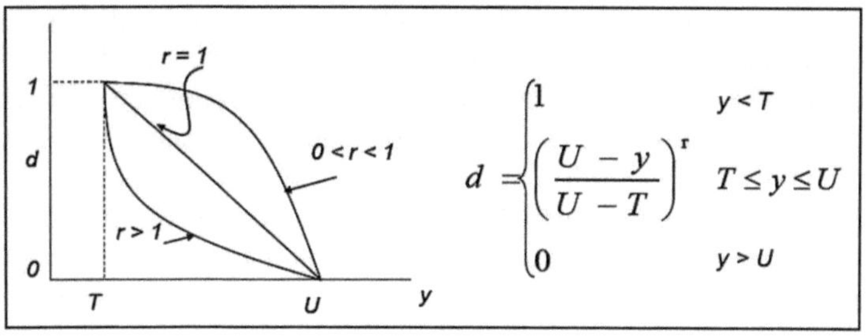

Fig. 10.13 Minimum value case—desirability function [12]

The first step (FR3131) is to group processes with similar characteristics. For this, an unsupervised learning approach using an agglomerative hierarchical clustering algorithm (DP3131) has been used to group candidate processes, based on common process characteristics [25].

In the second step (FR3132), through a desirability function analysis [26], we determine an overall desirability based on the weighted product of the individual desirability indices [27]. This step ascertains the individual desirability functions for each process characteristic (d_i) [28]. This is approximating the shape of the function relative to the importance of the factor. Based on the scoring system used, the model fits the "Minimum Value Case," since a lower score on the Likert scale is more desirable. Figure 10.13 illustrates the profiles of a desirability function for the minimum value case.

In addition, target value 'T' = 1, the Upper Limit 'U' = 5.05 is established, to ensure non-zero desirability values (Each process is scored on a scale of 1–5) [12].

Then, it is necessary to weigh the process parameters (FR3133) by selecting the shape parameters r. This could be done by an Analytical Hierarchy Process (AHP) as described by Saaty [11].

The shape parameter 'r' dictates the shape of the desirability function. For 'r' = 1 the function is linear. For 'r' > 1, the function is convex and places more importance on the parameter being closer to the target value. When $0 < r < 1$, the function is concave, with smaller values of 'r', denoting that the factor is less important as far as meeting the target. For this data set, assumptions were made to determine the shape parameters for each of the factors. The shape parameters were chosen based on a pair wise comparison of each factor.

For example the parameters with the highest importance were scored '5', while factors with the least importance were scored '0.5'.

The value of the shape parameter 'r' significantly impacts the overall desirability score of a cluster, and hence care must be taken to ensure that the values are chosen appropriately.

Overall Desirability ($D_i = (d_1 d_2 d_3 \ldots d_m)^{1/m}$) is the geometric mean of the individual desirability indices.

Finally, a cut-off point (FR3124) is established to aid in deciding which clusters are process efficiency conducive and which are not. The cut-off is at an overall desirability of 0.17 and is a function of the scaling parameter *'r'* used in the desirability indices. This cut was set based on clusters that scored at least 'X' (from 1 to 5 depending on the cutoff assigned to each process parameter).

Clusters above the cutoff tend to not have the best characteristics for process efficiency engagements. For these processes, alternate transformation options could include a change in the business model, policy, or even IT based infrastructure changes.

For the processes that are process-efficiency conducive, the next step is to identify specific projects to address the key performance indicators and the process parameters. Specific project could additionally address process simplification, process standardization, product quality, and process lead time. As a result, there could be multiple projects for each process that are process improvement conducive.

An effective tool for identifying the intervention areas (FR314 employs the following: Identify the intervention areas and the list of Potential Projects) on which to implement efficiency projects, using the Cost Deployment Framework (DP314), an innovative system for management and control of establishments that introduces a strong link between individualization of areas to be improved and the results of the performance improvements obtained, constituting a reliable means of program budgeting [29]. This framework is composed of seven steps, each of which uses a specific tool. Figure 10.14 shows the method and the tools used. DM314x is decoupled and satisfies Axiom 1.

A Cost Deployment Framework allows for defining improvement programs that have an impact in reducing losses or everything that can be classified as wastes or non-value added in a systematic way [30].

The foundation of the methodology is the systematic identification of waste and losses of the area under examination, its evaluation and transformation into values. This is possible because it relates waste and losses to their causes and origins, allowing a complete definition of the loss.

In addition, the framework guides the individualization of the best technical method to remove the cause and assess in detail the activity costs of removal and performance improvement [30].

After identifying a significant number of potential improvement projects, it is necessary to prioritize them (FR315).

First, it is necessary to select and evaluate projects according to the business goals and the constraints of the system. For this purpose, a Portfolio Optimization Model (DP3151) is used, which will provide both objective functions (such as Projects probability of success or Net Present Value of Saving) and constraints due to the system (Resources, Budget, mix of projects, etc.). Then it is necessary to prioritize the portfolio projects. For the purpose of this work, a Project Prioritization Matrix (DP315), which is a simple objective method widely used by many companies, will suffice (Fig. 10.15).

FR314/DP314	DP3141 Cost Allocation Matrix	DP3142 Loss Allocation Matrix	DP3143 Loss – Causes Matrix	DP3144 Costs - Losses Matrix	DP3145 Loss – Solutions Matrix	DP3146 Cost-Benefit Matrix	DP3147 Projecting and Budgeting
FR3141 Identify the costs of transformation and their nature and classify them in the appropriate categories	X	0	0	0	0	0	0
FR3142 Allocate losses and wastes	x	X	0	0	0	0	0
FR3143 Identify the possible causes of losses and wastes	0	x	X	0	0	0	0
FR3144 Calculate the cost of losses and wastes	x	x	0	X	0	0	0
FR3145 Identify methods for recovery losses and wastes	0	0	X	0	X	0	0
FR3146 Estimate the cost of the improvement and the corresponding reduction of losses and wastes	x	0	0	X	X	X	0
FR3147 Establish and Implement the improvement plan	0	0	x	x	x	X	X

Fig. 10.14 Design matrix DM_{314X}

FR315/DP315	DP3151 Portfolio Optimization Model	DP3152 Project Ranking Matrix
FR3151 Select the projects according to business goals and system's constraints	X	0
FR3152 Prioritize Portfolio Projects	x	X

Fig. 10.15 Design matrix DM_{315X}

10.4 Concluding Remarks

A Project Identification and Prioritization model is proposed in order to identify the parts of the Production System that are conducive to performance improvement: processes that use a system design are able to deploy the company strategy through singular operations and the reciprocal interactions [31] that are required for management.

The model, through characterization and process clustering, establishes the evaluation criteria that enable the identification of the business parts conducive for process effectiveness, a topic that is not yet widely addressed in literature. Through a Cost Deployment Framework and a project prioritization matrix, it is possible to identify a list of potential projects and prioritize them.

Three key elements of AD, adaptable to various manufacturing environments and extendible across industries, are:

1. Decomposition in design domains
2. Zig-zagging to create the design hierarchy
3. Independence Axiom

The decomposition includes functional and physical domains and provides the methodology for designing a Project Identification and Prioritization model. The decomposition facilitates the selection of new DPs (system designs) to meet new FRs. The zig-zagging process establishes a hierarchy of DPs at a higher level, determining the decomposition of FRs at lower levels through the FRs-DPs levels. The Independence Axiom drives the designer to select one and only one DP to satisfy an FR. Through the application of the Axiomatic Design method, we identify five steps within the model, suggesting necessary improvements to facilitate the implementation of Process Efficiency within a company. As evident in the Kaplan and Norton model, the connections between the legs are multidirectional. As a result, targeting efforts for improvement projects tend to maximize not only the strategic internal aspect of organizational processes, but the performances of the entire company. A culture of Continuous Improvement affects the Production Systems, but also the innovation approach and the competences needed; as a consequence, this also affects the customer perception of the product/service produced, the shareholders, and the financial domain of the organization. Given the possibilities for interdisciplinary use of this model and tools, further work and studies are required to test the proposed method in the actual industrial environment in order to validate and update its definitions.

References

1. M.L. Emiliani, Origins of lean management in America: the role of Connecticut businesses. J. Manage. Hist. **12**(2), 167–84 (2006)
2. J. Arturo Garza-Reyes, S. Eldridge, K.D. Barber, H. Soriano-Meier, Overall equipment effectiveness (OEE) and process capability (PC) measures: a relationship analysis. Int. J. Qual. Reliab. Manage. **27**(1), 48–62 (2010)

3. J.P. Zimmerman, J. Weiss, *Six Sigma's Seven Deadly Sins* (2005), pp. 62–66
4. A. Larson, *Demystifying Six Sigma: A Company-Wide Approach to Continuous Improvement* (AMACOM, New York, 2003). 191 p
5. T. Pyzdek, P.A. Keller, *The Six Sigma Handbook: A Complete Guide for Green Belts, Black Belts, and Managers at All Levels*, 3rd edn. (McGraw-Hill Companies, New York, 2010). 548 p
6. C.W. Adams, P. Gupta, C.E. Wilson, *Six Sigma Deployment* (Boston: Butterworth-Heinemann, Amsterdam, 2003), 290 p
7. M. Kelly, Three steps to project selection, in *ASQ Six Sigma Forum Mag*azine, vol. 2, no. 1 (2002), pp. 29–33
8. P.S. Pande, R.P. Neuman, R.R. Cavanagh, *The Six Sigma Way How GE, Motorola, and Other Top Companies Are Honing Their Performance (Internet)* (McGraw-Hill, New York, 2000) (cited 23 June 2015)
9. F.W. Breyfogle, J.M. Cupello, B. Meadows, *Managing Six Sigma: A Practical Guide to Understanding, Assessing, and Implementing the Strategy That Yields Bottom Line Success* (Wiley, New York, 2001). 272 p
10. A. Hira, T.W. Parfitt, *Development Projects for a New Millennium* (Praeger, Westport, Conn, 2004), 204 p
11. T.L. Saaty, Decision making with the analytic hierarchy process. Int. J. Serv. Sci. **1**(1), 83–98 (2008)
12. B. Duarte, D. Montgomery, J. Fowler, J. Konopka, Deploying LSS in a global enterprise—project identification. Int. J. Lean Six Sigma **3**(3), 187–205 (2012)
13. R.S. Kaplan, D.P. Norton, *The Strategy-Focused Organization: How Balanced Scorecard Companies Thrive in the New Business Environment* (Harvard Business School Press, Boston, Mass, 2001), 400 p
14. C.A. Brown, Decomposition and prioritization in engineering design, in *Proceedings of ICAD2011 The Sixth International Conference on Axiomatic Design*. Daejeon, 2011
15. N.P. Suh, *The Principles of Design* (Oxford University Press, New York, 1990), 401 p
16. R.A. Brualdi, *Introductory Combinatorics*, 3rd edn. (Prentice Hall, Upper Saddle River, New Jersey, 1999)
17. E.M. Rasiel, *The McKinsey Way: Using the Techniques of the World's Top Strategic Consultants to Help You and Your Business* (McGraw-Hill, New York, 1999), 187 p
18. M. Nordlund, D. Tate, N.P. Suh, *Growth of Axiomatic Design through Industrial Practice* (Tokyo; 1996), pp. 77–84
19. G. Arcidiacono, C.A. Brown, Axiomatic design of production systems for operational excellence, in *Proceedings of the 7th ICAD 2013 paper ICAD 2013-35*. Worcester, 2013
20. S. Kalpakjian, S.R. Schmid, *Manufacturing Engineering and Technology*, 7th edn. (Pearson, Upper Saddle River, NJ, 2014). 1180 p
21. G. Arcidiacono, R. Capitani, Guidelines for defining a design standard for an electrical steering system. J. Eng. Design **13**(4), (2002)
22. N.P. Suh, *Axiomatic Design Advances and Applications* (Oxford University Press, New York, NY, 2001), 503 p
23. D.S. Cochran, W. Eversheim, G. Kubin, M.L. Sesterhenn, The application of axiomatic design and lean management principles in the scope of production system segmentation. Int. J. Prod. Res. **38**(6), 1377–1396 (2000)
24. https://www.apqc.org/pcf. 30 June 2015
25. R. Likert, Technique for the misure of attitudes arch. Psycho **22**(140) (1932)
26. R. Xu, D.C. Wunsch, *Clustering* (Wiley, IEEE Press, Hoboken, N.J, Piscataway, NJ, 2009), 358 p
27. G. Derringer, R. Suich, Simultaneous optimization of several response variables. J. Qual. Technol. **12**, 214–9 (1980)
28. R.H. Myers, D.C. Montgomery, C.M. Anderson-Cook, *Response Surface Methodology: Process And Product Optimization Using Designed Experiments*, 3rd edn. (Wiley, Hoboken, N.J, 2009). 680 p

29. L. Massone, *Fiat Group Automobiles Production System: Manual do WCM, World Class Manufacturing: Towards Excellence Class Safety, Quality, Productivity and Delivery* (Ed. Fiat, Brazil, 2007)
30. H. Yamashina, T. Kubo, Manufacturing cost deployment. Int. J. Prod. Res. **40**(16), 4077–4091 (2002)
31. G. Arcidiacono, C. Calabrese, K. Yang, *Leading Processes to Lead Companies Kaizen Leader & Green Belt Handbook* (Springer, Dordrecht, 2012)

Chapter 11
Challenges in Designing and Implementing Large Systems (Overcoming Cost Overruns and Missed Project Schedules)

Nam P. Suh

Abstract People have developed many imaginative and innovative *systems* in all fields of human endeavor to satisfy human and societal needs. These systems were designed to satisfy a specific set of goals, or functional requirements (FRs), within a set of constraints (Cs). One of the major goals is to develop and deliver the final functioning system on time and within the original budget. Sometimes, achieving these goals at all times has been challenging. Many highly publicized projects, such as the new Berlin Airport, the F-35 fighter airplane, Boston's underground highway ("The Big Dig"), the new Berlin Airport, and the US healthcare system, have missed their original cost estimates and delivery schedules. These projects have been staffed with some of the most experienced, skilled, and intelligent engineers and managers, who put in countless hours to ensure their success. Yet they failed in terms of two important metrics: budget and delivery time. While there were most likely non-technical factors such as continuously changing requirements and constraints that affected the outcomes of these projects, in many cases, the basic root cause may be attributed to coupling of FRs. This paper supports this conclusion by using the systems the author has designed in the past as examples. Some of these systems are technical, and some are non-technical. Finally, a theorem is presented on one of the root causes of cost overruns and missed schedules.

Keywords Axiomatic Design · Large systems · Cost overruns · Missed schedule · Coupled design · Complexity

Some of the materials presented was the work done at KAIST (Korea Advanced Institute of Science and Technology), Korea, where he was the president for seven years from 2006 to 2013.

N.P. Suh (✉)
M.I.T, Cambridge, MA, USA
e-mail: npsuh@mit.edu

© Springer International Publishing Switzerland 2016

A.M. Farid and N.P. Suh (eds.), *Axiomatic Design in Large Systems*,
DOI 10.1007/978-3-319-32388-6_11

11.1 Introduction

Human civilization has been built on a series of imaginative and innovative solutions that people devised or designed to satisfy human and societal needs in many fields: energy, electric power generation, food, transportation, health care, education, information technology, banking, defense, environment, communications, and materials. Many of these innovative artifacts were in the form of *systems* that were designed to satisfy a specific set of functional requirements (FRs) and constraints (Cs). Some of these systems were large physical systems that consisted of with many subsystems. These systems varied in physical size, information content, materials used, complexity, etc. Some systems were time invariant, satisfying the same set of FRs at all time. However, some satisfied different sets of FRs as a function of time, i.e., time variant. People have designed amazing artifacts in the past. In contrast to physical systems, some systems are sociological and organizational systems such as universities and governments, where the issues are different but face the same problems, often manifesting in inefficient costly systems. Notwithstanding the advances made in creating and using systems, still there are many intellectual challenges in designing systems, especially large systems. Even today, many systems are still created through trial-and-error processes based on past experience and company-specific design processes, which may result in poorly functioning systems that require a series of expensive revisions or redesigns.

Systems are designed to produce certain outputs using a set of input parameters within a set of constraints. Therefore, we should be able to design all these systems, both physical and organizational, using the same methodology rather than a variety of ad hoc approaches, although the specific nature of the problem (or goals), physical principles, data, and acceptable variations are field and organization specific. However, the reality is that the system design and implementation are mostly ad hoc. In fact, there are many famous systems that could not be delivered on time and within the original budget. On the other hand, there are systems that were completed as originally intended within budget and on schedule. The basic question is: "what is the basic difference between these two cases? Can we design and implement systems that satisfy FRs and Cs to be on time and within budget at all times?" Furthermore, "how will they perform in the field?"

This paper addresses two challenging issues involved in designing and commissioning large physical systems: cost overruns and missing the original schedule. We will assume that the systems that exceed the original budget and miss the delivery time are poorly designed systems, although in the end they may satisfy a compromised set of FRs and constraints (Cs). In general, experience shows that cost overruns and missed schedules are a consequence of having introduced functionally coupled designs in the system. This happens when we design a system primarily based on intuition, past experience, and trial-and-error processes without the rigor of design principles. Modeling and simulation are necessary tools in system development, but they are not the best means of identifying causality of cost overruns and missed schedules. Experience with large system design reinforces the view that it is the coupling of FRs, i.e., coupled designs,[1] that is primarily responsible for cost overruns and missed schedule. In some cases, the people in charge of the project may compromise (or modify) the FRs and constraints in order to accept the product they ended up producing, not the product that satisfies the original set of FRs and Cs. Each delay and each change in FRs and Cs would incur additional cost.

One of the interesting and extremely profound questions is: "why are so many designers create or choose coupled designs if they are indeed responsible for cost overruns and missed schedules?" The behavior of designers will not change unless we can answer this question. In some ways, the root cause of this problem is the engineering design education, which does not teach the fundamental principles of design.

The assessment of these problems in sociological and organizational design is much more difficult, because the determination of the outputs is more difficult and subjective, although the same methodology has been used. However, exceptional cases are cited where noticeable changes were made, which qualitatively assess the design and performance of organizations. Eventually, as more case studies are made, and as the number of institutions that are designed based on these methodologies increases, more definitive assessment may be made.

This paper presents the importance of *System Architects* in preventing the creation of a coupled design. The function of the System Architect is to systematically monitor the design process to be certain that it does not inadvertently create coupled designs. An example will be given later. The rationale is that when the coupling of FRs is identified early, the design team can alter the design to come up with uncoupled designs rather than finding out later when the system does not function as intended after the hardware is made. Design changes can be made more easily and inexpensively when a coupled design is identified as early as possible in the project or design cycle.

Finally, this paper will conclude with a theorem on cost overrun and missed deadlines in completing the project and with comments on how industry, universities, and government can play their roles in overcoming the shortcomings encountered in designing large systems.

11.2 Major Projects with Cost Overruns and Missed Schedules

The enormous cost, both financial and human, of making wrong design decisions can be discerned from highly publicized cases. Recently, Bloomberg Business (Joshua Hammer on July 25, 2015) published a story on the fiasco involving the new airport being built for Berlin, Germany, under the heading of "How Berlin's futuristic airport became a $6 billion embarrassment." Its scheduled opening in

[1]In this paper, "coupling" implies "functional coupling" not physical coupling.

2012 has been delayed to 2017 with its project cost exceeding the original budget by tenfold, apparently due to the poor design of its fire alarm and communications systems. It appears that the managers of this project were chosen for their experience, and they relied on their experience rather than instituting a rational system for design, development, and execution of the project. Unfortunately, this is not an isolated case in many industries.

Another example of large systems project that exceeded the original cost estimate by a wide margin was the Boston's underground Central Artery, known as the "Big Dig"—the largest and the most expensive highway project in the history of the USA. This underground and undersea highway has improved the skyline of Boston and significantly improved the traffic in the greater Boston area. However, this project exceeded its initial cost estimate, from $2 billion to $18 billion, and took 30 years of planning and 12 years to construct. Why did it cost so much and took it so long to finish? Only the deep pockets of the Federal government of the USA and the Commonwealth of Massachusetts could have prevented a complete fiasco.

Another well-known large system that has exceeded the original cost and the development schedule is the F-35 Joint Strike Fighter airplane developed by Lockheed Martin for the US Department of Defense. According to David Francis (The Fiscal Times; July 31, 2014), "The F-35 Joint Strike Fighter is the most expensive, and possibly the most error ridden, project in the history of the United States military. But DOD has sunk so much money into the F-35—which is expected to cost $1.5 trillion over the 55-year life of the program—that the Pentagon deemed it 'too big to fail' in 2010. … American taxpayers will pay between $148 million and $337 million per jet plane in 2015, depending on the model." Some blame the ever-changing requirements, etc., but perhaps there are more fundamental reasons related to design and poor execution.

Lockheed Martin was aware of the need to adopt a more systematic approach in developing large systems in order to avoid cost overruns and project delays similar to that incurred in executing F-35 project. When Lockheed Martin got the contract to develop the Orbital Space Plane (OSP)[2] from NASA, its visionary program manager, Robert Ford, had decided to use a system design methodology in developing OSP. In 2002–2003, he invited the author to teach system design based on Axiomatic Design to his lead engineers and designers of the OSP program. About 250 engineers were taught over a period of about six months primarily in Denver, Colorado. Young engineers learned it relatively quickly, but the experienced engineers had difficult times learning a new way of thinking about system design. They were used to starting out with a physical embodiment of what the system should look like without initiating the design process for the system based on thorough definition and decomposition of functional requirements (FRs) and constraints that must be satisfied.

[2]Robert Ford, *Orbital Space Plane (OSP) Program at Lockheed Martin*, American Institute of Aeronautics and Astronautics, September 2003, Long Beach, California, U.S.A.

There are many other examples of ad hoc approaches to system development that had created major problems and sometimes led to a complete failure. In 2014, the US government launched the national healthcare system (so-called Obamacare).[3] The enrollment in Obamacare could not begin when it was originally scheduled to begin because of the glitches in its large software system. The system had many operational glitches and took intensive effort to correct the problem. This poor design of the software system put the entire healthcare system in jeopardy. It may be a good example of a large software system not performing its functions due to poor design. Many universities had encountered similar problems when they try to replace their disparate administrative processes with computerized software systems they had purchased from vendors. It took a significant sum of money and time to overcome the problems, partly because the original software was not designed to deal with academic affairs.

Other countries also had notable failures of large systems. One of the most catastrophic failures is the demise of Japan's Fukushima nuclear power plants. The tsunami that hit the coastal area of Sendai, Japan, on March 11, 2011, was caused by an underwater earthquake with a magnitude of 9.0, the biggest recorded earthquake to hit Japan. The Fukushima nuclear power plants were completely destroyed, contaminating a vast area with radioactive materials. It will take many decades for the Fukushima region to recover from this disastrous natural disaster and man-made failures. The human toll and the cost of ameliorating all the damages done in the region can hardly be overstated. According to Nakao et al. [8] and Hatamura et al. [1], it was the coupling of functions of its electrical systems with those of mechanical systems that led to the failure of the entire system. When the tsunami water came into the ground floor of the reactor building, all the electrical systems were submerged in water, which led to the failure of mechanical systems.

Some large organizations such as universities, state governments, and financial institutions also fail to achieve their goals. First, many institutions do not go through the rigorous design process to clearly articulate their institutional goals and means of achieving the goals. Second, the rules and regulations that were created in the past prevent them from achieving their new institutional goals. Even after the new mission and the accompanying plans have been clearly established, many defend and persist to use the old familiar system. Thus, some government agencies are bound by past practices and regulations that are not compatible with their current mission, goals, and programs.

---

[3]The official name for "Obamacare" is the Patient Protection and Affordable Care Act (PPACA), or Affordable Care Act (ACA) for short, which was signed into law to reform the healthcare industry by President Barack Obama on March 23, 2010, and upheld by the Supreme Court on June 28, 2012.

11.2.1 Why Cost Overruns, Missed Schedules, and Poor Performance?

The engineers and managers who worked on the Berlin Airport project, the F-35 Joint Strike Force airplanes, and the Fukushima nuclear power plants must have been intelligent and capable people, since these advanced technology projects attract some of the most experienced and brightest engineers and scientists. These important projects also receive the generous financial support. It is most likely that they had good intentions to create great systems within the budget, on schedule, and indeed to exceed the original expectation. It is also possible that non-technical factors have had the progress of these projects. If we exclude these non-technical factors, what is then the fundamental reason for the cost overruns and missed schedules?

Even those who led the project or participated in making decisions related to the systems might not know the true fundamental cause of their problems, which culminated in cost overruns and delays. They may know the *symptoms* of the failures but not *the cause*, i.e., basic design decisions that led the project to failure. At this time, the only way to delineate causality of system failures is to compare them with projects that were finished in time and within the estimated cost. I have found that the successful projects were the ones that adhered to the principles of the Independence Axiom. Based on these limited case studies, it may be concluded that projects that have encountered problems are those with coupled designs. The designers of these systems, intentionally or inadvertently, introduce coupled designs at the system level as well as at lower levels of the design hierarchy. These couplings of FRs lead to cost overruns and time delays. Unless they had a systematic means of checking for coupling, it would have been difficult to identify and avoid the coupling. Companies want to use the same methodology and system that is yielding a good product after the system has been debugged over many years of refinement, but when a completely new product has to be developed, such copying process does not work well.

This conclusion was reinforced at Lockheed Martin when about 250 lead engineers, working on a major national project, were taught in groups of 25 engineers for about six months. They were experienced engineers, many with advanced engineering or science degrees. It was more difficult to teach those who had years of experience in doing design and development through a repetitive cycle of design/build/test. Some of them were used to the practice of coming up with a physical embodiment first before explicitly stating the FRs.

11.2.2 Lessons Learned: Development of Technological Systems

Axiomatic Design was established as a result of my attempt to establish a new strategic direction for the newly established MIT Laboratory for Manufacturing and Productivity in 1976. The goal of LMP was to create the science base for design and manufacturing, since much of the work in these fields was largely empirical and specific to a given situation. To achieve this goal, we extracted the common elements that were present in good designs. These common elements were used to create axioms. This approach was orthogonal to algorithmic approach used for specific design and manufacturing operations.

The discovery of the importance of the Independence Axiom (i.e., "maintain the independence of FRs") and the Information Axiom (i.e., "minimize the information content") stems from the projects I worked on in industry and MIT. It began with my first industrial job after my junior year at MIT. The job initiated me into the world of systems and system design, without my being aware of its significance at the time. Since then, I was fortunate to have had the opportunity to design and implement a number of products, manufacturing systems, technological systems, and organizational systems in several different fields. In nearly all cases, I was successful in producing products and the manufacturing systems in relatively short times and at a minimal cost, because I adhered to principles of Axiomatic Design.

11.3 Industrial Journey and "Data" on the Design of Systems[4]

11.3.1 Design of Products and the System for Mass Production of Foam/Straight Plastic Laminate Plates, Dishes, etc.—Journey in Guild Plastics, Inc.

My first industrial job was with Guild Plastics, Inc. In 1958, I joined the company after my junior year at MIT. I was very fortunate to get the real industrial job with major responsibility, although I had not received my formal degree yet. I worked at the company full time during the summer of 1958 and 24 h a week during my senior year while finishing up my undergraduate education at MIT. The job was not only interesting and fun but also paid well, almost three times the minimum wage I had been paid at MIT. After graduating from MIT, I worked at Guild Plastics full time during the summer of 1959 and then again part-time until January 1960.

Guild Plastics was a small company that made disposable ice-cream dishes and cups by vacuum forming extruded impact-grade polystyrene sheets. Samuel Shapiro, a grandson of a Russian immigrant who set up an ice-cream cone factory, started the plastics business in a warehouse where they also made ice-cream cones. I learned about plastics and polymer processing at Guild Plastics while working there. Many years after I left the company, it merged with Sweetheart Plastics, a large company. In 1958, I was the only engineer in the company. My title was development engineer, although I did not have an engineering degree yet. I worked

[4]Some of the cases presented in this paper were discussed in Suh [11, 12].

with and learned from many experienced machinists and production foremen about the real-world practice.

Samuel Shapiro gave me the task of developing a product that will enable Guild Plastics to compete with "Styrofoam cup" that had good thermal insulation and stiffness, but was too thick to replace the paper cup used in vending machines (note the wall thickness determines the height of stacked cups). The FRs for the product were the following: FR1 = provide rigidity, FR2 = provide thermal insulation, and FR3 = provide strength.

The manufacturing process I created was to laminate straight plastic to a sheet of foamed plastics, form it to the desired shape by "match molding," and punch out the finished product. Thus, I invented the product (i.e., a laminated foam/straight plastic cup) and the entire manufacturing system for mass-producing the product, i.e., a "manufacturing system for making composite products." Guild Plastics manufactured "high-end" disposable plates, dishes, and cups using the process. The manufacturing system included an extruder, die for extrusion of polystyrene foam tube, calendaring rolls, lamination process, match molding, punching press, recycling the scrap, etc. We received a US patent on this manufacturing system and the laminated foam/straight plastic product (see Fig. 11.1).

The project was completed in about *one and a half years*, working full time during the summer and part-time during the academic semesters—three days a week at the company and the rest of the week attending classes. During the fall semester of 1958, I took a graduate-level heat transfer course, since the process I was developing included conductive, convective, and radiant heat transfer. More than 30 years later, when I visited one of the manufacturing plants of Sweetheart Plastics located in Summerville, Massachusetts, they were still producing laminated dishes and plates using the same manufacturing system. There were a lot more identical machines, producing many products continuously.

It should be noted that the manufacturing system I created at Guild Plastics was a *decoupled design*, which satisfies the Independence Axiom. It should be noted that the implementation of decoupled design must follow a specific sequence. At the time, I just developed the product and processes without thinking about any theoretical implications. This industrial process has provided me with a data point in developing the theory for design of large systems many years later when I decided to pursue an academic career.

11.3.2 High-Pressure USM Molding Process for Composite Shoe Soles

After I finished my master's degree at MIT, I accepted a job at the world's largest shoe machinery manufacturing company, USM Corporation, in Beverly, Massachusetts, although the offer I got from Guild Plastics, Inc., was substantially better in terms of compensation. I joined this large company, because I was concerned that I might become "a big frog in a little pond" at Guild Plastics, Inc. My

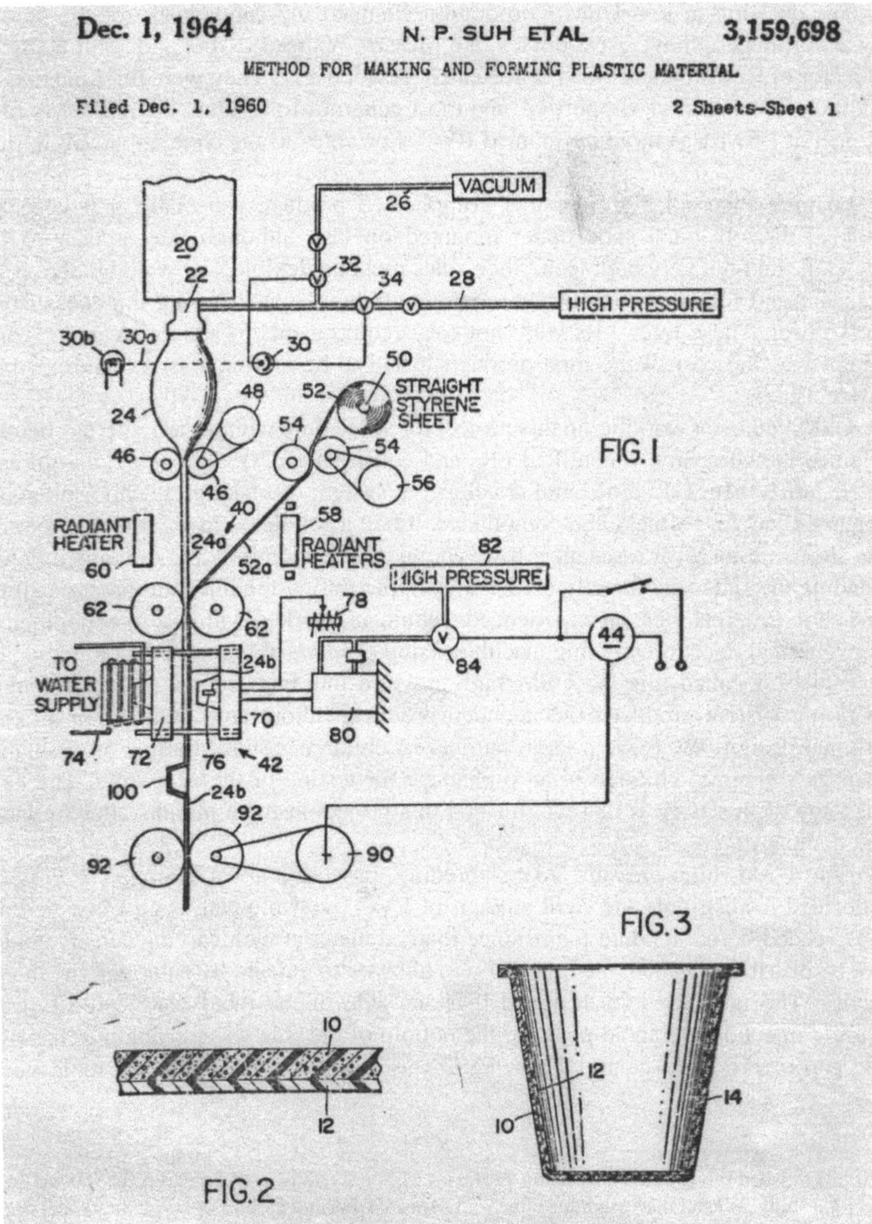

Fig. 11.1 Plastic/straight plastic laminated cup and the manufacturing system

goal was to learn how big companies operate, conduct their product development, and deal with engineering. Although Guild Plastics had initiated me into the world of engineering and industry, there were no metrics and guidance for my professional development and performance.

The decision to join USM Corporation changed my career path for the better. I was fortunate to have great bosses like Messrs. Walter L. Abel and John Hollick, Director of Research and Head of Research, respectively. They were big thinkers, in addition to being kind, supportive, and most generous to a young engineer. My first project at USM was molding foamed PVC shoe soles to the shoe upper, all in one step.

Foamed shoe soles are a rather complicated product, especially if it is to be molded directly on a shoe upper mounted on last, although they appear to be mundane and not very high tech. Shoe soles must be flexible, lightweight, and wear resistant and follow a complicated contour of shoes, with varying thickness from toe to heel. These three FRs were not easy requirements to satisfy. A major constraint was the cost of the final product: It had to be less expensive than existing products.

USM had been working on this project for some time without any success before I joined the company. I identified FRs and constraints (Cs), although I did not use those terms then. I designed and developed a system of making the shoe sole with foamed PVC for softness and compliance. It had a solid skin layer at the bottom of the shoe sole for wear resistance. It was done by injection molding of foamed PVC, bonding the shoe sole directly on the shoe upper during the injection process, all in one shot. Ralph Hobbs, an experienced technician, worked with me. We modified a conventional injection molding machine, using *expandable mold* with cold movable sole plate. I called this the USM high-pressure molding technology. The manufacturing system produced the intended product without any modification of the original design. We made a large number of children's shoes by this process and then gave them to children in an orphanage for testing of the new soles. The key message of this story is that we finished this project in three months after I joined the company.[5]

The USM high-pressure foam molding process was a *decoupled design.* I decided to distribute the right amount of PVC (with blowing agent) to where it was needed before it could foam, since foamed material with varying density could not be distributed. Then, the material was allowed to expand by enlarging the mold cavity. The mold was so designed that one side of the mold plate could move. During injection, the mold plate for the bottom of the sole was moved in to narrow the gap of the mold cavity. Then, PVC was rapidly injected into the mold with

---

[5]After I finished this project, my mentor Professor Milton C. Shaw asked me to come to CMU for doctoral study. When I discussed this offer with Messrs. Abel and Hollick at USM, they came back with a counter-offer to finance my doctoral education paying my full salary, all the research cost, etc., an unbelievable kind offer, with no strings attached. The only condition was that they would do it only for two years. After working for USM for a couple of years after finishing my doctorate, my former officemate at CMU asked me to join his university to teach. Again, my bosses were most understanding. They gave me a leave with a research contract to work on for USM at the University of South Carolina. Even after leaving USM, I maintained a close working relationship with Wally Abel for more than 20 years, who became vice president for research at USM and later at Emhart Corporation. I was most fortunate to have had Mr. Abel as my boss and a lifelong friend.

narrowed gap. The narrow gap and the high-speed injection of molten PVC raised the pressure of the melt, preventing it from foaming during injection. After the completion of the injection of PVC, the movable sole plate was retracted to expand the mold volume and let the PVC with foaming agent expand to fill the space created. The surface of the movable plate was maintained cold throughout the process so as to prevent the PVC adjacent to the plate from foaming, thus creating a wear-resistant skin layer. The process was later used to make automotive components, furniture panels, etc.

11.3.3 Coated Tungsten Carbide Tools (Surftech Corporation)

After I joined the MIT faculty, one of the research projects we worked on was the wear of carbide tools. Cutting tools must satisfy three FRs: stiffness, toughness, and wear resistance. Cemented a tungsten carbide tools satisfied the first two FRs well, but had high wear rates. We showed that the high wear rate is associated with the chemical instability of WC in cutting steel, i.e., WC dissolves in iron at the high temperature. Dr. Bruce M. Kramer wrote a doctoral thesis on the topic, which is the seminal work showing how chemical solubility of various carbides in iron controls the wear rates. His theoretical prediction of the relative wear rates of various carbides was right on the mark!

As part of this research, Bruce Kramer came up with a way of coating the surface of WC with other carbides by chemical vapor deposition of pure metals that reacted with the substrate WC, which created a thin layer of other carbides such as HfC and TiC on the surface of the substrate. Using these tools coated with HfC, etc., we prevented the wear of the substrate, i.e., cemented WC. He showed that the HfC-coated tools lasted about nine times longer than uncoated tungsten carbide tools. His doctoral thesis provides the theoretical basis for choosing tool materials. MIT Development Foundation established Surftech Corporation and hired Dr. Jack Smith, the former president of Hampshire Chemical, in order to commercialize these coated tools. None of the domestic carbide tool manufacturers would sell Surftech the substrate (cemented tungsten carbide, which is WC powder bonded with cobalt through the formation of eutectoid) at reasonable price. Surftech ended up importing the substrates from Poland. However, we had problems with the quality of the imported product, mostly the toughness of the tungsten carbide/cobalt substrate.

This HfC-coated tool was an uncoupled design. The process of making the coated tool by chemical vapor deposition was a decoupled design, consistent with the Independence Axiom.

All these processes and products discussed so far were created before the advent of AD. Therefore, they did not benefit by AD, although the thought process was based on AD. These projects provided the data for creating the design axioms.

11.3.4 Unsuccessful Projects

There were some of the patented systems that I created at USM Corporation and MIT that looked promising at the time, but later were found to have shortcomings because they violated the Independence Axiom. One of these was the rapid processing of plastisol, [6] and the other was a system for processing polyurethane. At the time, I did not know that these ideas had basic flaws, which I recognized as such much later when the Axiomatic Design theory was advanced at MIT.

11.3.5 Axiomatic Design and LMP at MIT

I began to think about design of systems because of the establishment of the MIT Laboratory for Manufacturing and Productivity (LMP). In 1976, Professor Herbert H. Richardson, head of mechanical engineering at MIT, asked me to organize a major interdisciplinary effort in the field of manufacturing. This action was to follow up on the recommendation of the task force on manufacturing,[7] which was established by Dean Alfred Keil and headed by Robert Lund. They had recommended that MIT establish a major effort in manufacturing. For a couple of years, the report sat on a bookshelf collecting dust until Herb Richardson took the initiative to create an interdisciplinary organization.

I was given \$20,000[8] as the seed fund for LMP and the privilege of working with two able younger colleagues, Professors Adam Bell and David Gossard, to launch LMP. David Gossard was just appointed as assistant professor upon finishing his doctorate, specializing in computer-aided design. Adam Bell, an untenured associate professor, was a control specialist with interest in design. They devoted 50 % of their time for LMP.

---

[6]Some polyvinyl chloride (PVC) parts are made from plastisol, which is a suspension of PVC particles in a liquid plasticizer. In PVC applications, plasticizers are liquid additives, most commonly phthalate esters. Plastisol flows as a viscous liquid. It can be poured into a metallic mold and heated to a set temperature, typically 177 degrees, to make solid PVC parts. Then, the plasticizer diffuses into the PVC particles, which then bonds with other PVC particles to form a solid article. Typically, plastisol is heated to around 177 degrees Celsius to dissolve PVC in plasticizer. Some PVC parts are made by heating plastisol in a heated mold. I found out that it takes a long time to make solid parts out of plastisol in a mold, because of the low thermal diffusivity of PVC and plasticizer. An idea was proposed to overcome this problem, but it had major limitations.

[7]This task force was created in response to the general consensus that the USA was no longer competitive in manufacturing as indicated by the trade deficit started in the early 1970s. It was acknowledged that MIT had made a mistake of de-emphasizing these fields after the launching of Sputnik by the Soviet Union.

[8]This sum was equivalent to the academic year salary (i.e., nine-month salary) of a full professor at MIT.

As the director of LMP, the first question I had to answer for myself was: "why should MIT start a new research activity in manufacturing?" In the late 1950s and the early 1960s, MIT's leadership had tried to eliminate the field of manufacturing and design from MIT so as to concentrate primarily on engineering sciences as a response to the Soviet Union's launch of Sputnik in 1957. Prof. Gordon Brown, then dean of engineering at MIT, spearheaded this effort, joined by many professors, including those in the Department of Mechanical Engineering. It was a turbulent period in the history of MIT and the department. Until then, the Department of Mechanical Engineering had major research activities in manufacturing headed by Professor Milton C. Shaw, my thesis advisor and a world-renowned scholar. He did not agree with MIT's leadership on eliminating his field of specialty and joined Carnegie Institute of Technology (now Carnegie Mellon University) as head of the Department of Mechanical Engineering.

The research done in the 1950s and 1960s in the field of manufacturing, including design, was largely empirical, which could not be generalized. Some of the papers published in these fields by many researchers were often descriptions of experimental observations and explanations that were applicable only to specific situations and could not be generalized. Furthermore, most of the research results in these fields were "algorithmic," i.e., led to right results if one followed a specific sequence, but not generalizable. We convinced ourselves that the field of manufacturing needs a "science base" for design of processes and systems in order to improve engineering education and industrial productivity.

To fulfill the long-term goal of LMP, we decided to establish an axiomatic approach to design and manufacturing. We wrote a major research proposal to NSF, asking for significant funding to achieve this goal. The program director at NSF, Dr. Bernard Chern, a thoughtful scholar, was extremely helpful. He decided that the only way such a proposal could be evaluated fairly was for me to explain the proposed idea directly to a review panel, because it would be difficult to evaluate it through the normal mail review process. He organized a site visit to MIT with 12 reviewers (six from industry and six from academia). One day before the site visit, I was visiting the West Virginia University to give a talk on tribology at a coal fluidization conference. David Gossard called me to inform me that one of the reviewers he met at a conference wanted to know how one would actually create such axioms.

Our research proposal had described the research goal, but had not mentioned how we were going to create such design axioms. I had less than 24 h before the site review meeting at MIT to prepare an answer to the most important question raised. To come up with a plausible answer, I decided to create hypothetical axioms based on the things that I did in industry and MIT, including those presented in the preceding section, which had yielded successful designs. I came up with 12 hypothetical axioms during the lunch break at the coal fluidization conference. After returning to MIT from West Virginia, Adam Bell, David Gossard, and I discussed the 12 hypothetical axioms and reduced the 12 hypothetical axioms to six hypothetical axioms and six corollaries. The next day at the NSF review meeting, we presented our approach to creating design axioms and had a lively discussion.

We received a strong positive review. After we received the NSF funding for Axiomatic Design project, we reduced the hypothetical axioms to two axioms: the Independence Axiom and the Information Axiom and four more corollaries.

Once we formalized these axioms, I finally realized why we could finish some of our projects so quickly without having to make corrections, etc. Those designs done at Guild Plastics, Inc., USM Corporation, and MIT were either uncoupled or decoupled designs that satisfy the Independence Axiom. Furthermore, having taught many engineers in industrial firms in the USA, Europe, and Asia, I surmised that some of the projects these companies were working on were coupled designs and sometimes took them a long time to create their products or systems to correct and overcome the coupling of FRs created during the course of the system development.

11.3.6 Lessons Learned from the Systems I Invented in Industry and MIT

What I learned from my industrial experience and my academic research at the University of South Carolina and MIT may be summarized as follows:

- Clear understanding of needs.
- Importance of defining goals (i.e., functional requirements—FRs).
- Finding means (DPs) of satisfying FRs.
- Changing multi-input/multi-output systems into a set of one-input/one-output systems. (the Independence Axiom: Maintain the independence of FRs, i.e., avoid coupled systems.)
- Minimizing complexity (the Information Axiom).
- The design of all systems can be "normalized."
- Understanding that good designs, regardless of their physical nature and the domain of applications, share the same characteristics.

11.4 Foundations of Axiomatic Design

The two axioms of Axiomatic Design, which were deduced by identifying the good design features that were always present in good designs, were the following [11]:

The Independence Axiom: Maintain the independence of functional requirements.
The Information Axiom: Minimize the Information content.

Functional requirements (FRs) represent the design goals. FRs are, by definition, independent from each other. To satisfy these FRs, we choose design parameters (DPs). The Independence Axiom states that we must choose DPs so that FRs remain independent from each other at all times. There are a large number of theorems that are generated from these two axioms [12]. For example, one theorem states that in an ideal design, the number of FRs and DPs is the same. Another theorem states that when two or more FRs are dependent on each other, those FRs constitute a single FR. The chosen FRs must be independent from each other by definition. If one DP satisfies all of them, the FRs are not independent from each other, i.e., they are related to each other. Then, the design becomes one FR–one DP design problem, i.e., satisfying one of the FRs, and the rest of FRs must be satisfied at the same time. There are many other theorems.

There have been a large number of papers written on Axiomatic Design as well as several books on topics related to AD, including the three books authored by Suh [11–13]. At the 9th International Conference on Axiomatic Design (ICAD), a large number of interesting and intellectually stimulating papers were presented.

These axioms were confirmed by a large number of new projects undertaken after the introduction of the design axioms. Some of these projects will be reviewed next. The major point to be highlighted in this paper is that when the designs satisfy the design axioms, they can be developed quickly, often leading to the lowest cost for system development.

11.4.1 Large Technology Systems Created Using the Design Axioms

To answer the question of why so many large projects sometimes fail to meet the original schedule and exceed the estimated cost, I would like to review three moderately large projects—Mixalloy, microcellular plastics (MuCell), and On-Line Electric Vehicle (OLEV)—that were designed and executed based on Axiomatic Design. The important point to be highlighted through these technological projects is that although these are highly innovative and technologically successful products and processes, *they were completed comparatively quickly at low cost.* Every project may have its reasons for cost overruns and delayed execution, but what we learned from these projects may be equally applicable to such projects as F-35 fighter planes and the Berlin Airport problems.

When there is a cost overrun and when the project cannot achieve its goal on time, it is often difficult to identify the fundamental causality. Often we identify symptoms rather than the real cause. I learned about the problems associated with coupled designs by actually executing large projects that included a coupled design. We knowingly allowed the coupled design, because the majority of the senior staff thought that it would make the execution of the project a lot easier, i.e., I was outvoted. After trying to make the coupled system work for about six months, we replaced it with an uncoupled design in time, but that one wrong decision still incurred additional cost and delayed the project by about six months. This incident was a good lesson for all of us.

To avoid such mistakes inadvertently creeping in, we appointed "System Architects" in executing OLEV (On-Line Electric Vehicle) and Mobile Harbor

(MH) projects at KAIST. Their job was to spot any coupling of FRs introduced by any one of the designers working in the large projects. As a result, we commercialized OLEV in two years from the start of the project and the demonstration of the Mobile Harbor was also done in two years as well. Coupling of FRs is the major source of time delays and cost overruns. This conclusion is equally applicable to non-technological projects (e.g., National Science Foundation, MIT, KAIST), which will be discussed in a later section. The cost overruns and delays in well-publicized large projects may also be due to coupled designs introduced during the project design and execution. Unless they have a systematic means of checking for coupling, they would not have known that they had made a mistake. The responsibility of System Architects at KAIST was to monitor design decisions made by all engineers and designers engaged in these large projects by constructing the design matrix (DM) for their project.

From a technological point of view, the three projects—Mixalloy, On-Line Electric Vehicle (OLEV), and microcellular plastics (MuCell)—were successful and innovative. We received many patents and recognitions for these inventions. From the commercial point of view, one has become a successful business, the second is in the early stage of commercialization, and the third failed as a business venture. The nature of these three projects will be briefly described here with descriptions of the important lessons learned.

11.4.1.1 Mixalloy

This project was remarkable in that we went directly from the concept for the new product and process of making Mixalloy to production without the benefit of preliminary laboratory demonstration to test the feasibility of the invention, because we could not perform small-scale experiments to verify the basic concept of Mixalloy.

Mixalloy is a dispersion-strengthened copper alloy made up of a pure copper matrix phase with a plethora of nanoscale TiB_2 particles dispersed throughout the matrix in order to strengthen the alloy without sacrificing other properties such as electrical conductivity, thermal conductivity, formability, ductility, and fracture toughness. This alloy and the process of manufacturing the alloy were invented to meet the industrial need for copper alloys with high strength, high ductility, and high conductivity at high temperatures. The first application for the material was for spot welding tips used in assembling automobile bodies.

The invention of Cu/TiB_2 alloy (i.e., Mixalloy) and the process for making it were possible, because we had worked in two different fields, i.e., polymer processing and metal physics, in addition to design. However, we could not have done it without the outstanding people who developed the technology and operated the business. Both the alloy and the manufacturing system were designed based on Axiomatic Design.

The physical basis of dispersion-strengthened Cu/TiB_2 alloy is as follows: The FRs of this alloy are high conductivity, high strength, and high toughness. We can obtain high conductivity by controlling the purity of the copper matrix phase and the high strength by controlling the spacing between TiB_2 particles and obtain the fracture toughness by controlling the size of the particles. The smaller the spacing, the higher the yield strength. To obtain high toughness, we have to make particles smaller than a critical size.[9] If we can achieve such a microstructure, we have a decoupled design that will enable us to vary these three properties independently from each other by controlling the stoichiometric ratio to obtain high-purity Cu, the velocity of impingement, and the solidification rate of the liquid Cu with dispersed TiB_2. It is a decoupled, not an uncoupled, design, because the spacing between the particle size is affected by the particle size, which was controlled by the solidification rate of the molten Cu/TiB_2 solution.

The processing method we devised for the Cu/3 % TiB_2 alloy was to impinge a stream of liquid solution of copper (Cu) with 3 % titanium (Ti) against another stream of liquid solution of Cu with 6 % boron (B) at high enough velocity to create a nanoscale turbulent eddies in the impingement-mixing chamber. Then, when Ti and B are in close proximity, they react to form TiB_2 particles, lowering the overall free energy of formation of the mixture. Initially, the particle size is of nanoscale, but they grow unless the mixture of Cu and TiB_2 is cooled quickly. This idea of impingement mixing was learned while working in the field of polymer processing because General Motors was interested in this process. To make polyurethane, we must intimately mix polyol with di-isocyanate. While working on this impingement process for polyurethane, Tucker and Suh [15] performed dimensional analysis to show that the turbulent eddy size, δ, is proportional to −3/4 power of Reynolds number, i.e., $\delta \propto (Re)^{-3/4}$, where $Re = \rho VD/\mu$. For fine mixing of liquids, we want to make the Reynolds number high. In the case of liquid metals, it is rather easy to achieve a high Reynolds number because of the high density and low viscosity of molten metals. We need to cool the mixture of liquid Cu and solid TiB_2 particles quickly to prevent the coalescence of TiB_2 particles.

After conceiving the idea for the Cu/TiB_2 dispersion-strengthened alloy, we had to manufacture the alloy to test the viability of the basic concept. However, we could not verify the merit of the idea by performing small-scale experiments in laboratory. Only a large-scale testing would enable us to reach a steady-state, isothermal flow of molten metal solutions at 1200 °C in order to verify our processing idea and make the alloy for testing of their properties. However, we could

[9]Dislocation theory for metals states that if we put tiny particles in pure element with narrower spacing between them (of the order of 100 nm), the shear stress τ required to extrude dislocations through between the particles is inversely proportional to the spacing λ of the particles as $\tau = Gb/\lambda$, where G is the shear modulus and b is the Burgers vector ($\sim$lattice constant). Thus, the narrower the spacing, the stronger the alloy. The question has been how we can form a plethora of such small particles in pure metal matrix. One of the known ways of making dispersion-strengthened alloy, before the Mixalloy process was introduced, was to dissolve a small amount of aluminum in pure copper by melting them together to form a solution of Al/Cu and then expose the alloy to oxygen atmosphere at high temperature. Then, oxygen diffuses into copper, since the free energy decreases, forming Cu/Al_2O_3 microstructure. We felt that this “internal oxidation” process of manufacturing dispersion-strengthened copper was expensive.

not build the full-scale production equipment from the outset purely based on our theoretical reasoning and then make the alloy for the first time to determine its properties. Therefore, this idea for Mixalloy was put on a shelf for a couple of years.

Then, a sponsor came along who was willing to commit a sizable sum of money to commercialize Mixalloy. We designed and built the production equipment and all the auxiliary machines. Concurrently, we also built the factory that housed this production machine, which was about 20 ft high. In about three years, we were in production of mix alloys with the first machine we designed and built without the benefit of laboratory-scale experiments. We sold the products to both domestic and international companies. We only made a minor change to the original equipment to go into production, which we built for the first time.

We built a unique production-scale manufacturing system from scratch and made the alloy for the first time using the machine, the only machine of its kind in the world. The project was finished in about three years, which included designing and building the production equipment, making samples of Mixalloy, testing them, building the 27,000 square feet plant for commercial operation, working with potential customers (i.e., Chrysler, GM), hiring people, and working with hot isostatic extrusion company, and we start selling the product to the automobile companies in Detroit, Japan, and Korea. We lost about six months or so because we knowingly tried a coupled design for the conduit for the flow of molten copper alloys. The lesson learned is that although a coupled design may appear to be attractive, resist the temptation.

Just to provide a perspective to the whole commercialization process, typical industrial firms might have taken ten to twelve years to do what we did in three years. One of the major reasons we could do it in three years was because we had very smart people (i.e., three brand new MIT PhDs[10]), taking on the leadership roles. The other reason for the rapid development of the commercialization of Mixalloy is the rational design of the large systems, including the design of alloy, the design of the manufacturing process, and design and fabrication of production equipment.

The FRs of the process equipment were three: Maintain the temperatures of the liquid metal solutions at 1200 °C, maintain the flow rates of two streams the same throughout the impingement process, and quench the mixture of copper and TiB_2 quickly before coalescence of TiB_2 to a larger size.

We did our best to make sure that we do not violate the Independence Axiom. In order to match the flow rates in each conduit precisely the same, and also to maintain isothermal flow from the molten metal reservoir to the impingement-mixing chamber, we had to control the flow rate and the temperature of the liquid precisely. It is relatively easy to melt copper in a crucible with alloying elements and keep it at 1200 °C. The question we struggled with was how to control the temperature of the conduit, i.e., pipe, at 1200 °C in order to control the flow rate

[10]J. H. Chun, PhD in mechanical engineering at MIT; L. Sanchez, PhD in mechanical engineering at MIT; and A. Lee, PhD in materials science and engineering at MIT.

and maintain isothermal flow to the impingement chamber. Our team had two different ideas. The first idea was to wind a ceramic tube with resistance heating wire and control the temperature by means of electric power input to the heating element. Another idea was to use a tube with a larger diameter and let the molten metal solution flow through the ceramic tube and let some of the metal freeze on the wall rather than heating the conduit. The second idea was based on the argument that it costs money to make a heated tube by winding a ceramic pipe with resistance heating elements. However, this seemingly simple idea of using unheated large tubes never worked, since the temperature of molten metal solution varied from run to run (as well as within a given run) and the flow rates in these two tubes were difficult to control because it was non-isothermal flow. The temperature and the flow rate affected all other functional requirements, including the temperature of the mixture, the ratio of titanium to boron, and the flow rate. It was a coupled design. After trying to make this non-isothermal unheated tube work for about six months, we switched to the original idea of isothermal tube (i.e., tube with the electric heating wire), which made the system uncoupled. Soon thereafter, we were able to manufacture Mixalloy commercially and shipped the products to companies that made spot welding tips for automakers in the USA, Japan, and Korea.

We could have gone into commercial production much sooner had we not violated the Independence Axiom. This one wrong decision had diverted the development process by at least six months and wasted a significant financial resource of a small start-up company. My guess is that the developers of F-35 fighter plane, the managers and engineers of the Big Dig project, and managers and engineers who designed the Berlin Airport might have made similar mistake, but they might not even be aware of the source of their problems, i.e., coupling of their FRs.

11.4.1.2 OLEV (On-Line Electric Vehicle)

When I was at KAIST, we decided to solve global problems associated with EEWS (energy, environment, water, and sustainability) as a strategic goal of the institution. There were a large number of projects undertaken under the EEWS program.

One of the projects undertaken under the EEWS program was the OLEV project to reduce CO_2 emission by replacing automobiles that use internal combustion (IC) engines with electric vehicles that receive its electric power wirelessly from underground electric cables. We initiated the OLEV (On-Line Electric Vehicle) project in 2009, and two years later in 2011, OLEV buses were running commercially in Korea. The basic technology for transmitting high electric power wirelessly to moving vehicles, which is used in OLEV, is SMFIR (shaped magnetic field in resonance). Following the basics of Axiomatic Design, we made sure that SMFIR is an uncoupled design [14].

The short period for the design and development of OLEV is attributed to the dedicated leadership of Professor D. H. Cho, the director of the OLEV project, outstanding people who worked on the project, and the rigorous design reviews that prevented coupling of FRs.

We had a "System Architect" who monitored all the design decisions made using a design matrix to be sure that coupling of FRs was not introduced inadvertently [2, 5]. They made sure that the design, which involved a large number of engineers and designers, did not create coupled designs.

At the peak, about 200 people worked on the OLEV project. We had weekly project reviews with all the key staff members, which were often presided by the president of KAIST. The budget for the project was about $50 million ($25 million per year), and we finished the project within the budget.

11.4.1.3 MuCell

MuCell is the trade name for microcellular plastics,[11] which was invented in the early 1980s as part of the MIT-Industry Polymer Processing Program.[12] A number of PhD theses have been written on processing of microcellular plastics at MIT and at other universities. Processing of microcellular plastics is still an active academic research field with many active industrial applications. Most of the automobile companies in the world use this process in order to improve properties of polymer components, increase productivity, and reduce the cost.

The idea for microstructure of microcellular plastics was created while the author was having a lunch with Mr. Gordon Brown, a thoughtful executive of Eastman Kodak Company. Gordon Brown represented Eastman Kodak in the MIT-Industry Polymer Processing Program. Each year, member companies suggested research topics to the MIT Program. During the luncheon meeting to choose a project for Eastman Kodak, Mr. Brown stated that if the MIT Program can come up with an idea for reducing plastics consumption at Eastman Kodak, it could be a major contribution to his firm, since many of the products of Eastman Kodak were made of plastics. The constraints imposed on any solution were that the technology must allow the production of the parts without changing the geometric shape of their product and without sacrificing physical properties. To satisfy these FRs, the author proposed the idea of putting in a large number of microvoids into the plastic. The three FRs of the microcellular plastics are as follows: introduce voids, control the size of the void, and control the geometry of the part. The corresponding design parameters (DPs) were as follows: a plethora of tiny voids in polymer matrix (more than a billion bubbles per cm^3), the size of the void, and molding polymers with

---

[11] Trade name created by Trexel, Inc., a licensee of MIT.

[12] MIT-Industry Polymer Processing Program was established in 1973 with a major five-year grant of the National Science Foundation (NSF) to demonstrate how university and industry can collaborate to promote innovation. Up to 14 industrial firms participated in the program paying large membership fees to promote innovative research in polymer processing. Members shared the results of the research and cooperated in selecting research topics. NSF has created the Industry–University Cooperative Program at a large number of universities to replicate the MIT Polymer Processing Program. The success of this program has been used as a model in creating Industry–University Cooperative (IUC) Program at NSF.

bubbles into desired shape. In other words, we came up with three DPs that can fulfill three FRs, an uncoupled design.

After trying a few ideas, we demonstrated that we can make microcellular plastics on a batch process [7]. The microcellular plastics technology consisted of dissolving gases (CO_2 or N_2) in polymers by increasing the pressure of the gas at a temperature higher than the glass transition temperature of the plastic and then suddenly lowering the pressure to change the thermodynamic state of the polymer/gas solution quickly to create a two-phase material consisting of the gas phase and the matrix phase (i.e., the polymer). That is, after creating a solution of the polymer and gas, if the thermodynamic state of the solution is suddenly changed to a state of lower solubility, the dissolved gas cannot stay dissolved in the matrix as a solution and thus must diffuse out of the solution. However, the gas cannot suddenly diffuse out, i.e., a martensitic transformation, and thus, polymer/gas mixture forms a two-phase material by forming tiny bubbles in the polymer matrix. However, this batch process was not amenable for mass production,[13] which was required for industrial use.

Four PhD students[14] and I developed a continuous process of manufacturing microcellular plastics in about three years (which is a typical length of doctoral research), making sure that we do not violate the Independence Axiom of AD. The basic design for screw plasticating polymer processing techniques and machines had to be developed for mass production of microcellular plastics. This technology was taken over by Trexel, Inc., which developed an industrial version of the MIT extrusion process as well as extending it to injection molding. It took relatively a long time to develop the customer base and become profitable. A complete description of the microcellular plastics technology, including both the continuous process and batch process, is given in Wong et al. [16].

The lesson learned here is that three bright and able students could design and implement the process and system of manufacturing microcellular plastics for the first time by following Axiomatic Design, in relatively short period of time in an academic institution.

11.4.2 Likely Root Causes of Technical Systems Failures

There are many different reasons for not meeting the original schedule and performance criteria, including non-technical business reasons, when industrial firms

[13]Since then, Professor V. Kumar and his students at the University of Washington, Seattle, Washington, modified and improved this batch process for mass production.

[14]C.B. Park, D. Baldwin, V. Kumar, and S. Cha.

undertake ambitious projects that involve the creation of innovative large systems.[15] However, the projects discussed in the preceding section and others, including the design of large organizations (briefly discussed later in this paper), indicate that coupling of FRs is the root cause of cost overruns and project delays. Then, the question is why it is so difficult to instill a *discipline of avoiding coupling* in designing systems. The following is a partial list of reasons engineers and managers create so many coupled designs:

- Not knowing Axiomatic Design.
- Too much reliance on past experience.
- Not recognizing a coupled design.
- Starting with a physical solution even before FRs are defined.
- Incomplete or wrong specification of FRs.
- Creation of coupled designs by having more FRs than DPs.
- Poor decomposition of FRs, DPs, and PVs that lead to coupled designs.
- Lack of a "System Architect" who can rigorously oversee system architecture
- Choosing wrong DPs.
- Confusing between typical engineering requirements established in some industrial firms and the FRs and constraints that are the foundations of AD.
- Erroneous idea that physical integration of parts is always a good thing to do.
- Temptation to come up with a magic solution that solves the problem in one shot, satisfying many FRs with one DP.
- Not knowing the theorem that the number of FRs and DPs must be the same in an ideal design.

11.4.3 Important Roles of System Architects

In order to avoid typical coupling problems encountered by many large projects that involve a large number of designers and engineers, there should be a "System Architect" to monitor the design decisions made in all branches of the design team.[16] Their task is to catch wrong design decisions that couple the FRs at any

[15]Notwithstanding the statement of the previous paragraphs, it should be noted that there are many diverse views on system architecture [4]. Computer programming, aerospace engineering, chemical engineering, and other fields have developed their specific systems engineering approaches. Many involve analysis and simulation of the designed systems to find root causes of system failure and devise means of improving systems. Some treat systems engineering as an optimization problem. Another view is that "system architecting" is what a smart and experienced system engineer does, using heuristic approaches and simple rules to develop system architecture [10]. In contrast to these approaches, in AD, the main function of the system architect is to monitor the creation of coupling by constructing the design matrix for the entire design. Lee and Park [5], Hong and Park [2].

[16]Comments of Professor Stephen Lu: "Industry uses the architecture of previous system changing only the values of the parameters. That is reason they use Hierarchical Target Cascade (HTC) method." See Liu and Lu [6].

level.[17] Sometimes the design decisions made at lower levels in different branches of decomposition can couple the highest level FRs. In executing the OLEV (On-Line Electric Vehicle) and MH (Mobile Harbor) projects, we had two teams led by outstanding System Architects—professors T. S. Lee and G. J. Park—monitored the design of projects [2, 5]. They constructed the design matrix by getting the design information from engineers and designers working in different branches of the design decomposition in order to spot possible coupled designs. When they identified coupling, they made suggestions for appropriate changes. The design matrix for OLEV is shown in Fig. 11.2. The section of the design that could have created coupling is circled. Once the possibility of coupling is identified, we can modify the design to eliminate the coupled design.

Many designers and engineers engaged in a large project with many FRs and DPs, and many layers of decomposition might not have the relevant information on how the decisions made by other designers working on other parts of the system design might be affecting their part of the system. Conversely, they may not be able to assess how their own design decisions are affecting other parts of the system other designers are working on. It is the responsibility of the System Architect to assess the interactions between various FRs and DPs so as to prevent coupling of FRs working with various groups. This is best done by constructing design matrix [DM].

It should be noted again that regardless of the size of the system, if the design is an uncoupled design or decoupled design, a large system with many FRs and many DPs (i.e., multiple inputs and multiple outputs (MI/MO)) can be treated as if it is a series of a single-input/single-output (SI/SO) problem. In this case, the analysis of the system behavior is simple, becoming almost trivial. Uncoupled design leads to simple mathematical relationship between FRs and DPs.

11.4.4 Observations on System Design

Observation #1: Often designers begin the design process with physical embodiment (i.e., solutions) rather than defining FRs and C first. This is a major mistake.

In 2003, Mr. Robert (Bob) Ford, an executive of Lockheed Martin Company, in charge of the Orbital Space Plane (OSP), a NASA project, asked me to teach "system design" to their lead engineers. The goal of OSP was to replace the space shuttle that had been used in operation since 1981. The goal of the Lockheed Martin

[17]There are many different views of system architecture, depending on the nature of the system and specific fields of engineering [4, 10]. Computer programming, aerospace engineering, chemical engineering, and other fields have developed their specific systems engineering approaches. Instead of designing the system right, many conduct analysis and simulate the systems designed to find possible causes for system failure or optimize the performance. Some people regard a system architecting is what a smart and creative engineer with experience does using heuristic approaches and simple rules. In contrast to these approaches, AD specifically emphasizes proper system design, decomposition, and elimination of coupling of FRs by selecting new DPs.

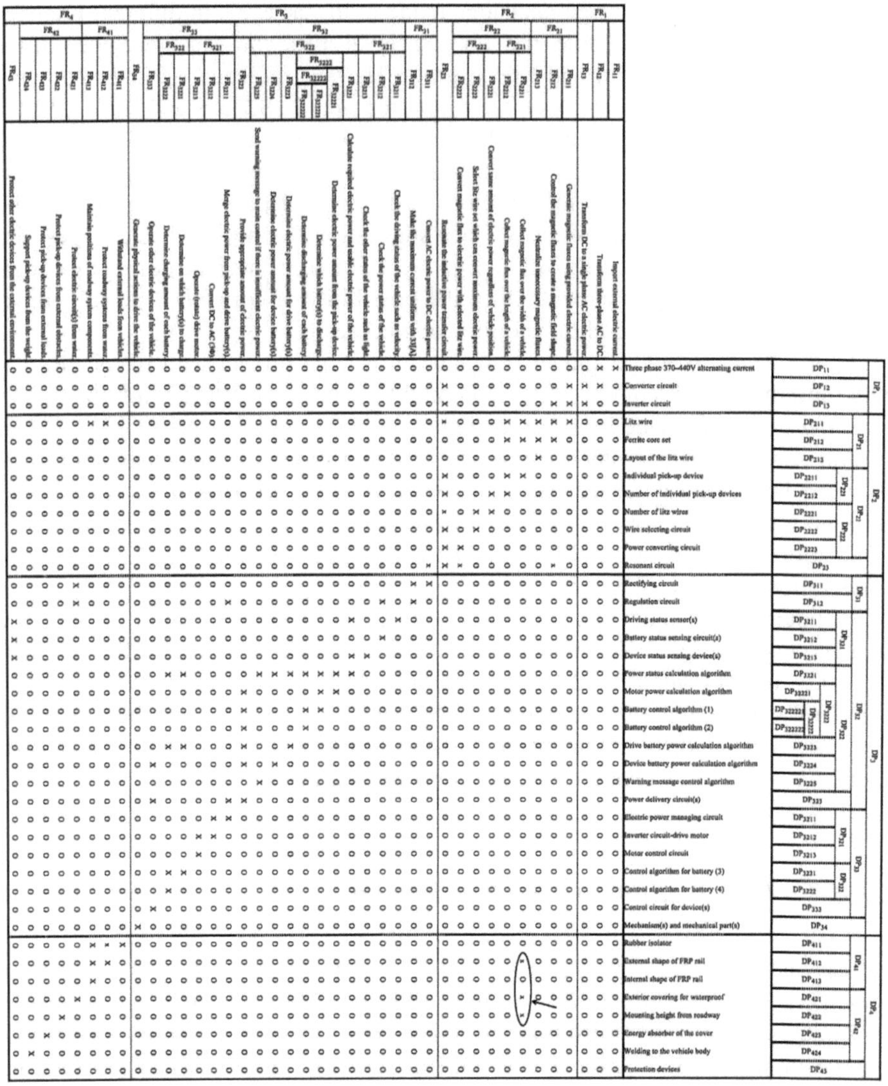

Fig. 11.2 The design matrix for OLEV. The circled elements of the design matrix may cause coupling. Lee and Park [5], Hong and Park [2]

project was to deliver their final product on time and within the budget by minimizing the trial-and-error processes often used in such large projects. Bob Ford was a visionary industrial leader. I taught a total of about 250 engineers, 25 engineers in each group, for about six months mostly in Denver, Colorado, and some in Alabama. One thing I found out teaching these experienced aerospace engineers and rocket scientists—very smart and many with advanced degrees—is that like many other experienced engineers in other fields, some start their project first thinking about the physical embodiment, i.e., the solution, rather than starting out with customer needs, FRs, and DPs. Often experienced engineers are used to, and depend on, the idea of developing complicated products by repeating the cycle of design/build/test/redesign/build/test. Therefore, it is often more difficult to teach experienced engineers than young engineers who are not yet imbued in industry tradition and not encumbered by their past experiences. Every one of us in the team was excited to work on the OSP project that would ultimately replace the space shuttle. However, suddenly a political decision was made by the Bush administration to discontinue this project. Now the USA must use Russian system to ferry astronauts and supplies to the International Space Station!

Observation No. 2: Failure to transform Multi-Input/Multi-Output (MI/MO) design to One-input/One-output (OI/OO) design by following the Independence Axiom (i.e., maintain the independence of functional requirements).

If we satisfy the Independence Axiom, even a large design project with many FRs can be treated as a set of one-input/one-output (OI/OO) project, a real advantage in system design. Modeling, simulation, and optimization become much simpler to implement, and actual physical implementation can avoid the trial-and-error process. However, more engineers are working on optimization of coupled MI/MO designs, which has a low probability of success.

Observation No. 3: Not able to recognize and build in "Functional Periodicity" into a large system.

Many natural and man-made systems have "functional periodicity," the best known being the circadian cycle. Machines that repeat the same functions throughout its life can accumulate errors. To stabilize such a system, it should be "reinitialized" periodically at the beginning of each cycle, somewhat similar to rebooting computers. This rebooting feature can be hidden from the users of the system by building in reinitialization as part of the software design. The use of reinitialization of FRs at the beginning of new cycle simplifies the design of large systems.

Observation No. 4: The physical size of the system or the number of parts in the system or the number of lines of code in software system is not a true measure of complexity.

Complexity is defined in many different ways depending on the field. Professor Seth Lloyd[18] of MIT once stated that there are as many definitions of complexity as the number of people working in the field of complexity.

In AD, complexity is defined as *"a measure of uncertainty in achieving FRs."* Therefore, a functionally coupled system is much more complex and an uncoupled system (note the difference between functional coupling and physical integration). The physical size or the length of software code or the number of parts in a system is not a measure of system complexity of designed systems, according to this

[18]Professor Seth Lloyd is the Nam Pyo Suh professor of mechanical engineering at MIT. The professorship was made possible by the generous gift of Mr. Hock E. Tan, CEO of Avago, Inc.

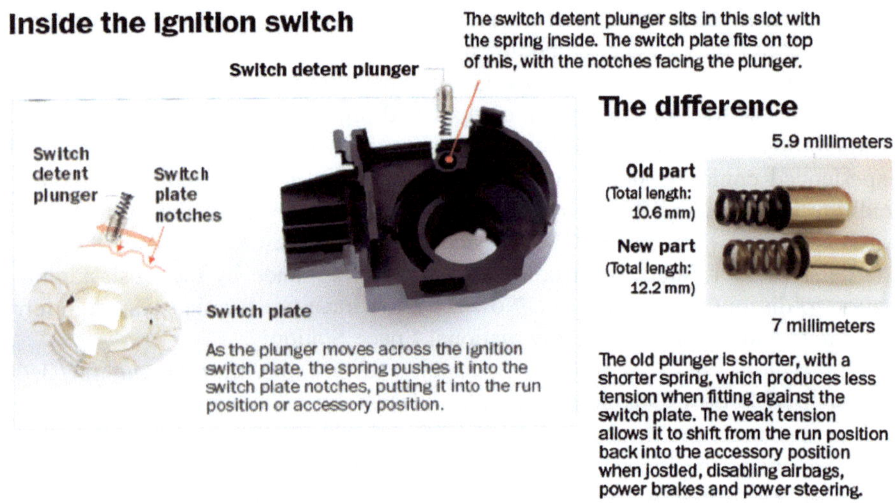

Fig. 11.3 An example of complex small system: GM ignition key

definition. A system with a larger number of parts is more *complicated* but not necessarily more *complex* than a system with a smaller number of parts. This definition of complexity is, in a way, counterintuitive.

An example of a small system becoming complex is the GM ignition key [9]. As shown in Fig. 11.3, the major part of the switch is small with a limited number of parts. *Yet it is complex*, because it cannot satisfy its FRs at all times. It has two FRs: (1) Move the plunger across the ignition switch plate to enable spring-loaded plunger to lock into the notch of the plate for either ignition or the accessory position. (2) Lock the plunger in the preset notched position. It is a coupled design, since there is only one design parameter (DP, the spring) for two FRs. Therefore, the DP requires fine-tuning of the spring force to satisfy both of these FRs. Sometimes, when one of these FRs cannot be satisfied, the vehicle will malfunction. That is the reason for the fatal accidents that bedeviled GM for so many years.

11.4.5 Design of Large Organizations

The foregoing presentation on design of large systems used technological designs as examples. The most important finding was that the coupling of FRs (violating the Independence Axiom) was the major cause for cost overruns and missed schedules. The question is whether the same can be said of organizations. In the case of organizations, the issue can be less clear because typically there is no delivery schedule as such and it is difficult to measure "cost overruns" in an organizational sense. In large organizations, these things appear as inefficiencies of organizations, leading to bloated organizations, where the number of personnel is increased to diffuse cost overruns and missed schedules. Thus, many leading research universities have a lot more administrative personnel, although the size of the faculty and students has not changed.

In order to deal with these issues related to large organizations, the Axiomatic Design methodology has also been used to design and operate large organizations. Often it is more difficult to deal with organizations, because FRs are not clearly defined and the DPs could be intelligent people with their own ideas and goals. Furthermore, sharing the same set of FRs among diverse groups of people in an organization is a challenge, especially if the new FRs are substantially different from prior goals and past practices. Often organizational design involves *more* DPs than FRs, which is classified as a redundant system Suh [11, 12].

The AD thinking was applied to several organizations: the MIT Laboratory for Manufacturing and Productivity (LMP), Engineering Directorate of the National Science Foundation (NSF), MIT Department of Mechanical Engineering, and the Korea Advanced Institute of Science and Technology (KAIST). Although the author may not be the best judge of the results of AD-based reorganization of these leading institutions, these organizations have transformed for the better after applying AD thinking. The result has been remarkable and positive, when measured in terms of the achievements of these organizations. The results of the implementation of FRs will be briefly described.

11.4.5.1 MIT Laboratory for Manufacturing and Productivity (LMP)

The reason for establishing LMP was fully discussed in a preceding section. The specific FRs were as follows:

FR1 = Develop a science base for manufacturing and design
FR2 = Create interdisciplinary research effort
FR3 = Emphasize innovation of new processes, new systems, etc.
FR4 = Promote collaboration with industry
FR5 = Raise significant research fund
FR6 = Attract talents

Results:

1. Created Axiomatic Design as well as 3D printing and others.
2. Faculty from four departments participated.
3. Raised most research fund among all units of the Department of Mechanical Engineering (about 30 % of the total department research volume of 65+ faculty members in the department).
4. Became the model for the NSF's Industry–University Cooperative (IUC) Program.
5. Helped to change national policy (The Stevenson-Wydler Technology Innovation Act of 1980)
6. Produced a large number of outstanding graduates with advanced degrees.

11.4.5.2 National Science Foundation (NSF)

Since its establishment in 1950, NSF had served and promoted traditional disciplines of natural sciences such as physics, biology, chemistry, and mathematics. The support for engineering was rather meager. When I was invited to join NSF by the White House in 1984, it was clear that many people in Washington had felt that the engineering group at NSF needed strengthening and changes, especially in view of the international challenges in manufacturing and technology. I was the first engineer and first (as well as the last) presidential appointee to head up the NSF Engineering Directorate.

Prior to my appointment, the Engineering Directorate was organized very much like a college of engineering of a typical university, i.e., Division for Mechanical Engineering, Chemical engineering, Electrical engineering, and Civil engineering. Furthermore, the NSF program directors were not encouraged to support research in new emerging fields or bold new ideas unless the peer review yields certain scores. They were simply reinforcing the existing activities of the engineering colleges, which, in turn, discouraged universities from moving into new fields since NSF did not support research in new or emerging fields. In fact, the fields that received more proposals received a larger budget, which meant that new emerging fields did not get budget allocation.

To my surprise, I found out that NSF's engineering group defined engineering as "applied science," reflecting their survival instinct in the political capital dominated by scientists. Therefore, they did not actively support any research in such engineering fields as design, manufacturing, optics, and MEMS. Also in spite of the importance of emerging technological fields for future economic development of the USA, NSF did not support any research in emerging and critical engineering fields. Also many of the program directors of the NSF Engineering Directorate thought that their job was to receive unsolicited proposals and have them evaluated through peer review and give out grants to those who received higher scores without any consideration of relative strategic importance of various engineering fields. In short, NSF forgot the original reasons for establishing NSF in 1950. To overcome these shortcomings, the Engineering Directorate of NSF had to be completely reorganized during the first three months of my tenure at NSF.[19]

Some of the important FRs for the NSF Engineering Directorate in 1985 were as follows:

FR1 = Satisfy NSF Act of 1950: advance science and engineering; secure national defense; and provide health, welfare, and prosperity
FR2 = Support advances in engineering science
FR3 = Create science base for fields that do not have science base

[19]These changes—both organizational and budgetary—required the approval of Office of Management and Budget (OMB) of the White House and the approval of the US Congress. Many thought that the proposed changes could not be achieved during my tenure, since I had agreed to serve at NSF only for one year.

FR4 = Support emerging technologies
FR5 = Support critical technologies
FR6 = Support minority institutions

Results:

1. Reorganized to be more effective in satisfying FRs, whereas the original organization was similar to the college of engineering of typical engineering schools.
2. Attracted highly qualified talents from universities and industry to NSF.
3. Created new programs such as the Engineering Research Centers (ERC) program, which are still in operation after 30 years.
4. Raised the budget.
5. Affected the perception of engineering in Washington and at universities.
6. Changed the financial resource allocation to emphasize the original mandate of NSF.
7. Affected engineering education and research in the USA.

11.4.5.3 MIT Mechanical Engineering Department

When I was appointed as the head of the Department of Mechanical Engineering at MIT, the department had always been ranked No. 1 in the USA. The department had many well-known professors, who had made important contributions to various fields of mechanical engineering that were important in the first half of the twentieth century. The emphasis of its educational and research programs was mostly in those topics related to automobiles, power generation (i.e., heat engineering), manufacturing, design, control, and mechanical aspects of bioengineering. Some of these fields had matured with no major technological challenges remaining in the twenty-first century. Therefore, most progress was incremental. To deal with the problems of the twenty-first century, the disciplinary base of the department had to be broadened. The department lacked expertise in such fields as information technology, nanotechnology, nanomanufacturing, semiconductors, life sciences, and materials that were outside of the traditional department of mechanical engineering. Since advances in engineering are often made at the interface between and among traditional disciplinary boundaries, the department had to be transformed.

In order to prepare for the twenty-first-century engineering needs, the FRs for the Department of Mechanical Engineering were established as follows:

FR1 = Open up new frontiers of knowledge for the twenty-first century
FR2 = Recruit the best faculty from fields that will be important in the twenty-first century
FR3 = Create financial resources
FR4 = Upgrade facilities
FR5 = Provide multidisciplinary (or cross-disciplinary) education

Results:

1. Changed the direction of the department research and education to deal with important issues of the twenty-first century.
2. Hired many outstanding professors who could deal with important issues of the twenty-first century.
3. Strengthened inter- and multidisciplinary research.
4. Now almost 50 % of the mechanical engineering faculty has doctorate degrees in fields outside of traditional mechanical engineering.
5. Renovated facilities, laboratories, and acquired additional space.
6. Eliminated in-breeding of the faculty to facilitate intellectual diversity.
7. Made the curriculum more flexible to allow students to combine traditional mechanical engineering subjects with those of other disciplines, e.g., biology and computer science by strengthening Course 2A (non-traditional mechanical engineering program).
8. Increased student enrollment.
9. Secured a significant sum of financial gifts.

11.4.5.4 KAIST (Korea Advanced Institute of Science and Technology)

Around 1970, the Republic of Korea decided to transform its industry from labor-intensive business such as apparel and shoe manufacturing to heavy industries such as shipbuilding, automobile, steel making, and electronic. In order to support this major transformation of Korean industries from labor-intensive to "heavy" industries, the Korean government established KAIST in 1971, with the help of the US government, in order to produce engineers and scientists with advanced graduate education.

From the beginning, KAIST received special support from the Korean government under a special legislation. It received full support for faculty salary, full support for graduate students, cost of infrastructure, etc. Faculty members were paid about three times the prevailing salary of professors in other universities in Korea to entice Korean PhDs in residing in the USA to Korea. Students were deferred from military service, in addition to receiving free education, room, and board. They also received spending money. As a result of these benevolent policies, some of the best students came to KAIST for their masters' and doctorate degrees. The quality of the faculty was very high, too. KAIST produced a large number of key technical leaders of major companies such as Samsung and others. About 10 % of professors in science and technology of Korean universities are KAIST graduates. By 2006, when the author assumed the presidency of KAIST, it has become one of the best two universities in Korea. Internationally, its ranking was 196th. By any measure, it was a success story.

The author took the presidency of KAIST in order to make KAIST one of the top ten research universities of the world. In order to achieve the goal, KAIST had to be transformed into a globally competitive university from the one that nurtured and had grown under the protection of the government. Also, KAIST no longer received exceptional financial support from government. There were many issues that had to be solved. The first issue was related to the faculty—its size, productivity, diversity, and attitude. In 2006, it had about 400 professors and 6000 graduate students. The faculty size was too small to take care of the large number of students in a research university. To be competitive with world's leading research universities, KAIST needed to double the faculty size. It was also an aging faculty: The median age was about 55 with a few professors under age of 45. They could not hire new professors because they depended on the budget provided by the government. They could hire a new professor only to replace a retired professor. Too much in-breeding—in intellectual sense—took place, which discouraged outstanding scholars in new emerging fields from seeking faculty positions at KAIST. Although they produced many papers, a limited number of these papers were cited by others, indicating that they were not at the cutting edge of the field. Professors were mostly depending on their students for scholarly outputs. Senior professors were taking advantage of younger colleagues to take care of chores of the department. Graduate students were not productive because they were not challenged. Furthermore, they stayed in school for too many years to finish their degrees. All that time, their financial support was guaranteed by the government, which did not provide incentives to finish their degrees as soon as possible. KAIST also lengthened their stay at KAIST as students by insisting that the student should publish a paper in leading journal before they could graduate. KAIST's physical infrastructure was falling apart. They could not run any experiments at night in winter because KAIST did not heat the buildings to save money. They accepted all these as norm, which had to change to be competitive with leading universities of the world.

FR1 = Instill a sense of "Can Do" attitude
FR2 = Solve important problems for humanity
FR3 = Lower the barriers between disciplines
FR4 = Adopt merit-based system
FR5 = Increase faculty size, especially women faculty and international faculty
FR6 = Globalize
FR7 = Improve physical facility
FR8 = Measure real intellectual impact, not the number of papers published
FR9 = Increase productivity of faculty, students, and staff.

Results:

1. The ranking of KAIST jumped from 196 in 2006 to 10 in 2015 in the world's ranking of innovative universities. Just for comparison, the highest ranked university in the UK was Imperial College which was ranked 11th in this Thomson Reuters survey [3].
2. Increased the faculty size from 400 to 625 without significant government support by hiring 350 new young professors.
3. Increased the budget by a factor of 2.7 in seven years.

4. Raised the standard for tenure.
5. Globalization: Adopted English as the language of instruction, hired many international faculty members, recruited more international students, and promoted collaboration in research with institutions outside of Korea.
6. Built 14 new buildings for research, education, dormitories, sports complex, etc.
7. Increased it endowment.
8. Received major gifts (a difficult task in Korea).
9. Gave more opportunities to those students without financial means and limited opportunities.
10. Created major interdisciplinary research efforts.
11. Emphasized research in EEWS (energy, environment, water, and sustainability).
12. Created new interdisciplinary departments.
13. Created new teaching paradigm (Education 3.0).
14. Invented On-Line Electric Vehicle and Mobile Harbor to present bold concepts for the future development of society.
15. Absorbed a private university to strengthen KAIST in the field of information technology.

11.4.6 Theorem on Development of Large Systems

Based on the foregoing discussion of large technical systems with coupled FRs, cost overruns and missed schedules, the following theorem may be stated:

Theorem on Cost Overrun and Project Delay:

The coupling of the functional requirements (FRs) of the system under development is the root cause of cost overruns and project delays.

Proof of the Theorem:

Consider two designs—Design A and Design B—with an equal number of FRs and DPs. Then, they are both ideal designs per Theorem 4 [12]. Assume that Design A is a coupled design and Design B is an uncoupled design. Then, the information content of Design A, I_A, is greater than the information content of Design B, I_B. That is, $I_A > I_B$. Therefore, the design parameters of Deign A must have tighter tolerances than those of Design B. Furthermore, Design A works only when there is a unique solution (i.e., all DPs associated with coupled FRs must have the exact values requiring tight tolerances for the system to satisfy its FRs). In such a design, all FRs are functions of other FRs. In Design B, each DP satisfies one specific FR and thus, the tolerances can be much larger. That is, Design B can be treated as a series of one-output one-input systems with larger tolerances for DPs. Therefore, the cost of Design B, $\$_B$, is less than the cost of Design A, $\$_A$, i.e., $\$_B < \$_A$. Similarly, the time taken to complete the project A, T_A, is longer than the time taken to complete Design B, since each task does not have to be repeated. Thus, $T_A > T_B$.

11.4.7 Why Is the Coupled Design so Attractive and Enticing to Some?

People are attracted to the idea or the "thing" that is beautiful, elegant, and simple. They often look for a simple solution to a complex problem. Designers and engineers, like their compatriots, are equally fascinated and enticed by what appears to be an elegant simple solution that solves a multitude of problems all at once. They look for one magic knob that satisfies all the requirements. It may be part of human DNA. That may be the reason why some people are attracted to coupled designs. Sometimes a coupled system appears to take care of so many problems all at once, simply and elegantly. Then, when they actually try to implement it, they find that the system does not simply work the way they had expected, so they begin the process of changing this or trying that, sometimes forever. In the end, they accept a compromised design that they can sort of justify, coming out with rationale as to why the deficient solution should be equally acceptable, although they do not quite satisfy the original FRs and constraints.

To those who have found the beauty of uncoupled designs, uncoupled designs are much more elegant and simple to implement than coupled designs. They surpass coupled designs in their functionality and in many other ways, e.g., in satisfying the FRs and constraints, in reducing the cost of development, and in meeting the original schedule. Uncoupled designs are easy to develop and implement. Most of all, they perform reliably for a long time. Furthermore, as we work with uncoupled designs, we get more fascinated by their enduring performance that is beyond our imagination. The production system at Sweetheart Plastics is still running after a half century of operation. It is their simplicity and elegance that makes them to work so well without going through a continuous cycle of changes or modifications. We often take the performance of uncoupled systems for granted. To create uncoupled or decoupled designs, we simply must know from the very beginning what our objectives are in terms of FRs and constraints. Then, we need to come up with appropriate DPs that generate an uncoupled or decoupled design! To be proficient in creating uncoupled or decoupled designs, we have to go through the experience of creating an uncoupled design and discover how well it works.

Sometimes in academia, an elegant uncoupled design is often less appreciated. Unlike coupled designs that require lengthy mathematical derivations in order to obtain a unique solution, uncoupled design is much easier to analyze, because the design can be treated as a one-input one-output system, regardless of the total number of FRs and DPs involved in the uncoupled or decoupled system. However, when people write papers with lengthy mathematical derivations because the system is coupled with multiple FRs and DPs, academics tend to be more impressed. When one presents an elegant uncoupled design that can be solved with simple mathematics, they wonder why the paper is worth publishing. Coupled systems with several FRs and DPs may have only a unique solution without the freedom to change any one FR without affecting other FRs, whereas uncoupled or decoupled systems do not have a unique solution because FRs are independent from other FRs.

This attitude of the academia is more than a problem for academics. The issue at hand is also more than simply a matter of elegance and simplicity. Our inability to deal with design rationally affects the industrial development negatively. It increases the cost of research and development, wastes valuable human resources as well as natural resources, and even deprives those whose quality of life depends on our ability to produce high-quality goods at a lower price and conserve resources. The ultimate goal of intellectual endeavor should be, and is, for the betterment of humankind.

Universities and other organizations can be run more efficiently if they are designed properly, i.e., uncoupled designs for organizations and strategies. The latest trend of the operating cost escalation of many universities and ineffectual policies may be attributed to poor design, i.e., coupled design or no design.

11.4.8 Dilemmas and Action Items

11.4.8.1 Dilemmas

(1) **Universities' dilemma in dealing with large systems**

Many people are aware of the need to teach students on how to design and implement large systems. However, universities have not done much to meet this need. Some of the difficulties may be as follows:

- Difficult to teach design of large systems because it is resource-intensive.
- Many schools need to improve their design curriculum.
- Large project execution is difficult under the current academic structure.
- It is nearly impossible at most universities to secure financial support and personnel for execution of large systems projects, e.g., OLEV and Mixalloy.
- Lack of professors with experience in dealing with large systems.
- Universities design and operate their universities based on their experience and trial-and-error processes.

(2) **Industries' dilemma in dealing with large systems**

Industry is always interested in improving its productivity, and innovating and developing new products, especially for large capital goods. However, they do not know what tools they can use to reduce the cost and shorten the development time. They simply may assume that the current state of their operation is the best that can be done. Also, many firms may not be aware of the fact that the root cause of long development times and high cost is associated with coupled designs. They may not even be aware of the havoc that is created by coupled FRs. They may accept the status quo as being reasonable.

The following is a list of issues confronted by industrial firms:

- Most engineers they hire were educated in universities that do not teach large system design.
- They depend too much on experience.

- They tend to design large systems by trial and error.
- Program managers operate purely based on their experience. (Few industrial leaders think like Robert Ford.)
- No systematic approach.
- They tend to copy their old successful designs, which is not conducive to real innovation.
- Many industrial leaders are expected to "act and produce" rather than "think before jumping and produce."

11.4.8.2 Action Items

(1) **What should universities do to teach large systems design?**

Perhaps the most important thing for universities to do is to recognize the importance of the field of design for both technological and organizational systems. Until now, design and synthesis have received only a limited (or marginal) attention at most universities, since a majority of the scholarly contributions recognized by the scholarly community has been in the domain of analysis and scientific discoveries. Synthesis of large systems was largely left to industry and is mostly carried out in a highly ad hoc manner. This has been the case even in non-technological fields such as economics and public policy, although the design of systems affects many societal and economic outcomes. Major policies (e.g., health care, economic policies) would benefit if they were founded on more rigorous analysis and creative synthesis.

The following is a partial list of actions universities can undertake to strengthen the field of design and synthesis:

- Try to get large ambitious projects funded.
- Create a simulation program to teach system design. We can teach how large system should be designed and executed on computer, i.e., how to define FRs, how to select DPs, how to decompose a higher-level FRs and DPs until the design is finished, and how to monitor coupling introduced by someone in another branch of the project.
- Create "System Architects" to monitor the simulation of a hypothetical system development.
- Create large projects with a hierarchical structure for project execution rather than one professor working alone with his or her graduate students.
- Practice AD in operating their universities.

(2) **What should industries do to deal with large systems design?**

Industrial firms have a large stake in reducing the cost of designing and implementing large systems. Their competitiveness depends on shortening the time and reducing the cost of such systems. In order to achieve such goals, industrial firms may take the following actions:

- Educate young design engineers on fundamentals of design and systems.
- Educate, train, and create professional System Architects.
- Mandate that managers must learn about the systems design and systems management.
- Create System Architects.
- Pay these people well.
- Review their own organizational design for coupling of FRs of their companies.

(3) **What should government do to encourage innovation in designing large systems?**

Governments are directly or indirectly the ultimate beneficiary effective system development. They procure their weapons and create the infrastructure. They also collect taxes from their industries for the products manufactured and sold by their industries. The competitiveness of their nation ultimately depends on productivity of their industry and universities. Therefore, governments should nurture the ability of their institutions and people to be more effective in designing and implementing large systems.

- Fund more goal-oriented large projects.
- Do not confuse fundamental research in large systems with one PI-initiated project. Large system projects are also made up of many "one PI-led" projects, except that they must be rationally organized and integrated to achieve the overall goals of the system. The intellectual issues in developing large system are as fundamental as one PI-led projects.
- Balance the support for research in basic sciences and for research in system development.
- Evaluate the merits of problem definition of proposed research as much as the details of proposed approaches.

11.5 Conclusions

- Many problems humanity must solve involve design and development of large systems. The current process of developing these systems generates many unexpected problems, increasing the cost of development and prolonging the development period.
- "Coupling of FRs" is one of the primary reasons for cost overruns and missed schedules in developing large systems. A theorem is presented.
- Industry must replace the repetitive design/build/test cycle with a better methodology to reduce the cost and time taken for development of large systems.
- Academia must expand and develop more rigorous teaching methods in the field of design of large systems.

- Academia should re-examine their university design to determine the root cause of the cost increase that has been faster than the general inflation rate.
- To promote innovation, government should fund more system-related research and development projects.
- This matter of the increased financial cost and the unnecessary expenditure of valuable human and natural resources because of our inability to design it "right" from the beginning must be addressed through educational processes and changes in industrial practice.

Acknowledgements Several colleagues read the manuscript and made useful comments and suggestions. In particular, I would like to thank Professors Amro Farid, Paul Citti, Stephen Lu, Kate Thompson, Chris Brown, and Mitchel Tseng for their comments.

References

1. Y. Hatamura, S. Abe, M. Fuchigami, N. Kasahara, *The 2011 Fukushima Nuclear Power Plant Accident (How and Why It Happened)* (Woodhead Publishing, Elsevier, 2015), p. 21
2. E. Hong, G. Park, Design information management of an on-line electric vehicle using axiomatic design. SAE Int. J. Mater. Manufact. **3**(1), 133–141 (2010). doi:10.4271/2010-01-0279
3. http://mobile.reuters.com/article/idUSKCN0RG0CX20150916
4. A. Kossiakoff, W.N. Sweet, S.J. Seymour, S.M. Biemer, *Systems Engineering—Principles and Practice* (Wiley, 2011)
5. T. Lee, G. Park, Managing system design process using axiomatic design: a case on KAIST mobile harbor project. SAE Int. J. Mater. Manuf. **3**(1), 125–132 (2010). doi:10.4271/2010-01-0278
6. A. Liu, S.C.-Y. Lu, A new coevolution process for conceptual design. CIRP Ann. Manuf. Technol. (2015)
7. J. Martini, The production and analysis of microcellular foam. S. Thesis, Mechanical Engineering, MIT, 1981
8. M. Nakao, K. Kusaka, K. Tsuchiya, K. Iino, Axiomatic design aspect of the Fukushima-1 accident: electrical control interferes with all mechanical functions, in *Proceedings of the 7th International Conference on Axiomatic Design*, ed. by M.K. Thompson. ICAD-2013-17 (2013), pp. 113–118
9. H. Oh, Private communication (2014)
10. E. Rechtlin, System Architecturing, *Creating and Building Complex Systems* (Prentice Hall, Englewood Cliffs, NJ, 1990)
11. N.P. Suh, *The Principles of Design* (Oxford University Press, New York, 1990)
12. N.P. Suh, Axiomatic Design, *Advances and Applications* (Oxford University Press, New York, 2001)
13. N.P. Suh, *Complexity: Theory and Applications* (Oxford University Press, New York, 2005)
14. N.P. Suh, D.H. Cho, Electrification of ground transportation systems, to be published in 2016
15. C.L. Tucker III, N.P. Suh, Mixing for reaction injection molding II. Impingement mixing of fiber suspensions, Polym. Eng. Sci. **20**(13), 887–898 (1980)
16. A. Wong, H. Guo, V. Kumar, C.B. Park, N.P. Suh, Microcellular Plastics. Encycl. Polym. Sci. Technol. (2016)

Appendix
Axioms, Corollaries and Theorems in Axiomatic Design

Abstract Beginning with the original statement of Axiomatic Design's two Axioms, the theory has developed to include many corollaries and theorems. These are now collated in this appendix for the convenience of the interested reader.

1. Axioms

Axiom 1 (*The Independence Axiom* [1]) Maintain the independence of FRs.
Axiom 2 (*The Information Axiom* [1]) Minimize the information content.

2. Corollaries

Corollary 1 (Decoupling of Coupled Designs [1]) *Decouple or separate parts or aspects of a solution if FRs are coupled or become interdependent in the designs proposed.*
Corollary 2 (Minimization of FRs [1]) *Minimize the number of FRs and constraints.*
Corollary 3 (Integration of Physical Parts [1]) *Integrate design features in a single physical part if FRs can be independently satisfied in the proposed solution.*
Corollary 4 (Use of Standardization [1]) *Use standardized or interchangeable parts if the use of these parts is consistent with the FRs and constraints.*
Corollary 5 (Use of Symmetry [1]) *Use symmetrical shapes and/or components if they are consistent with the FRs and constraints.*
Corollary 6 (Largest Tolerance [1]) *Specify the largest allowable tolerance in stating FRs.*
Corollary 7 (Uncoupled Design with Less Information [1]) *Seek an uncoupled design that requires less information than coupled designs in satisfying a set of FRs.*
Corollary 8 (Effective Reangularity of a Scalar [1]) *The effective reangularity R for a scalar coupling "matrix" or element is unity.*

© Springer International Publishing Switzerland 2016
A.M. Farid and N.P. Suh (eds.), *Axiomatic Design in Large Systems*,
DOI 10.1007/978-3-319-32388-6

3. Theorems in General Design

Theorem 1 (Coupling Due to Insufficient Number of DPs [1]) *When the number of DPs is less than the number of FRs, either a coupled design results, or the FRs cannot be satisfied.*
Theorem 2 (Decoupling of Coupled Design [1]) *When the design is coupled due to the greater number of FRs than DPs (i.e.,* $\mathbf{m} > \mathbf{n}$*), it may be decoupled by the addition of new DPs so as to make the number of FRs and DPs equal to each other, if a subset of the design matrix containing* $\mathbf{n} \times \mathbf{n}$ *elements constitutes a triangular matrix.*
Theorem 3 (Redundant Design [1]) *When there are more DPs than FRs, the design is either a redundant design or a coupled design.*
Theorem 4 (Ideal Design [1]) *In an ideal design, the number of DPs is equal to the number of FRs.*
Theorem 5 (Need for New Design [1]) *When a given set of FRs is changed by the addition of a new FR, or substitution of one of the FRs with a new one, or by selection of a completely different set of FRs, the design solution given by the original DPs cannot satisfy the new set of FRs. Consequently, a new design solution must be sought.*
Theorem 6 (Path Independency of Uncoupled Design [1]) *The information content of an uncoupled design is independent of the sequence by which the DPs are changed to satisfy the given set of FRs.*
Theorem 7 (Path Dependency of Coupled and Decoupled Design [1]) *The information contents of coupled and decoupled designs depend on the sequence by which the DPs are changed and on the specific paths of changes of these DPs.*
Theorem 8 (Independence and Tolerance [1]) *A design is an uncoupled design when the designer-specified tolerance is greater than*

$$\left(\sum_{j \neq i; j=1}^{n} (\delta \mathrm{FP}_i / \delta \mathrm{DP}_j) \Delta \mathrm{DP}_j \right) \tag{1}$$

in which case the nondiagonal elements of the design matrix can be neglected from design consideration.
Theorem 9 (Design for Manufacturability [1]) *For a product to be manufacturable, the design matrix for the product,* **[A]** *(which relates the* **FR** *vector for the product to the* **DP** *vector of the product), times the design matrix for the manufacturing process,* **[B]** *(which relates the* **DP** *vector the* **PV** *vector of the manufacturing process), must yield either a diagonal or triangular matrix. Consequently, when any one of these design matrices, that is, either* **[A]** *or* **[B]***, represents a coupled design, the product cannot be manufactured.*
Theorem 10 (Modularity of Independence Measures [1]) *Suppose that a design matrix* **[DM]** *can be partitioned into square submatrices that are nonzero only along the main diagonal. Then, the reangularity and semangularity for* **[DM]** *are*

equal to the products of their corresponding measures for each of the nonzero submatrices.

Theorem 11 (Invariance [1]) *Reangularity and semangularity for a design matrix* [**DM**] *are invariant under alternative orderings of the FR and DP variables, as long as orderings preserve the association of each FR with its corresponding DP.*

Theorem 12 (Sum of Information [1]) *The sum of information for a set of events is also information, provided that proper conditional probabilities are used when the events are not statistically independent.*

Theorem 13 (Information Content of the Total System [1]) *If each DP is probabilistically independent of other DPs, the information content of the total system is the sum of information of all individual events associated with the set of FRs that must be satisfied.*

Theorem 14 (Information Content of Coupled vs Uncoupled Designs [1]) *When the state of FRs is changed from one state to another in the functional domain, the information required for the change is greater for a coupled process than an uncoupled process.*

Theorem 15 (Design–Manufacturing Interface [1]) *When the manufacturing system compromises the independence of the FRs of the product, either the design of the product must be modified, or a new manufacturing process must be designed and/or used to maintain the independence of the FRs of the products.*

Theorem 16 (Equality of Information Content [1]) *All information contents that are relevant to the design task are equally important regardless of their physical origin, and no weighing factor should be applied to them.*

Theorem 17 (Design in the Absence of Complete Information [3]) *Design can proceed even in the absence of complete information only in the case of a decoupled design if the missing information is related to the off-diagonal elements.*

Theorem 18 (Existence of an Uncoupled or Decoupled Design [3]) *There always exists an uncoupled or decoupled design that has less information than a coupled design.*

Theorem 19 (Robustness of Design [3]) *An uncoupled design and a decoupled design are more robust than a coupled design in the sense that it is easier to reduce the information content of designs that satisfy the Independence Axiom.*

Theorem 20 (Design Range and Coupling [3]) *If the design ranges of uncoupled or decoupled designs are tightened, they may become coupled designs. Conversely, if the design ranges of some coupled designs are relaxed, the designs may become either uncoupled or decoupled.*

Theorem 21 (Robust Design When the System Has a Nonuniform pdf [3]) *If the probability distribution function (pdf) of the FR in the design range is nonuniform, the probability of success is equal to one when the system range is inside the design range.*

Theorem 22 (Comparative Robustness of a Decoupled Design [3]) *Given the maximum design ranges for a given set of FRs, decoupled designs cannot be as robust as uncoupled designs in that the allowable tolerances for DPs of a decoupled design are less than those of an uncoupled design.*

Theorem 23 (Decreasing Robustness of a Decoupled Design [3]) *The allowable tolerance and thus the robustness of a decoupled design with a full triangular matrix diminish with an increase in the number of functional requirements.*

Theorem 24 (Optimum Scheduling [3]) *Before a schedule for robot motion or factory scheduling can be optimized, the design of the tasks must be made to satisfy the Independence Axiom by adding decouplers to eliminate coupling. The decouplers may be in the form of a queue or of separate hardware or buffer.*

Theorem 25 ("Push" System vs. "Pull" System [3]) *When identical parts are processed through a system, a "push" system can be designed with the use of decouplers to maximize productivity, whereas when irregular parts requiring different operations are processed, a "pull" system is the most effective system.*

Theorem 26 (Conversion of a System with Infinite Time-Dependent Combinatorial Complexity to a System with Periodic Complexity [3]) *Uncertainty associated with a design (or a system) can be reduced significantly by changing the design from one of serial combinatorial complexity to one of periodic complexity.*

4. Theorems Related to Design and Decomposition of Large Systems

Theorem S1 (Decomposition and System Performance [3]) *The decomposition process does not affect the overall performance of the design if the highest level of FRs and Cs is satisfied and if the information content is zero, irrespective of the specific decomposition process.*

Theorem S2 (Cost of Equivalent Systems [3]) *Two "equivalent" designs can have substantially different cost structures, although they perform the same set of functions and they may even have the same information content.*

Theorem S3 (Importance of High-Level Decisions [3]) *The quality of design depends on the selection of FRs and the mapping from domain to domain. Wrong selection of FRs made at the highest levels of design hierarchy cannot be rectified through the lower-level design decisions.*

Theorem S4 (The Best Design for Large Systems [3]) *The best design for a large flexible system that satisfies m FRs can be chosen among the proposed designs that satisfy the Independence Axiom if the complete set of the subsets of FRs that the large flexible system must satisfy over its life is known a priori.*

Theorem S5 (The Need for a Better Design [3]) *When the complete set of the subsets of FRs that a given large flexible system must satisfy over its life is not known a priori, there is no guarantee that a specific design will always have the minimum information content for all possible subsets, and thus, there is no guarantee that the same design is the best at all times.*

Theorem S6 (Improving the Probability of Success [3]) *The probability of choosing the best design for a large flexible system increases as the known subsets*

of FRs that the system must satisfy approach the complete set that the system is likely to encounter during its life.
Theorem S7 (Infinite Adaptability vs. Completeness [3]) *A large flexible system with infinite adaptability (or flexibility) may not represent the best design when the large system is used in a situation in which the complete set of the subsets of FRs that the system must satisfy is known a priori.*
Theorem S8 (Complexity of a Large Flexible System [3]) *A large system is not necessarily complex if it has a high probability of satisfying the FRs specified for the system.*
Theorem S9 (Quality of Design [3]) *The quality of design of a large flexible system is determined by the quality of the database, the proper selection of FRs, and the mapping process.*

5. Theorems for Design and Operation of Large Organizations

Theorem M1 (Efficient Business Organization [2]) *In designing large organizations with finite resources, the most efficient organizational design is the one that specifically allows reconfiguration by changing the organizational structure and by having flexible personnel policy when a new set of FRs must be satisfied.*
Theorem M2 (Large System with Several Subunits [2]) *When a large system (e.g., organization) consists of several subunits, each unit must satisfy independent subsets of FRs so as to eliminate the possibility of creating a resource-intensive system or a coupled design for the entire system.*
Theorem M3 (Homogeneity of Organizational Structure [2]) *The organizational structure at a given level of the hierarchy must be either all functional or product-oriented to prevent duplication of effort and coupling.*

6. Theorems Related to Software Design

Theorem Soft 1 (Knowledge Required to Operate an Uncoupled System [3]) *Uncoupled software or hardware systems can be operated without precise knowledge of the design elements (i.e., modules) if the design is truly an uncoupled design and if the FR outputs can be monitored to allow closed-loop control of FRs.*
Theorem Soft 2 (Making Correct Decisions in the Absence of Complete Knowledge for a Decoupled Design with Closed-Loop Control [3]) *When the software system is a decoupled design, the FRs can be satisfied by changing DPs if the design matrix is known to the extent that the knowledge about the proper sequence of change is given, even though precise knowledge about the design elements may not be known.*

7. Theorems Related to Complexity

Theorem C1 (Complexity of an Uncoupled System with Many Interconnected Parts [4]) *Complexity of an uncoupled system with many interconnected parts is not necessarily greater than that of a system with fewer interconnected parts unless the interfaces between the interconnected parts of the uncoupled system increase uncertainty by reducing the overlap between the system range and the design range.*

Theorem C2 (Complexity of a Decoupled System with Many Interconnected Parts [4]) *Complexity of a decoupled system with many interconnected parts is not necessarily greater than that of a system with fewer interconnected parts unless the interfaces between the interconnected parts of the decoupled system increase uncertainty by reducing the overlap between the system range and the design range.*

Theorem C3 (Complexity of a Coupled System with Many Interconnected Parts [4]) *Complexity of a coupled system with many interconnected parts is not necessarily greater than that of a system with fewer interconnected parts unless the interfaces between the interconnected parts of the coupled system increase uncertainty by reducing the overlap between the system range and the design range.*

Theorem C4 (Complexity of an Uncoupled System with Complicated Arrangement of Parts [4]) *Complexity of an uncoupled system with complicated arrangement of parts is not necessarily greater than that of a system with less complicated arrangement of parts unless the interfaces between the parts of the uncoupled system increase uncertainty by reducing the overlap between the system range and the design range.*

Theorem C5 (Complexity of a Decoupled System with Complicated Arrangement of Parts [4]) *Complexity of a decoupled system with complicated arrangement of parts is not necessarily greater than that of a system with less complicated arrangement of parts unless the interfaces between the parts of the decoupled system increase uncertainty by reducing the overlap between the system range and the design range.*

Theorem C6 (Complexity of a Coupled System with Complicated Arrangement of Parts [4]) *Complexity of a coupled system with complicated arrangement of parts is not necessarily greater than that of a system with less complicated arrangement of parts unless the interfaces between the parts of the coupled system increase uncertainty by reducing the overlap between the system range and the design range.*

Theorem C7 (Imaginary Complexity of a Decoupled System with Complicated Arrangement of Parts [4]) *The time-independent imaginary complexity of a decoupled system with complicated arrangement of parts can be large if the design parameters (DPs) are not changed in the sequence given by the design matrix.*

Theorem C8 (Complexity of Sociopolitical–Economic Systems [4]) *The complexity of sociopolitical–economic systems increases with the number of entities (i.e., organizations or individuals) that can affect the ultimate outcome.*

Theorem C9 (Reduction of Complexity of Sociopolitical–Economic Systems [4]) *If all the constituents of a social system can agree on the common set of FRs and if the FRs can be satisfied independently, the complexity of the decision-making process can be reduced when the final decision is made by a single entity after understanding and taking into account the uncertainties introduced by other constituents of the system.*

Theorem C10 (Reduction of Complexity of Sociopolitical–Economic Systems through Reinitialization or Redesign [4]) *When a sociopolitical–economic system is moving into a chaotic state because of time-dependent combinatorial complexity, the system should be reinitialized or redesigned to reduce complexity.*

References

1. N.P. Suh, *The Principles of Design* (Oxford University Press, 1990)
2. N.P. Suh, Design and operation of large systems. J. Manuf. Syst. **14**(3) (1995)
3. N.P. Suh, *Axiomatic Design: Advances and Applications* (Oxford University Press, 2001)
4. N.P. Suh, *Complexity : Theory and Applications* (Oxford University Press, New York, 2005)

Index

© Springer International Publishing Switzerland 2016
A.M. Farid and N.P. Suh (eds.), *Axiomatic Design in Large Systems*,
DOI 10.1007/978-3-319-32388-6